Reconfigurations of the Bildungsroman

American Frictions

Volume 2

Gonçalo Cholant

Reconfigurations of the Bildungsroman

─────

Taking Refuge from Violence in Kincaid, Danticat, hooks, and Morrison

DE GRUYTER

ISBN 978-3-11-163114-1
e-ISBN (PDF) 978-3-11-075275-5
e-ISBN (EPUB) 978-3-11-075284-7

Library of Congress Control Number: 2022933874

Bibliographic information published by the Deutsche Nationalbibliothek
The Deutsche Nationalbibliothek lists this publication in the Deutsche Nationalbibliografie;
detailed bibliographic data are available on the Internet at http://dnb.dnb.de.

www.degruyter.com

Acknowledgments

I would like to thank the editors of American Frictions, especially Julia Roth, for the opportunity to contribute to this outstanding research endeavor. I am extremely grateful for all the support given by Professor Isabel Caldeira, my former supervisor and lifelong inspiration, without whom the present work could not have been done. I would also like to thank Professor Caldeira for her observations in the foreword.

I must thank the Portuguese Foundation for Science and Technology (FCT) for granting me the scholarship (SFRH/BD/103174/2014), which allowed me to pursue the investigation during my Ph.D., which has allowed me to produce a major part of the research that can be found in this volume, the Faculty of Humanities of the University of Coimbra (FLUC), and The Center for Social Studies (CES).

https://doi.org/10.1515/9783110752755-001

Foreword

"Race has been a constant arbiter of difference, as have wealth, class, and gender – each of which is about power and the necessity of control." (Toni Morrison, *The Origin of Others* p. 3). In these words by Toni Morrison I find the main structure of meaning that is exposed in Gonçalo Cholant's book, *Reconfigurations of the Bildungsroman: Taking Refuge from Violence in Kincaid, Danticat, hooks, and Morrison.* Through an insightful confluence of theoretical concepts such as coloniality (Quijano; Mignolo; Lugones), the Ecology of Knowledges and the Sociology of Absences (Santos), subalternity (Spivak), intersectionality (Crenshaw) and the need of its epistemological revision (Roth), Cholant dismantles the logic of systemic oppression and violence that marks the everyday experience of subaltern subjects and deprives them of their rights of citizenship. This rich theoretical framework provides the opportunity to unveil how colonialism continues to leave a mark on the experience of African-American and African diasporic subjects, and promotes a decolonial rethinking of hegemonic patterns of knowledge (Mignolo; Maldonado-Torres).

The focus of *Reconfigurations* is contemporary literature by a few women authors of African descent, from the United States and the Caribeean, "as [l]iterature offers a setting where the process of unearthing histories is possible, and while doing so, it brings to the forefront the complexities derived from the erasure fabricated by colonial historiography." (Cholant 33). The author's selection follows this order: Jamaica Kincaid (*Lucy*, 1990); Edwidge Danticat (*Breath, Eyes, Memory,* 1994); Toni Morrison (*God Help the Child,* 2015); and bell hooks (*Bone Black – Memories of Girlhood,* 1996). Mostly fiction, but also texts fluidly entangled with the autobiographical, these powerful narratives remake the connection between the present, history and memory. The writers are either from United States or chose the country as their destiny of immigration, therefore their narratives reveal an interesting transnational paradigm that connects the islands and the metropole, the immigrant experience and the diasporic one.

The circumstance of trauma, since it frequently derives from the experience of violence, direct and indiret, physical and pshychological, individual and collective, is a common trace in these literary pieces and offers the author a superb opportunity to display the logic of coloniality and intersectional oppression. As Cholant argues, "[t]he imaginative work has transformative possibilities over trauma, revitalizing the event, transposing its literal character and eventually surpassing it." (16) Besides displaying and analysing the traumatic processes in each story, the author underlines the way each protagonist works through trauma and builds her own strategy of resistance. The act of speaking out to

https://doi.org/10.1515/9783110752755-002

"break through silences to speak the truth of their lives, to give testimony" (hooks, *Sisters of the Yam* 35) such as the act of writing are an integral part of a struggle against invisibility for women in their plight against multiple oppressions of race, class and gender. For Cholant, the investigation that led to the production of this volume was also a political act and an act of ethical responsibility (cf. Butler, *Precarious Lives*) "trying to identify in the texts tendencies of emancipatory practices and knowledges." (59)

A common pattern in the texts selected is the form of the *Buildungsroman*, or coming-of-age story. However, against the aforesaid background – historical, social and individual – the *Buildungsroman* could only be "reconfigured", as the author chose to characterize the process of reacessment and revision of the canonical form. Through this process of reconfiguration the writers, as counter-hegemonic subjects, have searched for a more adequate means to convey experiences preyed by poverty, racism, sexism, and other kinds of violence and discrimination. Thus, as Cholant argues, we are exposed to a "type of text [that] ends up reverting the unitary discourse into a communitarian discourse that was broader, expressing not only the concerns of the self, but also the collectivity, the support network that makes survival possible, turning writing into a space for contestation against hegemonic patterns" (31). The author presents literature as the ground on which trauma and violence are negotiated, made sense of, and ultimately resisted, working as a way of denouncing experiences that have been made invisible, as well as providing a safe point of reference to dealing with traumatic experiences.

Awareness about the causes of trauma and the different ways violence (direct and structural) act upon subjectivities – when many variants such as sex, race, class, origin, and sexual orientation, among others, complicate one's experience – opens the space for Cholant's analysis of the different ways the selected writers unsettle the traditional *Bildungsroman* in their coming-of-age stories of Black women in the Americas. Cholant also points out to the reconfiguration of the autobiographical in the case of hooks' beautiful but so far quite neglected memoir. As he claims, new formulations of the genre are achieved

> [...] complicating notions of narrative closure, questioning the status quo instead of abiding to it, rethinking sexual identities and gender roles, defying hegemonic culture, redefining the meaning of belonging to the United States and to the Caribbean. They also subvert the genre, demonstrating how porous and flexible the limits that separate autobiography from fiction are, offering readers narratives that dismantle aesthetic/literary patterns, and consequently, through the subversion of the conventions, undo constructs already established about identity and the women's condition. (Cholant 39 – 40)

The reader of the present work can find not only a thorough and sensitive reading of the literary texts, but also a questioning of the possibility of representation, when the subject of analysis involves violence and trauma. But, as the author eloquently observes, "the possibility of representation of unbearably violent and traumatic events is not simply referential, but a transformative space where the resignification of the experience takes place. The remapping of experiences carried out by literature might be one of the first steps towards the mitigation of violence as a structural reality." 39

Gonçalo Cholant is a young scholar and this is his first book, but reading through it we are led into a world of sociological and cultural significance of very topical issues, grappled with with a bold political stance and a mature awareness of inequalities in our contemporary world and our common humanity in times of deep global instability. As he writes, "reading literary fiction boosts the empathy capacity of subjects, since it forces them to deal with the perspectives of others, expanding the knowledge about them, helping to recognize similarities among readers and the stories found in literature." (44) Academic research can also be a field that transforms us individually and collectively because we learn but also learn to unlearn.

To conclude, I believe that Gonçalo Cholant's study is a noteworthy contribution to the fields of Inter-American and Studies of the literature of the African diaspora, also adding value to recent scholarship on the reframing of the *Bildungsroman* (LeSeur; Rishoi; Bolaki).

Isabel Caldeira. University of Coimbra

References:

Bolaki, Stella. *Unsettling the Bildungsroman: Reading Contemporary American Ethnic Women's Fiction*. New York: Rodopi Press, 2011.

Butler, Judith. *Precarious Life: The Powers of Mourning and Violence*. London and New York: Verso, 2004.

Crenshaw, Kimberl. "Demarginalizing the Intersection of Race and Sex: A Black Feminist Critique of Antidiscrimination Doctrine, Feminist Theory and Antiracist Politics." *University of Chicago Legal Forum*, vol. 1989 issue 1, 1989, pp. 139–167, chicagounbound.uchicago.edu/uclf/vol1989/iss1/8. Accessed 3 September 2010.

Crenshaw, Kimberl." Mapping the Margins: Intersectionality, Identity politics, and Violence Against Women of Color." *Stanford Law Review*, vol. 43 no. 6, 1991, pp. 1241–99.

hooks, bell. *Sisters of the Yam: Black Women and Self-Recovery*. New York and London: Routledge, 2015.

LeSeur, Geta. *Ten Is the Age of Darkness: The Black Bildungsroman*. Columbia: University of Missouri Press, 1995.

Lugones, Maria. "The Coloniality of Gender." Worlds and Knowledges Otherwise, vol. 2, Duke University, Spring 2008, pp. 1–17.

Maldonado-Torres, Nelson. "Outline of Ten Theses on Coloniality and Decoloniality" http://caribbeanstudiesassociation.org/docs/Maldonado-Torres_Outline_Ten_Theses-10. 23.16.pdf. Accessed 26 February, 2018.

Mignolo, Walter D. "DELINKING". *Cultural Studies* 21.2 (2007): pp. 449–514. 13.02.2014. DOI: 10.1080/09502380601162647. Accessed 24 August 2019.

Mignolo, Walter, ed. *Género y descolonialidad.* Colección Pensamiento crítico y opción descolonial. Buenos Aires, Argentina: Ediciones del signo, 2008.

Morrison, Toni. *The Origin of Others.* Harvard University Press, 2017.

Quijano, Anibal. "Coloniality of Power, Eurocentrism, and Latin America." *Neplanta: Views from South* 1.3 (2000): pp. 533–580. Accessed 20 October 2003.

Rishoi, Christy. *From Girl to Woman: American Women's Coming-of-age Narratives.* Albany: State University of New York, 2003.

Roth, Julia. "Entangled Inequalities as Intersectionalities: Towards an Epistemic Sensibilization." DesiguAldades.net Working Paper Series no. 42, Berlin: DesiguALdades.net Reseach Network on Interdependent Inequalities in Latin America, 2013.

Santos, Boaventura de Sousa. "Para uma sociologia das ausências e uma sociologia das emergências." *Revista Crítica de Ciência Sociais*, vol. 63, Coimbra: Centro de Estudos Sociais, 2002, pp. 237–80.

Santos, Boaventura de Sousa. "Para Além do Pensamento Abissal: Das Linhas Globais a Uma Ecologia de Saberes." *Revista Crítica de Ciências Sociais*, vol. 78, Coimbra: Centro de Estudos Sociais, 2007, pp. 3–46.

Spivak, Gayatri Chakravorti. "Can the Subaltern Speak?" *Marxism and the Interpretation of Culture.* Eds. Nelson and Grossberg. Urbana: University of Illinois Press, 1988, 271–308.

Contents

1 Introduction

1.1 American Studies, Black Women, Literature and the Right to Say Something About All of These

"As students of the American nation in the world [...], Americanists must decide at all times which side they are on."
Maria Irene Ramalho

Firstly, I believe it is important to be aware of the situatedness of the following work, and in this initial moment, to state how, where, why, and by whom such study is proposed. Like Toril Moi affirmed in her introductory work on feminist literary theory *Sexual/Textual Politics* in 1985, "[o]ne of the central principles of feminist criticism is that no account is neutral" (XV). The present work has a clear agenda of emancipation, and it aims at a form of knowledge production that is engaged with social change and the raising of awareness about violence, racism, sexism, and other forms of oppression that are responsible for the generation and perpetuation of social inequalities. The present work aims at discussing the works of African diasporic female authors – from the United States and the Caribbean, more specifically from Haiti and Antigua, in their entanglements with a colonial past and a capitalist transnational neoliberal present.

Producing this work in Portugal, we are at a very specific position in Europe and at a vantage point when looking at the postcolonial reality. Portugal played an important role in the inception of modernity, due to its exploration of new territories that eventually became its colonies, and its involvement in the slave traffic. Portuguese history is permeated by the constant contact with other peoples and places and the awareness of such relation's influences. It is also worthy mentioning Portugal's geo-political position in the semi-periphery of contemporary Western European power play and its relation to the North-South inequality paradigm. Most importantly, the present work reflects a contemporary concern with making visible the global North-South relations, so as to deconstruct them.

Regarding the researcher proposing the present work, on the one hand, I write/speak from a position of marginality when it comes to the core matter of analysis: I am neither a woman, nor black; I am neither from the United States, nor from the Caribbean. However, I am not marginal to the experience of living under the capitalist order in a Southern country (Brazil and Portugal), neither to the dealing with violence and its representations in the 20th century. I am also not marginal to the process of growing up in times of uncertainty, in addition to being culturally trained/educated in the western tradition/culture.

https://doi.org/10.1515/9783110752755-003

Underneath it all, I am Portuguese citizen with Brazilian roots, an ex-colonial subject who came to the European metropolis, a trope that resonates in the stories under analysis.

1.1.1 Claiming American Studies

But who has the right to claim American Studies? Who does the field belong to? Maria Irene Ramalho asks these questions in "Who Owns American Studies? Old and New Approaches to Understanding the United States of America", while tracing the trajectory of the field since its inception during the Cold War until the outset of what are now called the "New Americanists".

American Studies emerged from the disciplines of History and English and started in the 1940s, significantly after the War, as a project to grasp the integrity of United States' culture and society *as a whole* and present it to the world as a model. American Studies have also been perceived as a movement rather than a field of studies, a movement composed by several different disciplines that deal with the United States and its culture, creating a varied composite of investigation that reflects the attempt to fully understand the complex and multifarious reality of the nation-state.

However, the completeness of this *whole* – integral, unitary and homogenous – object becomes problematic when the traditional understanding of America as an entity invariably embedded in a narrative of progress and exceptionalism started to be questioned. "Old Americanists" were concerned mainly with the definition of the American literary canon embedded in the Western tradition, forging the principles and the method that should guide any understanding and study of what constituted such *Americanness,* stressing the values that are commonly connected with the American ideology of individualism, democracy, simplicity, innocence, enterprise, liberty, exceptionalism, among others. Henry Louis Gates Jr., in "Writing 'race' and the Difference It Makes", published in 1985, describes the canon of Western literature, stressing its transcendental character:

> [...] the canonical texts of the Western literary tradition have been defined as a more or less closed set of works that somehow speak to, or respond to, "the human condition" and to each other in formal patterns of repetition and revision. And while most critics acknowledge that judgment is not absolute and indeed reflects historically conditioned presuppositions, certain canonical works (the argument runs) do seem to transcend value judgments of the moment, speaking irresistibly to the human condition. [...] The question of the place of texts written by the Other (be that odd metaphorical negation of the European defined as African, Arabic, Chinese, Latin American, Yiddish, or female authors) in the proper study of

"literature," "Western literature," or "comparative literature" has, until recently, remained an unasked question, suspended or silenced by a discourse in which the canonical and the noncanonical stand as the ultimate opposition. (2)

The revision of the American Studies field, and its subsequent shift in paradigm, is related to the recognition of minorities (racial, national, sexual, among others), as part of the American experience and as producers of culture and knowledge. New Americanists have enlarged the scope of research, which comprises not only literature and the subsequent opening of the canon, but any field of study, any topic related to America, from Imperialism to exceptionalism, from racial to gender studies, from migration to food, just to name a few. Shelley Fishkin addresses the American Studies Association in 2004, referring to the possibilities of American Studies, focusing on the diverse possibilities that derive from the field: "[t]here probably are as many definitions of American studies in this room as there are scholars; indeed, one of the reasons many of us were attracted to American studies in the first place was its capaciousness, its eschewal of methodological or ideological dogma, and its openness to fresh syntheses and connection" (19). This eschewal might be read as the combat and rigorous vigilance against ideas that perpetuate the earlier myth of the innocent and exceptional America, which also camouflaged its imperial endeavors. Ideology and politics played an important part in the revision, demonstrating how such tools of analysis were fundamental for the starting of an understanding of American studies and its multifaceted objects of inquiry as a complex set of identities and desires, who happened to share the fictional common ground of such *Americanness*. The stories of "the other America" began to occupy the mainstream of what Americanness might mean. Isabel Caldeira reflects in "Who Has the Right to Claim America?", while dealing with the notions of Eric Hobsbawn, that the memory of a nation is a key element in the exercise of rituals and traditions that foster its sense of continuity, and therefore of belonging. This sense however is lost to those who have been systematically excluded from the nation, experiencing instead a sense of disruption. She states:

> For peoples or groups who have been subjected to radical displacement, deep cultural deprivation, racial discrimination and other kinds of violent abuse to both their personal and national identities, there is a sense of disruption, not of continuity. These groups need to reestablish emotional ties with the past and to retell or rewrite a consistent narrative of history (Lyotard), carving out a sense of communion with "an immemorial past and [...] a limitless future [...]". ("Who Has the Right to Claim America?" 178)

The the Civil Rights movements, the anti-Vietnam War movement, and the second and third wave Feminist movements, are forces that have created a more complex realization of what it means to be American during the post-cold war era and further into the 21st century. The field has absorbed a quantity of diverse voices that have questioned the homogenized ideal of white/straight American-ism, creating thus a constellation of different perspectives, multiplying the ways of seeing and being America(n). African-Americans, Chicanos, Asian-Americans, Haitian-Americans, Latinos and so on, all now contribute to a more complex set of issues around national identity, as well as complicate notions of transnation-ality. This ideological revision goes hand in hand with the evolvement of Amer-ican Studies, which no longer sees the field as a contained unitary and cohesive body. In this respect, it is interesting to bring to the fore the perspectives of Aní-bal Quijano and Immanuel Wallestein, and their conceptualization of "Ameri-canity", in comparison to "Americanness", as it is an example of how American studies as a discipline shifted its perspectives. In their article "Americanity as a Concept, or the Americas in the Modern World System", published in 1992, the authors refer to Americanness as the state of being American, which is frequently understood solely as being a citizen of the United States; Americanity however, refers to the Americas as a set of territories that are distinct from the Old World, a "new" place used by colonizers during the period known as modernity as a test zone of several different modes of labor organization and forms of control never before exerted in such proportions. The concept of Americanity is related to the way in which the continent served as a stage for the implementation of a capital-ist system of exploitation and domination of different cultures, with brutal con-sequences to the indigenous peoples, and to slave work forces that were import-ed to boost the new economic order, in the United States and beyond.

Despite such diversity of topics, there is some sort of hegemony that runs through the current research tendencies established by the New Americanists. Ramalho states that:

> [...] [T]he multicultural, transnational, post-national, anti-imperialist, post-colonial, anti-discriminatory, comparative, and definitely secular agenda of the new Americanists quickly became hegemonic in the field. The rigorous questioning of the tradition and other givens proposed by Derrida's deconstruction, as well as the interdisciplinary, critical research de-rived from the Frankfurt School have been major influences. ("Who Owns American Stud-ies?" 8)

The present work converges at this tendency, since its objective is to study Afri-can diasporic expression of black women authors from the United States, Haiti, and Antigua, discourses that denounce and unveil violence in its various forms, and depict and criticize the coloniality of the structures that produce such violence. Haiti, Antigua, and the United States are seen here as axes from which cultures constantly reshape themselves, in exchanges influenced by language, economy, race, sex, sexuality, and so on. The transnational aspect is relevant as well, since the authors from the Caribbean to be studied here have constructed their stories (and their subjectivities) in these transnational contexts, as well as published their stories firstly in the United States, and in English, stories that are the result of the migration flows between the islands and the United States. In regard to the African-American authors, they write narratives about displacement, subalternity and disenfranchising in their own country. The present work, distancing itself from a perspective based solely on national constructs of identification, is concerned with understanding these transnational realities born of cultural exchanges and influences.

The geopolitics of identity become a rich framework for the revision of American Studies as a field. As Janice Radway states in her presidential address to the American Studies Association in 1998, the interaction between cultural spaces/geographies is the best way to the investigate human culture. In her speech, Radway questions the validity of the name of the association – the American Studies Association – and its sense of hegemony when facing contemporary challenges, speculating about different suitable options that could better represent the current work being developed. Radway questions if renaming the field as "The Society for Intercultural Studies" would offer a new perspective, fostering the study of the non-national and the transnational forms of identity construction. She states:

> A society that was not hemmed in by the need to peg cultural analysis of community and identity-formation to geography might better be able to attend to the full variety of cultural negotiations, negotiations that do not recognize national borders but flow across them to solicit the identifications of attentive and like-minded individuals. (22)

National identity based on geographical/ideological borders sufficed for a long time as a framework in which the field developed, mainly because of the Cold War strategy. The United States needed to see itself, more importantly, to project itself, as a cohesive unit with defined borders, and hence with a defined identity that sustained the ideological machinery of the war and could strongly face the Soviet ideology. The idea of American exceptionalism was fundamentally connected to such strategy. On the other side, the Soviet Union was constructed and perceived as the alterity force with which the Unites States was rightfully meant to compete with. Exceptionalism meant the right to bend the rules of international diplomacy, to live and perform outside of the laws that had been commonly agreed upon so far, and to justify such acts in the name of the enforcement of democracy, liberty, and all other American ideals, as Donald

Pease comments in his article "Re-thinking American Studies After US Exceptionalism":

> Throughout the Cold War, US dominance was sustained through the US's representation of itself as an exception to the rules through which it regulated the rest of the global order. But with the dismantling of the Soviet Union and the formation of the European Union, the US lost its threatening, socialist, totalitarian Russian Other as well as its destabilized, dependent European Other. (19)

After the weakening of the dichotomic understanding of the national identity of the United States, a transnational perspective seemed to be more adequate to perceive the position occupied by the country. The influence of the United States was now displaced as a network of multiple flows of interdependent relations. The uniqueness of the country's national identity would be preserved through the hegemonic presence of its exceptionalism, perceived then through a relational aspect. Pease comments:

> [...] With the disappearance of relations that were grounded in [...] macropolitical dichotomies, multiple, interconnected, and heterogeneous developments emerged that were irreducible to such stabilized oppositions. The demands of a newly globalized world order solicited an understanding of the US's embeddedness within transnational and transcultural forces rather than reaffirmations of its unique isolation from them. (20)

From then on, it is impossible to address the United States and it is even impossible for the United States itself to affirm its identity outside a circuit of interrelatedness with other nation-states, as well as to avoid its own intrinsic diversity of cultures. The 2004 presidential address delivered by Fishkin echoes these preoccupations, as she points out the numerous ways in which a transnational turn in American Studies would render the field richer in perspectives previously marginalized or made invisible. She states:

> As the transnational becomes more central to American studies, we'll pay increasing attention to the historical roots of multidirectional flows of people, ideas, and goods and the social, political, linguistic, cultural, and economic crossroads generated in the process. These crossroads might just as easily be outside the geographical and political boundaries of the United States as inside them. We will increasingly interrogate the "naturalness" of some of the borders, boundaries, and binaries that we may not have questioned very much in the past, and will probe the ways in which they may have been contingent and constructed. ("Crossroads of Cultures" 22)

The Unites States perception of itself also shifted from unitary to multifaceted, since there was a slow recognition of the complex set of different identities within the nation-state. The presence of the United States in the global world stage came to be understood as the result of the various relations of the United States with other nation-states, which made for a broader and more inclusive comprehension of the field[1]. Still, it must be asked: who are the ones producing these new diverse understandings? Julia Roth comments on this transnational turn in her article "Decolonizing American Studies – Toward a Politics of Intersectional Entanglements", emphasizing that such turn must encompass the perspectives of different agents in the conversation of what American studies are, namely when academia in its privileged position as the sole producer of valid hegemonic knowledge keeps undermining and erasing the possibilities and effective production of other voices. Roth states:

> However, the terms of the conversation are not changed by telling multiple stories, if these stories are told by the same storytellers (and regardless of their connectedness). The decolonization of received modes of doing knowledge production – and of American Studies respectively – rather requires listening to new and heretofore marginalized or silenced storytellers as well, and hence reflecting upon and scrutinizing the dominant positions of the power to define and represent, and to alter the theoretical frameworks, parameters, and the respective units of analysis. (137–138)

The transnational turn must accommodate not only the production of knowledge based in this interconnectedness, but also offer a praxis of investigation that really does pay attention to the other subjects involved in the transnational and transcultural conversation. The listening to stories that dismantle the hegemonic knowledge production can be found in many places, from grassroots political movements, to art, or literature. It certainly can be found in the voices present in stories that are engaged with telling the other side of Americanness.

1.1.2 Language, Literature, and American Studies

Since language is agency, the act of sharing stories that refigure/complicate the understanding of what America is, and consequently what American studies are as a field, reshapes the realities of this transnational endeavor. As Toni Morrison states in her Nobel Lecture (1993), through the creation of the parable of the

---

1 Robert Warrior critiques the transnational turn from a Native American studies perspective in "Native American Scholarship and the Transnational Turn", published in 2009, referring to a group of scholars to whom the concepts of the "national" and "nation" are fundamental in their emancipatory agendas, which goes in line with their critique of postcolonialism. These factors lead to a rejection of the terminology introduced by the centralization of the transnational in the field.

blind story-teller, language work (meaning the sharing of stories through the manipulation of language) is very much real, and shapes reality, at the same time it shapes itself, during the act of "doing language". Morrison describes her protagonist as someone who: "[b]eing a writer she thinks of language partly as a system, partly as a living thing over which one has control, but mostly as agency – as an act with consequences" ("Nobel Lecture"). These consequences may be seen as the dismantling/reinforcing of inequalities, which are fostered by the use of language. Isabel Caldeira comments on this notion in "Mourning for Citizenship in Morrison's Fiction", stating: "Morrison is able to respond to the issues she feels are important and significant in the world – in a critical venture that deeply implies, in her personal articulation, 'response-ability[2]' [...]" ("Mourning for Citizenship"). Morrison is aware that language is incapable of doing it all by itself, and like the character in her parable, she makes clear that the struggle to clearly depict the tensions, injustices and inequalities experienced by humanity is never fully successful, yet, the success of language lies in its exercise, in the attempt to tell. The author claims: "[i]t is the deference that moves her, that recognition that language can never live up to life once and for all. Nor should it. Language can never 'pin down' slavery, genocide, war. Nor should it yearn for the arrogance to be able to do so. Its force, its felicity is in its reach toward the ineffable" ("Nobel Lecture"). Though imperfect, language is the arena in which there is a space for the reconfiguration of the reality, and in which discourses and representations might be redressed, rejected, or remade. In "'What Moves at the Margins' as Vozes Insurretas de Toni Morrison, bell hooks e Ntozake Shange" Caldeira locates these concerns in the literature written by African-American women authors, stating:

> This is the concern that I find, for example, in the literature produced by African-American women, voices that rise up against racism, sexism, the cultural hegemony of the empire and the patriarchy that placed their lives and memories in spaces of subordination, which devalue and diminish them. From them we also receive the signs of resistance[3]. ("What moves" 146)

---

2 Regarding this terminology, used by Morrison in *Playing in the Dark*, published in 1992, Caldeira considers the plural implications of the word, emphasizing: "The deconstruction of the word responsibility underlines the ability to answer (respond to) the call for action. I relate this posture to Gayatri Spivak's in her essay on "Responsibility," (1994), where the scholar stresses the ethical responsibility, accountability, answerability that the intellectual bears to the Other (the subaltern)" ("Mourning for Citizenship").

3 Translation provided by the author.

The texts that are going to be analyzed in the present work tell stories that, fictionally or autobiographically, and in the interstices of these genres, deal with representations of violence and their subsequent traumas, and their author's response-ability. It is a given that such a task, the telling of trauma and violence, is damned from its onset, since pinning down these experiences through language is impossible, since they elusively escape the words that precisely try to represent them. The telling of violence and traumatic events precisely fits the paradigm of the limits of language; yet, the exercise of trying to communicate such experiences is valid simply for its effort, as it renders evident the social structures that underly the life experiences of the subjects who have gone through unspeakable things. António de Sousa Ribeiro in *Representações da Violência* (2013), while discussing the impossibility of the translation of the experience in the Nazi concentration camps into language, states that such impossibility is just an apparent paradox, since it is the very impossibility of translation that allows for the possibility of representation, and that the literary discourse is the realm where such (im)possibility exists (26).

> [...] the discourse of testimony can never aspire to a directly referential function; in fact, it is only in literary discourse that it is possible to realize the representation of this unlivable experience, the possibility of testimony resides in the literary dimension, that is, the transposition to another level of meaning allows to do justice to the density of the violent truth of the concentration and extermination camps [4]. (26)

The main objective of this work is focused on obtaining a broader understanding of how trauma and violence are present and how they may be represented in a set of coming-of-age stories from female authors of the African diaspora in the Americas, making evident how trauma and violence are related to the historical and cultural context of Americanity. Fiction and autobiography will be used as the modes of expression for this analysis, making possible a better understanding of the intersections of the two genres and the consequential implications to trauma representation. The present work will deal with four different authors and their coming-of-age narratives: Jamaica Kincaid's *Lucy* (1990), Edwidge Danticat's *Breath, Eyes, Memory* (1994), bell hook's *Bone Black – Memories of Girlhood* (1996) and Toni Morrison's *God Help the Child* (2015). The context of diasporic subjects of African origin in the United States, Haiti, and Antigua – but all of them writing in the United States – will expose the ways in which different types of violence and trauma exist and are perpetuated in the lives of these female protagonists, whose experiences may also be representative of collective re-

---

4 Translation provided by the author.

alities. The types of representation of the experiences will provide a better perception of how violence is constructed and dealt with, as well as making evident the ways in which resistance is performed by these subjects in their attempts to narrate trauma. This will be made throught the model of the *Bildungsroman,* as it allows for the revision of the development of the subject, revealing many anthropological realites, as stated by Rogério Puga in *O Bildungsroman: (Romance de Formação) Perspectivas:*

> [...] the representation of the inner and physical growth of the Self through retrospective and (auto)biographical narratives, allows (re)thinking the human condition and consequently anthropological issues, such as for example, multiculturalism, interculturality, human rights, social interaction, the male/female condition, the symbolic space of development (national, regional and colonial cultural, social and historical landscape), the singular determination of the protagonist, the surrounding environment, social pressures, the demand for a name, for an identity and a place in the world, gender and *Bildung,* in addition to individual and collective memory, among other themes and social issues[5]. (8)

Departing from the analysis of the selected works, my aim is to reach an intersectional understanding of the structures that perpetuate racism and sexism, comparing their historical and cultural backgrounds through narratives of development, to finally reveal and discuss the possible role of literature as a combative discourse in order to deconstruct such social practices and decolonize epistemic violence.

1.1.3 Violence and Trauma

> The Black female is assaulted in her tender years by all those common forces of nature at the same time that she is caught in the tripartite crossfire of masculine prejudice, white illogical hate and Black lack of power. The fact that the adult American Negro female emerges a formidable character is often met with amazement, distaste and even belligerence. (Angelou 209)

The concept of violence is known in sociological studies to be problematic, since it escapes objectivity in its definition. The different contexts in which the word might be used show the fluidity and evasiveness of the concept, making the attempt of defining its limits a theoretical endeavor on its own. Willem Schinkel, in *Aspects of Violence*, approaches this debate, dealing with the different definitions of violence and their applicability in social theorizing. The author describes

---

5 Translation provided by the author.

the potentiality of the concept to be not only something that objectively escapes the possibility of interpretation and the consequential categorization, but also the possibility of its misrecognition, since it is usually surrounded by different elements that make it less recognizable. Through etymological and semantic analysis of the German and Latin roots of the word, the author concludes that the use of physical force over something or someone characterizes violence in its most primordial sense. When facing an act of violence, namely when physical harm of the body is involved, it will rarely not be perceived as violent by the observer. However, it is possible to claim that the recognition of violence as such depends on the person who is making the interpretation, and not on the act/situation/representation itself. Violence may also be inflicted upon ourselves and not necessarily upon others, as well as animals and nature, making the range of options for a possible object of violence to be enlarged much more than simple physical violence. Psychological violence is also violence, which removes the physicality of the action but still inflicts force upon a victim's well-being and health. In the case of social injustice, the violent character of a state apparatus might either be or not be understood as such, since the implications of it over an individual subject are no longer clearly asserted, and they might be relegated to different sets of cause/effect that might not be inherently violent.

Schinkel points to three possible options open to the study of violence departing from a sociological approach: the first one consists of the empirical creation of a concept of violence, which would have the consequences previously mentioned. The second one is based on the delimitation of the possible readings of the concept, stretching its applicability, but diminishing its complexity in regard to the forms of violence that do not take part in the common sense. The third option would be not defining the concept, allowing the readers to be responsible for the comprehension of violence through their own paradigms. These strategies comprise some of the many developments of the field, reflecting the complexity of violence studies. What follows is a brief description of different ways of theorizing violence in plural understandings of what might constitute the concept. Johan Galtung, attempting to find a definition that is both broad and specific, defines violence as "the cause of the difference between the potential and the actual, between what could have been and what is" ("Violence, Peace, and Peace Research" 168). Or as the following: "I see violence as avoidable insults to basic human needs, and more generally to life, lowering the real level of needs satisfaction below what is potentially possible" ("Cultural Violence" 292). According to Shinkel, for Galtung, violence is an influence, something that limits the capacity of a human being to act and even his/her capacity to be. This definition enlarges the scope of a conception of violence which is based on the material/physical level, making it possible for different types of violence that were made invisible before to come to light. The author also names these social differences as "social injustice", an expression that better makes evident the idea that the social is responsible for the generation of differences of opportunities given to certain members of society.

Broader conceptualizations of violence, such as symbolic, cultural, institutional, and structural violence extrapolate the limits of physically direct violence. The impact of institutions rather than that of individuals characterize institutional violence. The state is an example of an agent of violence that acts though coercion by police forces and the army. Cultural violence, on the other hand, is diffused among all agents of society: racism, sexism, homophobia, and class oppression are some examples of cultural violence that might easily translate into physical violence, and institutional violence as well. In gender relations, symbolic violence presents itself in the legitimation of the masculine domination through the constant and ever present exercise of reproduction of symbolic patterns by men, women, and institutions, in which the androcentric logic is imperative, as stated in the works of Pierre Bourdieu[6]. António Sousa Ribeiro refers to the concept of cultural violence coined by Galtung, stating that it is related to Pierre Bourdieu's symbolic violence, which indicates a type of violence that is based on language or in symbolic systems, aiming at the naturalization and consequently at the legitimation that discretionary power exercises. (*Representações* 10). Schinkel also approaches symbolic violence, claiming that the definition of violence through language is already a form of violence, the violence that is inherent to language as it excludes meanings everytime a specific meaning is inferred. The choice of a specific terminology to define the concept of violence will directly reflect a set of beliefs and semantic power relations linked to historical definitions of violence and their applicability. So, the attempt to define violence through language is in itself an act of violence, considering that the power to define violence is responsible for the attributions and possible consequences of this categorization. Schinkel states that:

---

6 Bourdieu describes this kind of violence through the interactions between female and male subjects, dominated/dominant, stating that: "Symbolic violence is instituted through the adherence that the dominated cannot fail to grant to the dominant (and therefore to the domination) when, to shape her thought of him, and herself, or, rather, her thought of her relation with him, she has only cognitive instruments that she shares with him and which, being no more than the embodied form of the relation of domination, cause that relation to appear as natural; or, in other words, when the schemes she applies in order to perceive and appreciate herself, or to perceive and appreciate the dominant (high/low, male/female, white/black, etc.), are the product of the embodiment of the – thereby naturalized classifications of which her social being is the product" (35).

> The very existence of a certain concept of violence seduces us into thinking that there is no violence outside the denotation and connotation of that concept. The aptitude to misrecognition of violence can be seen as an intrinsic feature, an aspect of violence, so to speak as an illusion naturelle (Malebranche) or as a 'well-founded illusion'. (Bourdieu 33)

The plurality that is inherent to the concept of violence demonstrates that it ranges from the physical to the psychological, from the personal to the structural, violence which is manifest and which is concealed. Another relevant discussion regarding the definition of violence is the dual interpretation of one of its fundamental aspects: whether violence is a social construct, or an anthropological reality. Ribeiro states:

> [...] if violence is an anthropological constant, then it is ineradicable, it is a kind of human destiny and can, at best, be controlled and disciplined. If, on the contrary, violence is generated in contexts of interaction and within the scope of unequal social relations, if it depends on socialization practices specific to specific social contexts, then non-violence shows itself as a perspective that is not simply utopian [7]. (*Representações* 11)

Such discussion is important because it would eventually settle if it is possible to eradicate violence as a cultural practice, or if it will inherently be present as part of our shared humanity. Either way, efforts to mitigate violence as a reality that affects real subjects must be made. Ribeiro states that the need of a precise conceptualization of violence must be put at a second place, favoring approaches that, while dealing with broad definitions of violence, are contextually capable of unveiling violence precisely when its occultation constitutes a type of violence in itself.

Galtung argues that peace is the absence of violence, and "if peace action is to be regarded as highly because it is action against violence, then the concept of violence must be broad enough to include the most significant varieties yet specific enough to serve as the basis for concrete action" ("Violence, Peace, and Peace Research" 168). The author points out to the dimensions in which violence can be thought, indicating that dedication to this endeavor is more profiting to concrete action than trying to encapsulate the concept into a definition. According to Galtung, there are six different dimensions of violence, which help to better identify types of violence that might be overlooked. Thus, understanding violence in its different dimensions is profitable inasmuch it unveil types of violence that might be overlooked while trying to delineate a conceptualization that ultimately aims at the eradication of violence.

---

7 Translation provided by the author.

The first type is related to the differentiation between physical and psychological violence, the first being the obvious perceived violence that subjects the body, with its culmination at the death of the object of the violence; psychological violence on the other hand is the one which is inflicted on the mental state of the object of the violence, being exemplified with "[...] lies, brainwashing, indoctrination of various kinds, threats, etc" ("Violence, Peace, and Peace Research" 169). The second difference is related to positive and negative influence, this differentiation being of great interest to the present work, since it unveils how a capitalist order functions in the generation of violence and subsequent trauma. Negative influence is related to the usual understandings of violence as punishment. Regarding positive influence, Galtung's example is the consumer society, which positively influences people to participate in the market by rewarding them with instant satisfaction. These transactions work for an economy of manipulation, where people have fewer and fewer options to live outside the parameters sublty imposed, while the ones in power profit from them. The third distinction is whether there is an object of violence, opening questions about threatening as violence. If threatening is something that can decrease the potential of the subject to the actual realization of herself/himself, then it can be considered a kind of violence. The absence of an object of violence might imply that the action of the perpetrator is possible and likely close to happening, inhibiting possible objects of violence from action. This distinction works closely to the one related to psychological violence. The fourth dimension to be differentiated is related to the subject of violence and its existence. Acts of violence that have a subject are considered personal or direct violence, according to the author; on the other hand, when there are not any actors as subjects of violence, violence is then characterized as structural or indirect. "There may not be any person who directly harms another person in the structure. The violence is built up in the structure and shows up in unequal power and consequently unequal life chances" ("Violence, Peace, and Peace Research" 171). This kind of difference is seen in the colonial power structures that continue to define the range of possibilities of colonized subjects. The fifth difference regards the intentionality of violence. This difference is of great interest here as well, since unintentional violence and structural violence tend to intersect. The sixth and final difference pointed by the author is related to the levels of violence, being it manifest or latent. The first one is visible, and can hardly be denied, latent violence on the other hand is "[...] something which is not there, yet might easily come about" ("Violence, Peace, and Peace Research" 172). Racial tensions are a clear example of such dynamics, where one single act might trigger a torrent of violence that was building up silently. "It indicates a situation of instable equilibrium, where the level of realization is not sufficiently protected against deterioration by upholding mechanisms" ("Violence, Peace, and Peace Research" 172).

The concept of structural violence will be especially relevant to the analysis of the present work, since it makes evident the oppressive social structures, and thus, helps to dismantle them. Structural violence, having a less poignant character, is more easily not recognized as violence, the object of structural violence might not be aware of the existence of it, in opposition to the one involved in direct violence, who is generally conscious of the violent act. Galtung characterizes structural violence as silent and static, the kind that is naturalized both by the perpetrators and by the objects of violence. The structure is also responsible for the perpetration of violent acts against the individual and the society, becoming evident in the power asymmetries and consequently in the access to life opportunities. We might consider, for instance, the state as a perpetrator of structural violence – through the promotion and endorsement of politics that diminish the capacities of citizens, such as the biased distribution or resources and benefits, lower quality public education, or discriminatory legislation. It is important to think about how direct violence turns into structural violence, when the political and cultural systems become the oppressors. George Lipsitz, in *The Possessive Investment in Whiteness*, published in 1998, indicates how some measures (or their absence) are capable of inscribing racism, and consequently violence, within the social structure:

> Failing to enforce civil rights laws banning discrimination in housing, education, and hiring, along with efforts to undermine affirmative action and other remedies designed to advance the cause of social justice, render racism structural and institutional, rather than private and personal. (46)

The concept of trauma is also relevant to this discussion, since it frequently derives from the experience of violence, direct and indirect, and has real consequences in the lives of real people. According to Caty Caruth, the experience of trauma and its subsequent effect is unintelligible and incurable. What is possible is to find ways of coping with such an experience, relearning how to live around its presence. Trauma is not defined by the event, not even by its distortion of the subject's life; it is defined though by the involuntary recollection of that given moment that has not been (fully) processed. Caruth analyses the post-traumatic stress disorder, defining such as:

> [...] a response, sometimes delayed, to an overwhelming event or events, which takes the form of repeated, intrusive hallucinations, dreams, thoughts or behaviors stemming from the event, along with numbing that may have begun during or after the experience, and possibly also increased arousal to (and avoidance of) stimuli recalling the event. (193)

Trauma is then understood as a literal memory, in which the symbolic process did not take place, which hinders the subject from operating as usual, since the intrusion of this memory takes hold of her or his existence. In an attempt to deal with such a multifaceted subject Caruth points out the transdisciplinary aspect of trauma studies, claiming that psychoanalysis, psychiatry, sociology, and literature are called for the understanding of such phenomena. Márcio Seligmann-Silva claims in his article "Narrar o trauma – A questão dos testemunhos de catástrofes históricas", that: "[in] the witnessing situation past tense is present tense. [...] More specifically, trauma is characterized by being a memory of a past that does not pass[8]" (69). Narrating trauma is one possibility to overcome the isolation caused by the experience, which is particular and exclusive. Seligmann-Silva cites the works of Hélène Piralian about the representation of traumatic events, saying that the construction of a representation about the experience helps the subject to regain control over the unprocessed memory. "The linearity of the narrative, its repetitions, the construction of metaphors, everything works towards giving this new dimension to previously buried facts. Conquering this new dimension is like leaving the survivor's position to coming back to life[9]" (69). Through the act of telling, the subject starts to work the shock of trauma, giving it shape so that he or she might overcome this moment of temporal paralysis caused by the traumatic experience. The imaginative work has transformative possibilities over trauma, revitalizing the event, transposing its literal character and eventually surpassing it.

Literature then comes to the fore, serving as a context/process of enunciation that is fertile for the narration of trauma and the denunciation of violence. Literature, usually connected to its engagement with imagination and the fictional world, as well as with aesthetics, offers the subject a space in which the fragmentary character of trauma might be recreated, and consequently, (re)appropriated. However, critiques regarding the capabilities of literature a as a space for healing can also be found. Sabine Broeck comments in "Trauma, Agency, Kitsch and The Excess of The Real: *Beloved* Within The Field of Critical Response" that the reception of Morrison's novel has been mainly framed in more or less sophisticated versions of the PTSD perspective. She states that *Beloved* readings have "[...] oftentimes congealed into a rhethoric of trauma that verges on, and sometimes crosses over into kitsch" (202). By using the concept of "kitsch" the author points to a critical response that is seen somehow as demonstrative of an excessive sentimentality or garishness, in which critics have

---

8 Translation provided by the author.
9 Translation provided by the author.

read the novel either in terms of a redemption tale or as an "unframable"text characterized by "its very excess of the traumatic real". (203) Broeck criticizes the too smooth readings of trauma testimony in literature, questioning the reception of the text as an "[...] almost mythical, certainly naïve faith in the social and individual 'cure' the novel can preside over" (208), calling it a kitsch reception. The potential of the project of working through trauma successfully through literature is seen, when she states:

> As a poweful symbolic gesture *Beloved* extends an option to understand the recourse to trauma as sine-qua-non to negotiate subjectivity as sociality, to rescue trauma – as it were – from its pariah status o the shameful and secretively whithheld or suppressed" [...] *Beloved* enabled readerly containment of trauma as a sublime constituitive experience which may, finally, be absorbed rather that aggressively rejected. (209)

The realization that our reality is permeated by the overwhelming presence of violence and trauma is reflective of the kinds of power relations that have developed throughout history, and literature may be seen as space for the negotiation when violence and trauma meet culture. Violence and trauma are themes very much present in coming-of-age narratives, extending themselves from infancy to adulthood. According to Claudine Raynaud in "Coming of Age in the African American Novel", facing overt and direct racism (and the subsequent process of (non)symbolization of the traumatic experience) is a crucial moment in the development of the protagonist in their "education". The works about to be analized in this study are representative of the presence of violence and trauma during the identity development processes represented in coming-of-age stories, but first and foremost, one must investigate some of the causes of the presence of violence and trauma in the historical context of the selected works.

1.1.4 Globalization, Capitalism, Migration, and Social Inequality

Globalization plays an important part in the lives of contemporary subjects, and thus the telling of coming-of-age stories that are set during the second half of the 20th century depicts the impact of this circumstance during the growing-up process. The perspective of Zygmunt Bauman in *Liquid Times – Living in an Age of Uncertainty* (2007) will complicate readings regarding the individualism that dominates contemporary societies, in addition to global capitalism and its consequences. The implementation of a global economy with its transnational model of exploitation has made more evident the ideological danamics of the market, where a few profit in detriment of the many. The contemporary self is profoundly individualistic, embedded as it is in a social, economic, and political turn brought about by globalization. Bauman states that "[t]he new individualism, the fading of human bonds and the wilting of solidarity are engraved on one side of a coin whose other side shows the misty contours of 'negative globalization'" (*Liquid Times* 24). Regarding the topic, the author comments in *Liquid Love – On The Frailty of Human Bonds:*

> The fading of sociality skills is boosted and accelerated by the tendency, inspired by the dominant consumerist life mode, to treat other humans as objects of consumption and to judge them after the pattern of consumer objects by the volume of pleasure they are likely to offer, and in 'value for money' terms. [...] In the process, the intrinsic value of others, as unique human beings (and so also the concern with others for their own, and that uniqueness's sake), has been all but lost from sight. Human solidarity is the first casualty of the triumphs of the consumer market. (*Liquid Love* 75–76)

Bauman points to the capacity of the market to interfere in the dismantling not only of nation-states, but also of human relations, stating that "[i]n its present, purely negative form, globalization is a parasitic and predatory process, feeding on the potency sucked out of the bodies of nation-states and their subjects" (*Liquid Times* 24). Toni Morrison, similarly, in an interview with Sheila Foran, analyzes the changes in the relation of identity and capitalism in America, pointing to the dismantling of citizenship in detriment of the global market, stating:

> I remember being a girl and being called a citizen and this was important, [...] Sure I was a second-class citizen, but I was still a citizen. After World War II, [...] we were called American consumers not American citizens, and we are now called [American] taxpayers. This means our relationship with our country now is not the same as it used to be when being a good citizen was something important. ("Toni Morrison Reflects" 2011)

Global capitalism is a determinant factor in the life experience of the diasporic contemporary subjects who live structural and direct violence, who are forced to migrate, trying to escape poverty. They are confronted with new difficulties, seeing their civil rights cancelled in order to attend capitalistic demands of development, that regard them and their needs as subaltern. When inserted in new contexts, such subjects face power structures that force them down to the bottom of the social hierarchy, for the maintenance of the white/male/heterosexual/classist supremacy and its subsequent social hegemony. Audre Lorde specifies the diverse subalternities in a profit-oriented society in "Age, Race, Class, and Sex: Women Redefining Difference":

> Much of Western European history conditions us to see human differences in simplistic opposition to each other: dominant/subordinate good/bad, up/down, superior/inferior. In a society where good is defined in terms of profit rather than in terms of human need,

there must always be some group of people who, through systematized oppression, can be made to feel surplus, to occupy the place of the dehumanized inferior. Within this society, that group is made up of Black and Third World people, working-class people, older people, and women. (114)

This logic of systemic oppression can be applied to disenfranchised racialized and genderized citizens in the United States, who face discrimination in overt and subtle ways, remaining in systemic poverty, and made invisible inside their own country. African-American citizens have experienced such invisibilization for very long, and still suffer from it. Since the abolition of slavery until the Civil Rights movement gains, most of these citizens were condemned to poverty given the fact they could not access education and participate in the economy at equivalent positions with white citizens. It is also relevant to stress the position of immigrants in the United States, who continuously experience discrimination and are victims of structural forms of poverty and violence in their search for better lives in the "developed world". In an interview to PBS, Edwidge Danticat comments on the crisis experienced in Haiti during and after the earthquake of 2010, comparing it to the New Orleans crisis in the aftermath of the Katrina hurricane in 2005. More than two hundred thousand Haitians died in that catastrophe, and one and a half million are homeless since then. The proportions of the calamity were so great to an already historically impoverished country that recovering from such event is an ongoing struggle that is far from finished. Haiti is the poorest country in the Americas, with a GDP of US$ 846 per capita, as of 2014. The United States figures as the opposite in the spectrum, as the richest country in the Americas, with a GDP of US$53,042.00 per capita. Katrina occurred in a country where there are indisputably more resources and an overall structure to deal with such crisis. Yet, the people of New Orleans, which is predominantly African-American, received very little help from the government, who did not intervene to rescue its citizens as promptly as needed. More than one thousand deaths were registered during the event and in subsequent days, though numbers are not completely accurate. Danticat recalls that citizens of the United States, were made invisible in such time of crisis, referring that there was some perplexity regarding whether the victims affected by Katrina should be called refugees in their own country, stating:

[...]perhaps, these worlds are closer than we think and *maybe the poor inhabit their own country. And it seemed to me that what Katrina, that whole experience showed, the poorest amongst us are as invisible in the richest country in the world as they are in the poorest countries in the world"* [emphasis mine]. ("NEED TO KNOW" 2011)

The poorest are rendered invisible by a structure of power that does not recognize their needs and sees them as less than human, mainly because of their poverty and race, turning citizens into aliens. Their class status intersecting with their racial status is responsible for a displacement inside their own country. Citizens are systematically kept in poverty by the intersection of different factors, and finally are alienated and made invisible at the time they most need recognition[10] and support. The lack of solidarity is a trend, and it is ever present, from the richest to the poorest across the whole continent.

In *The Meaning of Freedom and Other Difficult Dialogues* (2012) Angela Davis claims that racism, as an institutional practice materialized in the racial segregation, was eliminated *de jure*. However, *de facto*, discriminatory policies are still in practice. Contemporary racism is much more elusive than the one which was practiced in the second half of the 20th century. She says that:

> [...] we tend to think racism was overt. [...] And now we tend to think that it's hidden. I wonder why. Maybe it's because we have again learned not to notice it, because we have been persuaded that the only way to eliminate it is by pretending that it doesn't exist, that the only way to eliminate racism is to pretend that race doesn't exist. [...] In the segregated South, the signs of racism, the literal signs, made us pay notice to it. But now that the signs are gone, discriminatory practices continue under the sign of equality. (*The Meaning of Freedom* 129 – 130)

Davis is claiming for the undoing of the invisibilization of social and institutional practices that continue to racially discriminate women and men. The Civil Rights Movement of the 1950s and 60s fought for and contributed immensely to the approval of laws that intended to end discrimination and segregation. Ula Taylor summarizes the ideas concerning the 1964 Civil Rights Act in her article "The Historical Evolution of Black Feminist Theory and Praxis":

---

10 Recognition is the first step towards action. In *Sister Citizen*, Melissa Harris-Perry approaches the Katrina disaster in a compelling argument about the different perception of white and black Americans regarding the aftermath of the disaster. Through the analysis of data collected by the Pew Research Institute, Harris-Perry claims that: "[w]hile 71 percent of blacks believed that the Katrina disaster showed that racial inequality remains a major problem in the country, a majority of whites (56 percent) felt that racial inequality was not a particularly important lesson from the disaster. Seventy-seven percent of African Americans believed that the federal government response was fair or poor, and although the majority of whites agreed, the percent was much smaller (55 percent). In a stunning racial disjuncture, 66 percent of African American respondents believed that if most of the victims were white, the response would have been faster. Seventy-seven percent of whites believed that race made no difference" (138).

The Civil Rights Act of 1964 was initially perceived as a major piece of protective legislation. Unlike the abolitionist movement and the first wave of feminism, political demonstrations against the federal and state governments in the 1960s were televised internationally. During the Cold War, government officials had a hard time positioning the United States as the home of democracy in light of the struggles initiated by African Americans for political and civil rights. Within this international context, legislative concessions had to be made. The Civil Rights Act of 1964 stipulated that the attorney general had to protect citizens against discrimination in voting, education, and public accommodations. (242)

These efforts, consolidated in the law, promised to bring the end of institutional practices that overtly perpetuated hierarchies in the United States. However, the discrimination practices are still present in the contemporary society, taking, however, different shapes. Racial equality seems like some kind of utopia, still far out of reach, at the same time that racial issues are made diffuse in perspectives that claim to be equitable. Color-blindness, or post-race are concepts that illustrate how the racial question is still a difficult one to be addressed, since they deny the claims of the victims of racism, by glossing over reality with discourses that state that the complaints of these subjects have already been overcome. It is important to realize that such discriminatory practices are still present and deeply systemic, that it is of great importance to identify them in their sophisticated strategies of opacity and occultation, so that it may be finally possible to combat them.

Literature was and still is (may be even more today) an important media in which the concerns with unveiling social inequalities are represented and discussed, in paths that move from silence to language, and hopefully into action.

1.1.5 A Minor Literature

Toni Morrison comments in "Unspeakeable Things Unspoken" that the process of classification of the production of those who do not belong to the canon often goes through a predictable patronizing process. It seems that the *status quo* perceives the possibility of opening up to these "minor" cultures as a threat. She states:

When the topic of third-world culture is raised, unlike the topic of Scandinavian culture, for example, a possible threat to and implicit criticism of the reigning equilibrium is seen to be raised as well. From the seventeenth century to the twentieth, the arguments resisting that incursion have marched in predictable sequence: (1) there is no Afro-American (or third-world) art; (2) it exists but is inferior; (3) it exists and is superior when it measures up to the "universal" criteria of Western art; (4) it is not so much "art" as ore – rich ore – that

requires a Western or Eurocentric smith to refine it from its "natural" state into an aestheti-
cally complex form. ("Unspeakeable Things Unspoken" 129–130)

The racial tension in this argument is palpable as Morrison makes explicit the
dynamics of invizibilization that are created by the normative culture and institutions, which are mostly white/European, evidencing that even when the production of minorities is deemed to be good, it needs to be "universal", meaning
that it must not explore the particular and the specific of a culture, but instead, it
should express values that are hegemonic, making possible for the rest of the
community to enjoy it. The last comment that sees this production as a rich
"source material" highlights that, according to this logic, there must exist the interference of those who "know" what "culture" is meant to be for the production
of the Other to be accepted. On this regard, Morrison concludes:

> Canon building is empire building. Canon defense is national defense. Canon debate, what-
> ever the terrain, nature, and range (of criticism, of history, of the history of knowledge, of
> the definition of language, the universality of aesthetic principles, the sociology of art, the
> humanistic imagination), is the clash of cultures. And *all* of the interests are vested. ("Un-
> speakeable Things Unspoken" 132)

Regarding the formation of canons and their opening, it is possible to say that
the African diasporic literary tradition(s) are part of what Gilles Deleuze and
Félix Guattari defined as a minor literature. The authors reached the definition
of this kind of expression when analyzing Kafka's texts (1986), claiming that
minor in this sense does not equate to inferior, but is connected to the idea of
a revolutionary literature. More specifically, the authors state: "[w]e might as
well say that minor no longer designates specific literatures but the revolutionary conditions for every literature within the heart of what is called great (or established) literature" (17). Here I argue that African-American and Afro-Caribbean literature may be read according to the three characteristics defined by
the authors as constituting a minor literature.

The first feature of a minor literature is concerned with the linguistic character, in which the expression of the subjects is made through a deterritorial-
ized language. Before having access to the written word, subjects brought from
West Africa through the Middle Passage to the Caribbean islands and the United
States already shared a tradition of storytelling, which was considered the most
important form of communal expression and socialization. In the shores of the
Americas, law denied slaves the right to literacy[11]. However, the protestant reli-

---

11 Henry Louis Gates Jr. reports a "[...] 1740 South Carolina statute that attempted to make it

gion gave the possibility of learning to read and write to many of them, so they could have access to the Bible. Those slaves who were literate in the United States were later encouraged by Abolitionists to start producing their own narratives in written language, appropriating the language of their masters, and telling readers about their side of the story/History ("Subjects in Time"). Using the English language, a colonial language, in opposition to the obliterated mother tongues of Africa, these subjects renegotiate meaning inside an oppressive language system, appropriating the master's language to be able to make public their concerns and interests, which originated a specific tradition of literature, the slave narrative.

The second characteristic is related to the political characteristic of the works attributed to a minor literature:

> In major literatures [...] the individual concern (familial, marital, and so on) joins with other no less individual concerns, the social milieu serving as a mere environment or background [...]. Minor literature is completely different; its cramped space forces each individual intrigue to connect immediately to politics. ("What is a Minor Literature?" 17)

The literature that will be explored in this study is fueled with political meaning, since its origins and subsequent developments were always linked to the interventionist character of the texts, which frequently were used as tools in the struggle for freedom and social justice. The representations of violence and trauma present here serve the purpose of connecting experience to knowledge production, intervening in the perceptions the readership might obtain regarding such issues. They reveal, unveil, and denounce experiences, creating opportunities for debate and the raising of awareness, consequently promoting the possibility for social change. Fiction and autobiography align with this precept, denouncing structural and direct forms of violence, as well as racism, sexism, class oppression, homophobia, just to name a few, all this through imagination and text in negotiation with the readership.

The third characteristic of a minor literature is that all of its aspects have a collective value. The issues explored by African diasporic authors tend to over-

---

almost impossible for black slaves to acquire, let alone master, literacy: 'And whereas the having of slaves taught to write, or suffering them to be employed in writing, may be attending with great inconveniences; Be it enacted, that all and every person and persons whatsoever, who shall hereafter teach, or cause any slave or slaves to be taught to write, or shall use or employ any slave as a scribe in any manner of writing whatsoever, hereafter taught to write; every such person or persons shall, for every offense, forfeith the sum of one hundred pounds current money'" ("Writing 'race'" 9).

flow the unitary and individualistic set of values found in major literatures; instead, they transmit the values and causes that permeate their collective experience, even if such homogeneity is only a collective imaginary. Contrary to canonic (white/male/heterosexual/cisgendered) literature, Black subjects never had access to thinking themselves unitarily, mostly because their experience of subalternity has conditioned them to develop a stronger sense of solidarity as a coping strategy of resistance, while their white counterparts have the privilege of imagining and representing themselves as single entities. This is much so mainly because the set of values that comprise the idea of Americanness are designed to value such indivisibility of the subject, who is the sole responsible for making it in the New World. Black subjects, on the other hand, have always been treated in terms of a collective identity, homogenized and decomplexified in the master narratives (in Jean-François Lyotard's terms). In addition to this, black people are affected collectively by a slave past and racial struggle for freedom and civil rights, and personal instances of racism reverberate directly on the lives of this imagined collective. This characteristic of making the personal something collective can be found in Black literature in the United States since its inception with the slave narratives. The abolitionist literature in which the slave narratives are included, is evidence of such trend, in which black subjects represent their personal stories, making clear the collective pain inflicted by the Peculiar Institution.

1.2 Race and Literature

1.2.1 African-America

> "The history of the American Negro is the history of this strife, – this longing to attain self-conscious manhood, to merge his double self into a better and truer self. In this merging he wishes neither of the older selves to be lost. He would not Africanize America, for America has too much to teach the world and Africa. He would not bleach his Negro soul in a flood of white Americanism, for he knows that Negro blood has a message for the world"
> W. E. B. DuBois (*The Souls of Black Folk* 3).

The definition of race, a useful category in biology, has been discredited by the humanities for a long time when dealing with the social. Gates clarifies: "[w]hen we speak of 'the white race' or 'the black race,' 'the Jewish race' or 'the Aryan race,' we speak in biological misnomers and, more generally, in metaphors" (4). These conceptions derive from "[...] dubious pseudoscience of the eighteenth and nineteenth centuries" (4), which served the purpose of inscribing the Other to the bottom of the social ladder, guaranteeing the supremacy of European powers in its colonial enterprize. The centrality of reason, in a Descartian perspective, would create the conditions to define black experience to be outside the circle of humanity, since the written word was not part of their civilizations.

> Blacks were "reasonable," and hence "men," if – and only if – they demonstrated mastery of "the arts and sciences," the eighteenth century's formula for writing. So, while the Enlightenment is characterized by its foundation on man's ability to reason, it simultaneously used the absence and presence of reason to delimit and circumscribe the very humanity of the cultures and people of color which Europeans had been "discovering" since the Renaissance. (8)

Several philosophers of the Enlightenment corroborated these perspectives, which equated reason with the written word, since it was through writing that one could best expose their intellect (the invention of the printing press also factors in as a promoter of the written word, which could then be more easily spread). The absence of a written language served as the explanation for the inequality between the African and the European, ultimately functioning as a reason for the conquest and enslavement of these racialized subjects, as explained by Gates:

> Since the beginning of the seventeenth century, Europeans had wondered aloud whether or not the African "species of men," as they most commonly put it, could ever create formal literature, could ever master "the arts and sciences." If they could, the argument ran, then the African variety of humanity and the European variety were fundamentally related. If not, then it seemed clear that the African was destined by nature to be a slave. (8)

The representation of black experience requires a set of values that can overcome the past of inhumane slavery imposed by the white telling of their lives. In the epigraph above W.E.B DuBois clearly outlines the path of African-American history, a history that is always duplicated in perspective seen through black and white eyes. However, referring to the African-American experience as solely masculine decries anachronism, as DuBois erases black women's particular experience in the journey towards self-consciousness[12]. DuBois is the author of *The Souls of Black Folk*, first published in 1903, in which he coined the concept of double consciousness, a definition that aptly describes the ambivalence of the black experience in the United States:

---

12 See *But Some Of Us Are Brave: All The Women Are White, All The Blacks Are Men: Black Women Studies*, edited by Gloria T. Hull, Patricia Bell Scott and Barbara Smith, published in 1993.

> After the Egyptian and Indian, the Greek and Roman, the Teuton and Mongolian, the Negro is a sort of seventh son, born with a veil, and gifted with second-sight in this American world, – a world which yields him no true self-consciousness, but only lets him see himself through the revelation of the other world. It is a peculiar sensation, this double-consciousness, this sense of always looking at one's self through the eyes of others, of measuring one's soul by the tape of a world that looks on in amused contempt and pity. One ever feels his two-ness, – an American, a Negro; two souls, two thoughts, two unreconciled strivings; two warring ideals in one dark body, whose dogged strength alone keeps it from being torn asunder. (2)

DuBois is referring to the moral division experienced by African Americans, based on a constant tension between two principles that seem to be at odds, being American and being black. Such tension results in a double perception of one's own subjectivity, one internal and one external. The unease described by DuBois is felt transversally among the African-American population, who seeks to make sense of the dynamics of national belonging. Constantly torn between two different cultural traditions, which seem to be at odds, African Americans must negotiate their identity construction within a hybrid composite. It is possible to articulate the notions of double-consciouness explored by DuBois with the idea of the third space, as described by Homi Bhabha in *The Location of Culture,* published in 1994. For Bhabha, all cultural statements and systems emerge in this third space of enunciation, which is characterized by ambiguity, dismantling any claim of cultural "purity", as both colonized and colonizer mutually contribute to the subjective construction of each other, although the power assimetries that are inherent to this process must be accounted.

> It is that Third Space, though unrepresentable in itself, which constitutes the discursive conditions of enunciation that ensure that the meaning and symbols of culture have no primordial unity or fixity; that even the same signs can be appropriated, translated, rehistoricized and read anew. (37)

The third space is understood then as the space of hybridity in which cultural meanings and identities invariably comprise some part of other identities. The ideas related to this third space might also be articulated with some relevant theories of poststructuralism, as pointed by Bill Ashcroft, Gareth Griffiths and Helen Tiffin in *Key Concepts in Post-Colonial Studies:*

> While Saussure suggested that signs acquire meaning through their difference from other signs (and thus a culture may be identified by its difference from other cultures), Derrida suggested that the 'difference' is also 'deferred', a duality that he defined in a new term '*différance*'. The 'Third Space' can be compared to this space of deferral and possibility

(thus a culture's difference is never simple and static but ambivalent, changing, and always open to further possible inter-pretation). (53–54)

Blackness in the Americas, and more specifically in the United States exists in this space of negotiation, as it is perceived as the difference within the culture by a white supremacist reality. Language is a powerful tool to address the erasures and assimetries, negotiating experience and representations, ultimately fostering change in real life contexts. In "The Transformation of Silence into Language and Action", a speech given to the Modern Language Association in 1997 by Audre Lorde, the author makes explicit the role of language in the deconstruction of oppressive silencing paradigms that are linked to racial matters as much as they are to class and sexual issues. Lorde claims:

> Each of us is here now because in one way or another we share a commitment to language and to the power of language, and to the reclaiming of that language which has been made to work against us. In the transformation of silence into language and action, it is vitally necessary for each and one of us to establish or examine her function in that transformation and to recognize her role as vital within that transformation. (43)

The power associated with language and the written word serves an activist role. At the same time the transition from passive silence to an active expressive subject position serves to rethink the language which originally was a tool belonging to the oppressive colonizer. Lorde echoes the precepts of a tradition in literature that started with slave narratives, which were used by the abolitionists as a tool to counter-attack the defense of slavery by Southern slavemasters. Through the written word the slaves could prove their own humanity, thus African-American women and men were no longer voiceless witnesses of their lives, a condition that had been imposed on them first by the slaveholding system and then by official historiography. It is this kind of writing, the voices from the "other side of history", that characterizes the tradition of the African diaspora to this day. Literature has been a rich source for the comprehension of racial relations in the United States, becoming a site in which one can perceive how subjectivities are constructed, reconsidered, accepted or rejected, and its study becomes indispensable as a tool for the understanding of reality. Gates concludes:

> We must, I believe, analyze the ways in which writing relates to race, how attitudes toward racial differences generate and structure literary texts by us and about us. We must determine how critical methods can effectively disclose the traces of ethnic differences in literature. But we must also understand how certain forms of difference and the languages we employ to define those supposed differences not only reinforce each other but tend to create and maintain each other. Similarly, and as importantly we must analyze the language of

contemporary criticism itself, recognizing especially that hermeneutic systems are not universal, color-blind, apolitical, or neutral. ("Writing 'race'" 15)

1.2.2 Slave Narrative

Women and men were taken from West Africa to the Americas and made into chattel slaves, mainly during the 17th and 18th centuries[13], and subsequently until the beginning of the 19th century. Slavery, though, was only officially recognized as such at the end of the 17th century through the register of the slave statute in the laws of Virginia and Maryland, in 1661 and 1664. The need for labor forces to explore the newly conquered territories fueled the market of slavery in numbers that were never previously seen. Estimates are that abour 12.5 million people were brought from Africa to the Americas during this period, mostly to Brazil and the Caribbean, approximations indicate that about 388 thousand arrived in the northern territories that would become the United States[14]. Destitute of national and cultural specificity, slaves were conceived as an amorphous mass, understood simply as the non-white Other, and therefore depleted from humanity and condemned to a life of servitude, violent coercion and exploitation.

The written word has been the key to the social transformation that has reshaped reality, and slave narratives emerged as the antidote to the reification and dehumanization of African subjects in the United States. Taking the written word as their weapon of attack, black people produced a series of discourses that proved their capacity of reasoning to a white audience, producing a sort of black

---

13 It is a fact widely stated that slavery existed before modernity and the colonizing of the Americas. However, slavery before such period was different in many aspects: people were made into slaves usually as a result of war in a transitional manner; in contrast with modern slavery, which was the result of a systematic practice of capture, shipping, and selling of forced labor that would become slaves for life. Modern slavery was much more permanent and emancipation was nearly impossible, though there are cases of freed slaves in America. During both stages, slaves could be liberated by their owners, or could also buy their freedom. In the Americas, children born out of slave mothers were automatically considered to be slaves too. On the matter, DeGruy states: "In most societies it was extremely rare for a slave population to reproduce itself through breeding, as did American slaves. Typically children born to slaves in places other than the Americas were born free. In many societies slavery was more akin to indentured servitude. There was a fixed amount of time a slave was held in service to his owner, after such time he would be granted his or her freedom. In America generations were born into slavery and died there" (DeGruy 2005: 48).
14 Data available at https://www.slavevoyages.org/

historiography that, though fragmented and dispersed, would mark their presence in the American narrative as subjects. Gates comments:

> Ironically, Anglo-African writing arose as a response to allegations of its absence. Black people responded to these profoundly serious allegations about their "nature" as directly as they could: they wrote books, poetry, autobiographical narratives. Political and philosophical discourse were the predominant forms of writing. Among these, autobiographical "deliverance" narratives were the most common and the most accomplished. ("Writing 'race'" 11)

Isabel Caldeira corroborates such claim in her doctoral thesis, *História, Mito e Literatura, A Cicatriz da Palavra na Ficcção de Toni Morrison,* stating:

> [...] [The] narrative of slaves comes to us as a text that invents its historical subtext, which, in opposition to other codifications of reality inherited and made official as the only truth, projects an alteration of 'reality' itself. It is this same subtext that will inscribe the tradition of Afro-American narrative to the present day[15]. (139)

Historiography is not clear regarding the first published slave narrative, given the divergences when it comes to works and dates. The African-American literary tradition started as a distinctively separate branch of literature in the United States, and progressively has become essentially an established expression of America in all its intricacies, complexities and anxieties. Gates comments on the issue:

> Black writing, and especially the literature of the slave, served not to obliterate the difference of race; rather, the inscription of the black voice in Western literatures has preserved those very cultural differences to be repeated, imitated, and revised in a separate Western literary tradition, a tradition of black difference. We black people tried to write ourselves out of slavery, a slavery even more profound than mere physical bondage. Accepting the challenge of the great white Western tradition, black writers wrote as if their lives depended upon it – and, in a curious sense, their lives did, the "life of the race" in Western discourse. ("Writing 'race'" 12–13)

African-American literature has played an important part in retrieving the humanity of African-American subjects, from the slave narratives that populated and fought for the abolitionist cause, to contemporary autobiography and fiction that denounce the oppression still felt by these subjects. While analyzing the thought of David Hume on writing as the utmost marker of difference between animals and humans, when black subjects performed the act of writing their

---

15 Translation provided by the author.

own narratives, Gates claims that "[s]imply by publishing autobiographies, they indicted the received order of Western culture, of which slavery was to them the most salient sign" ("Writing 'race'" 12).

Mary Jane Lupton, in *Maya Angelou – A Critical Companion*, presents some hypotheses to what might be considered the founding work in the African-American tradition. Some critics point to Briton Hammons's as the first printed narrative, published in Boston in 1760; while others believe it to be James Albert Ukawsaw Gronniosaw's *An African Prince*, of 1770 (43). While citing some other authors, through the perspective of the signifying monkey, Gates claims that the African-American tradition was created as a result of the criticizing of the chain of signs that denied their humanity, demonstrating the political aspect of this literary production and its emancipatory character:

> The writings of James Gronniosaw, John Marrant, Olaudah Equiano, Ottabah Cugoano, and John Jea served to criticize the sign of the chain of being and the black person's figurative "place" on the chain. This chain of black signifiers, regardless of their intent or desire, made the first political gesture in the Anglo-African literary tradition "simply" by the act of writing. Their collective act gave birth to the black literary tradition and defined it as the "Other's chain," the chain of black being as black people themselves would have it. ("Writing 'race'" 12)

Slave accounts had to be sponsored and authenticated by abolitionists or other white reliable mentors, which meant white men were responsible for attesting the veracity and authorship of such tales. The assistance of these actors resulted in a broader reach of the works, which often were published not only in the United States, but also in England, providing more efficient political articulations. Regarding white abolitionists, Lupton states that:

> What the slave wrote not only had to be 'true', but its truth had to be upheld or verified, in the preface or appendix, by conscientious white editors, publishers, and friends. Thus the *Narrative of the Life of Frederick Douglass* was authenticated in the preface by abolitionist leaders William Lloyd Garrison and Wendell Phillips. (*Maya Angelou* 35)

These accounts, however, have silences that strategically veiled the identities of former masters, places, and attitudes, so as to protect the writer from future reprimands and since they were usually fugitives, to protect themselves from being captured. These strategies in texts that were written to denounce the true cruelty of the slavery system characterize the adaptive character of black writers who needed to appropriate a white canonical form, so as to be heard and seen. The traditional autobiographical text, the tale of self-sufficiency and self-improvement, to be followed and upheld as a model of enlightened development

is changed into a search for escaping bondage and proving the author's humanity.

The scarcity of female voices during this period of African-American literature is noticeable. The undoing of paradigms of oppression during this era carries still the sexist *modus operandi* that oversaw the issues related to a female
kind of bondage. Male writers of the slave narratives often neglected questions
related to sexual violence, childcare, and stability of the family, for instance; instead, they emulated the white patriarchal social patterning of reality, oversha-
dowing these other questions.

Black women writers who were able to write their narratives do it differently.
Alison Easton, in her article "Subjects in Time – Slavery and African-American
Women's Autobiographies" affirms that: "[b]lack women are important adapters
of the genre, negotiating creatively with a form written largely by men (only
thirteen of the 115 personal narratives by former slaves published in the USA between 1760 and 1920 were by women)" (177). Easton points to the diversity of
texts written by slaves (and contemporary autobiographies written by black
women), and to their capacity of adapting the literary genres of the dominant
culture through a creative negotiation. The author cites as an example the
form of the canonical autobiographic text, with its individualistic character,
which is traditionally perceived as masculine, telling the story of an individual
who stands out from his environment to heroically triumph through his own efforts, without any help coming from the midst to which he belongs. When men of
African descent appropriated this genre, they enlarged its scope to tell the their
unitary accounts which would reflect the plight of enslaved subjects. However,
the predicaments suffered by enslaved women would frequently be absent
from such story. This type of text, when produced by an African American, especially when produced by a woman, ends up reverting the unitary discourse into a
communitarian discourse that was broader, expressing not only the concerns of
the self, but also the ones of the collectivity, the support network that makes survival possible, turning writing into a space for contestation against hegemonic
patterns, exposing matters that had previously been silenced, such as sexual violence, the unmaking of the black family structure, among others. As an example, Harriet Jacobs in *Incidents in the Life of a Slave Girl*, describes the way in
which she is able to escape her former master through the assistance provided
by her family. Jacobs would delay her escape, living seven years confined, so
she would be able to see her children through a peeping hole. J. M. Stover discusses in her article "Nineteenth-Century African American Women's Autobiography as Social Discourse: The Example of Harriet Ann Jacobs", how the autobiographical writing of this author is far from a self-referent individual account; on

the contrary, it transforms personal narrative into a social discourse, aimed at giving visibility to the humanity and integrity of all African-American subjects.

> The typical nineteenth-century black woman's autobiography is much more than a personal narrative that merely remarks on her personal growth; it is a social discourse that applies a unique black woman's voice to the interpretation and recording of her life experiences within a historical context that saw black Americans attempting to establish their humanity and self-worth in the eyes of a dominant white American society that granted them neither. Harriet Ann Jacobs's 1861 *Incidents in the Life of a Slave Girl, Written by Herself,* the black female slave narrative most often studied and anthologized, offers an excellent example of an African American woman's use of a revamped autobiographical genre as social discourse. (133)

It is important to highlight, however, that one of the most widely read works of the time, which contributed to the discussion regarding the abolitionist cause was written by a white woman, the abolitionist Harriet Beecher Stowe, in 1852. She is the author of *Uncle Tom's Cabin*, a very sentimental piece of literature that deals with the cruelties of the slavery system. It sold over 300.000 copies just after its publication, instigating the debate over slavery in the United States and in England. The work is seen as very influential, yet it is also problematic, since it depicts and reproduces African-American stereotypes, such a the Mammy, the ever present black woman responsible for child rearing, and the Uncle Tom, the subservient black man who lives in adoration of his master.

1.2.3 Writing as Agency – the Bildungsroman

> My silences had not protected me. Your silence will no protect you. But for every real word spoken, for every attempt I had ever made to speak those truths for which I am still seeking, I had made contact with other women while we examined the words to fit a world in which we all believed, bridging differences.
> Audre Lorde (*Sister Outsider* 41).

For very long the lives of women existed outside the literary canon, which created a silence about the life experience of these subjects, even more so when they belonged to an ethnic minority. However, with the advancements of the social movements, and especially the feminist movements, the voices of women multiplied over the last two centuries. The word is to Lorde an element that deserves attention given its social character, since it is capable of making short the distance between herself and other women, at the same time dismantling the linguistic sign to better shape it into her reality and the reality she aims at creating. The word, spoken or written, is the essential element for the creation of new social and (inter)personal realities, working for the subjects as an emancipatory tool. The preoccupations previously stated can be seen in the critical thinking of Patricia San José Rico, in her doctoral thesis, *The Call of the Past: Trauma and Cultural Memory in Contemporary African American Literature* (2013), which builds a parallel between the need for recuperating history as a political tool and the trauma that is present in the life experience of African-Americans represented in literature. The need to recover history must be central to a better understanding of the social past and to the promotion of political empowerment. The author states:

> Whenever history needs to be recovered, there is the possibility that all the documents are gone, as Gayl Jones writes in *Corregidora*. One may find that only the scars of history, like the chaotic writing of slavery on Sethe's back in Toni Morrison's *Beloved* remains. That scar needs to be as visible as blood in the process of bearing witness. It needs to be at the center of the willful effort to recover the past and start the process of working through, helping promote social understanding and political agency. If we traditionally hide scars from public view, the visibility of the scar becomes paramount in the process of giving visibility to the past. (308)

Literature offers a setting where the process of unearthing histories is possible, and while doing so, it brings to the forefront the complexities derived from the erasure fabricated by colonial historiography. It offers the possibility of the recreation of a past long lost, giving the readership a possibility to rethink history and its relation to contemporaneity. Working through trauma here is related not simply to personal accounts of unbearable situations, but mainly, about the working through of a cultural identity that needs to understand its background in order to overcome its present situation of oppression. The written word allows women to leave a space of subalternity and silencing, inscribing their histories in History, as pointed by Puga:

> [The *Bildungsroman*] comments and documents contemporaneity by focusing on the human being in development, critically questioning identities, stereotypes, obstacles and cosmovisions, questioning and transforming themselves in this process as well; hence it has also been used by feminist writers in the 20th century (female Bildungsroman) to fictionalize and comment on female pathways, spaces, traumas and obstacles[16]. (8–9)

The *Bildungsroman*, a German genre understood to be the original model for the novel of development, has been defined in several ways, as Geta LeSeur indicates in *Ten Is the Age of Darkness – The Black Bildungsroman*, stating that:

---

16 Translation provided by the author.

> Goethe's *Wilhelm Meisters Lehrjahre*, published in 1795, served as a model for the form as it was later seen in France, England, other parts of the European continent, and the United States. The form has been defined in various ways: the novel of development, novel of education (literal translation of *Bildungsroman*), "apprenticeship" novel, autobiographical novel, novel of childhood and adolescence, and the novel of initiation. (1–2)

Another term used in the United States to refer to this genre is the coming-of-age narrative, and in this work the terms mentioned above are going to be interchangeably used to refer to this specific form of novel and its modifications. Puga offers another work of German origin besides *Wilhelm Meisters Lehrjahre* as a founding model for the genre, in addition to presenting a definition:

> Broadly speaking, we can define the coming-of-age novel —of which Christopher Wieland's *Die Geschichte des Agathon* (1766–1767) and Goethe's *Wilhelm Meisters Lehrjahre* (1795–1796) are considered models and first examples—as a narrative fiction that represents the training path of a child or adolescent/young person until the adult stage of their life, as well as all the obstacles and tests that it overcomes, with the training process being predominantly informal, as opposed to formal or school education[17]. (10)

The *Bildungsroman*[18] is usually understood as a linear narrative of development in which the protagonist, generally a male, tends to conform to the social norm, offering the readership a narrative conclusion that is his assimilation to the *status quo*. It offers the feminist cause a space that is fertile for the questioning of identity and subalternity issues, since it is focused on a life period where certainty is scarce and fluidity is mandatory. Christy Rishoi in *From Girl to Woman – American Women's Coming-of-Age Narratives* claims: "[b]y focusing on adolescence, by definition a time of rebellion and resistance, and by foregrounding contradictory desires and discourses, the coming-of-age narrative provides a congenial form for women writers to successfully question the power of dominant ideologies to construct their lives" (9). This was not always the reality of the *Bildungsroman* for women, in which the protagonist would be taught to integrate into the *status quo* by learning about the limitations imposed on her sex (*O Bildungsroman* 25). However, when non-white women in the contemporaneity pro-

---

17 Translation provided by the author.

18 " The term *Bildungsroman* is coined by Karl von Morgenstern (1770–1852) in 1810, in a course he taught ('Ueber den Geist und Zusammenhang einer Reihe Philosophischer Romane') and in two communications entitled 'Ueber das Wesen des Bildungsroman' (1820) and "Zur Geschichte des Bildungsromans" (1824), and not by Wilhelm Dilthey in his biography of Friedrich Schleiermacher, Leben Schleiermachers (1870), as it was long thought, due to the greater projection of Dilthey's studies" (*O Bildungsroman* 10–11).

duce their discourse, these conventions are subverted[19]. These counter-hegemonic subjects restructure the literary narrative, since they share stories that end up affirming identity through difference, they question the ideologies that keep them in subaltern positions, and produce discourses of resistance that denounce their oppression. Such literary genre has, through the years, shown to be an interesting locus for the analysis of the social violence imposed on African-American and Caribbean women, since it offers a space where subjectivity construction is privileged, where social relations and their importance to survival and to the making of a healthy identity are demonstrated, at the same time as it offers flexibility to showcase individual expression.

In *Unsettling the Bildungsroman – Reading Contemporary Ethnic American Women's Fiction*, Stella Bolaki problematizes the matter of the *Bildungsroman* in the United States through new readings of Jamaica Kincaid, Maxine Hong-Kingston, Sandra Cisneros, and Audre Lorde, claiming that such authors[20] disrupt the conventions of this literary genre. Bolaki states:

> This study starts from the premise that the *Bildungsroman* is not an exhausted and outdated form but one that can be detached from its initial project and used productively across different historical periods and cultures. As a genre the novel of individual development may invoke concepts viewed with suspicion by the theoretically inclined literary critic, such as coherent identity, organic development, linear and teleological movement, and a closure that avoids openness (Felski, *Beyond Feminism* [...]) but exploring the category in new settings and through new perspectives reveals its usefulness for the representation of ethnic American and postcolonial subjectivities. (9)

The matters related to identity cohesion are deeply explored in the selected titles for this work, when questions of cultural/racial dominance are frequent, and the identification process takes place interstitially. Development is often arrested

---

19 Puga cites Lavobitz regarding the role of the patriarchy: "the role of patriarchy and its rejection in the heroines' quest for self is decisive [...]. As rebels, and feminists, the heroines of the female Bildungsroman challenge the very structure of society, raising questions of equality, not only of class, but of sexes as well [...]. Consequently, the female Bildungsroman is further defined by this most revolutionary characteristic" (qtd in Puga 28).

20 The writers selected for Bolaki's publication are considered to be "ethnic" by the author due to their distinctive non-white background which translates into the themes explored in their works, in addition to their heritage, which is a vital component in their experience, shaping the way the adjective *American* is developed in their narratives. They are all hyphenated subjects, claiming cultural allegiance to several nation-states, as well as stating their *Americanness*. They all share a history of immigration, one that differs from the hegemonic English migration in the United States, being perceived as part of the American whole, yet belonging to a non-white cultural minority.

due to the diverse forms of violence experienced by African-American and Caribbean subjects in the United States, and it shows in narratives that attempt to describe the formative periods, such as the *Bildungsroman*. The attention given by the authors to the experiences of their communities, rather than to tradition/genre conformity is evident in the development of narratives of subjects who extrapolate expectations of being part of an ideal America. The comparison between the traditional form of the genre and the modified post-colonial version is commented by Puga, as he states that the traditional *Bildungsroman* frequently works as a program of identification with the social order and society values by representing the assimilation of the thse scripts by the protagonists, while the ethnic or post-colonial *Bildungsroman* moves away from the standard traditional to deal with the processes that lead the individual to become aware of the difference/alterity and to identify with/reject the models that society imposes on them (18).

According to LeSeur, African-American writers are more interested in understanding their cultural production and group identity than in belonging to the American literary canon:

> Contemporary Black writers have [...] turned their attention inward, seeking to identify the traditions of their race by defining people individually, thus capturing a collective experience that is unique in terms of its circumstances of history and geography. They do not seek an entrée into the mainstream of European and American writing, but wish to explore the indigenous currents of these experiences – to communicate, often to educate, interpret and reveal the varied experience of four hundred years of suffering. (2)

Such description resonates with the terms postulated by Deleuze and Guattari concerning a "minor literature", namely the focus on collective identity instead of the unitary experience. The last four hundred years of disenfranchised living and structural violence are conveyed narratively to the public through examples that are personal and unitary, but also reflect the collective refiguring, reimagining what the cultural identity of the United States might be today. When looking inwards, black writers are seeking to understand themselves in their situatedness, no longer simply in relation to a white tradition/experience. And by doing so, they revisit their (personal) history, reassessing the narratives that constitute their identity as an imaginary collective.

Post-colonial authors of coming-of-age stories re-inscribe the historical discourse through narratives that, although particular are also universal. They end up reorganizing previous notions about race, class, and gender. In her study about the coming-of-age narrative, Rishoi points to the negotiations between the disenfranchised black subjects and coming-of-age narrative as a genre, stating the importance of the slave narratives to the development stories written by (black) women in America:

> Although the archetypical slave narrative is constructed by norms of male development, I suggest that it is the most direct literary ancestor of the type of women's narrative I am concerned with in this study. Concerned with demonstrating how 'a slave became a man' in Frederick Douglass's words [...], these narratives thematize the conscious and unconscious aspects of identity formation. (62)

Both types of narrative try to tell a story of overcoming, either the overcoming of the slave system, or the overcoming of the disadvantages of the development process that inevitably touches aspects such as identity formation, cultural affiliation, and in the cases to be analyzed in this study, trauma and violence. Rishoi claims that the slave narrative influenced the way in which coming-of-age stories developed in the United States, pointing out to the subjectivity constructions through the text, something that both kinds of text share in their stylistic predispositions, but only to a certain point, since each project is different in its finality. Rishoi reiterates the idea of sexual difference in the production of autobiographical texts by slaves, in which the male writers tend to glorify individualism, and the female writers tend to make explicit the interdependable network of people and communities as a factor for success in their path towards liberation. Such success might be here understood as the effective demonstration of humanity and the performance as an empowered subject. The coming-of-age stories written by women, and more specifically by African-American and Afro-Caribbean female authors, tend to showcase the tensions between the protagonists' self-determination and the influence of social and cultural factors in their potential opportunities. These authors are refashioning a genre that canonically belongs to the masculine normative experience, in which the *status quo* usually plays the part of enabler, or at least is a model to be followed. The coming of age narrative written by women tends to complicate the linearity of the traditional/canonical narrative, offering the opposite: a subject who perceives herself as external to its midst, who keeps herself marginal in the end of the narrative, refusing the perspective of assimilation to the social order. Rishoi clarifies: "[...] women's coming-of-age narratives often refuse closure, preferring instead an ambiguous textual ending that affirms the provisional nature of identity" (*From Girl to Woman* 63). This provisional nature of identity reiterates the transitional and post-modern status adopted by most of the authors, in which the construction of subjectivity is understood as an ever-changing process and not as a plan that can be accomplished. When these women write about growing-up histories/stories they are claiming their voice and empowerment, producing narratives that claim for recognition and denounce the oppressions they suffer. In addition, autobiography and fiction are elements that intermingle in this genre, making possible the creation and representation of imaginaries in which the protagonists are capable of resisting the hegemonic narratives of what it means to be a woman in their historical and geographical contexts. Puga comments that the narrative structure of the *Bildungsroman* facilitates the fictionalization of themes such as ethnicity, gender roles and changing social relations through the image that the characters convey of their own identity and that of others, the representation of gender is also associated with the description of power relations in patriarchal societies that are responsible for different forms of oppression to be overcome by women. Finally, the interaction between the characters thus reveals the hierarchy that exists within the same ethnicity or social group or gender (38–39).

Thus, it is noticeable that there is a tension between the traditional text and the expectations and the resignification process created by these writers who are not looking for agglutination to the social order but for an affirmation of subjectivity based on the idea of an empowered "outsider". Writers of Caribbean origin experience, for instance, some other issues in addition to their growing-up process and their subjectivity construction, such as the difficult tension regarding national identity in opposition to the colonial European domination, American imperialism, and the immigrant reality in the United States. The equation becomes even more complex given the constant flow these subjects experience, permanently moving in the direction of the United States, going back to the islands, searching for better work and life quality, visiting relatives, escaping violence, among many other motives. In such flow, lives are negotiated and identities are constructed, at the same time that they refigure memories of childhood, lived in the islands or in the host community. In their stories, they attempt to encompass this complexity, revealing the structures and conditions that regulate/influence this development. LeSeur claims that:

> The West Indian writer's concern is for the child who is born into an isolated community and grows up in a world influenced by European administrators. What happens to these children is the very subtle protest the authors project in their novels. Recent history can be seen through these records of childhood, and history is written into everyone's life. The impact of change, the clash of cultures, and the molding of communities are felt through these fictions, and the result is a learning about ourselves, our parents, and our children. (2)

The plurality of life stories, autobiographical, fictional, or even in between both registers, returns to the readership a possibility of complex interpretations that are opposed to the hegemonic narratives. The coming-of-age stories contribute even more effectively because they make public the perceptions of (fictional) children and teenagers. Especially because of this age factor, they are valuable witnesses who denounce the structures of violence in the making, in opposition to the hegemonic narratives (characterized by the white heterosexual middle-class male protagonist, usually an adult). LeSeur states that "[t]he perceptions of childhood are indispensable to any complete understanding of a community and its people, not only because any child is more honest than the most truthful adult, but because children are so often the forgotten camera in the corner (10).

1.2.4 A Child's Perspective and Subaltern Studies

The perspective of a child protagonist is of interest and is often used in this type of narrative, taking advantage of the temporal distance between the present of the character as narrator and the past experience of the narratee character as actor, since we are able to see the world through (allegedly) untrained eyes, making clear the encounters with long felt oppressions, before they are normalized by adults. This device gives the authors a vantage point of view, a way to express life in the making. LeSeur states:

> Children can play a part in emergent world literatures, because in cultures seeking Independence, children enjoy a natural, if precarious, enfranchisement. They provide a fresh point of view as Gullivers without fantasy, sojourners of the present, exploring the islands of manhood and womanhood, remaking the maps. (8)

It is possible to say that these voices in narratives of growing up enrich the readings on how different forms of violence happen and are naturalized in the lives of these subjects, who try to expose their personal and social and cultural contexts. These characters are responsible for removing from silence the structural elements of their social reality, taking advantage of their youth to see things as freshly new. Clearly, in this study, such unveiling is done through the representation of events in a literary form. However, as previously stated, the possibility of representation of unbearably violent and traumatic events is not simply referential, but a transformative space where the resignification of the experience takes place. The remapping of experiences carried out by literature might be one of the first steps towards the mitigation of violence as a structural reality.

The selected authors for this study demonstrate the potential to unsettle the genre of the *Bildungsroman*, complicating notions of narrative closure, questioning the *status quo* instead of abiding to it, rethinking sexual identities and gender roles, defying hegemonic culture, redefining the meaning of belonging to the United States and to the Caribbean. They also subvert the genre, demonstrating how porous and flexible the limits that separate autobiography from fiction are, offering readers narratives that dismantle aesthetic/literary patterns, and consequently, through the subversion of the conventions, undo constructs already established about identity and the women's condition.

Violence, direct and structural, shapes the growing up process for women in the United States and in the Caribbean. It is interesting to investigate what the role of violence in the identity construction of these subjects is, the way it is present in literary representations, more specifically in the coming-of-age narratives. What is the impact of globalization development in the generation of structural violence and trauma? Will the individualism preconized by the neoliberal model alter the interdependence practices that have been experienced so far? What are the consequences of the weakening of the community in the narrative of African-American and Caribbean women? In which ways does neoliberalism contribute to the maintenance and perpetuation of racism? How do the migration flows influence the development of these subjects? How do the authors deal with all these pressures while writing about (their) coming-of-age?

The coming-of-age story offers a privileged perspective for the observation of the presence of violence in the African-American and Caribbean experience in the United States, since the analysis of its narrative structure reveals what the consequences of violence and subsequent trauma in the development of the characters are/would be.

Subalternity is also a relevant concept to articulate in this understanding, since it is related to the right of non-hegemonic subjects to claim for history in their own terms. Originally coined by Antonio Gramsci, the term was first used to distinguish subjects who were submitted by the elites, such as peasants, workers, and the lower classes. The term has been appropriated by a south Asian group of postcolonial scholars who investigated the subaltern condition, producing a large corpus of theorization on the region, inaugurating a field of research known as Subaltern Studies. This group of researchers' aim was originally to address the academic production of the post-independence historiography of India, which focused on history lines that privileged the higher classes/castes, erasing subaltern subjects. The history of the ruling classes was widely accepted as the "official history", and subaltern subjects did not have access to power so as to control the means for their own representation, in addition to having much less access to the cultural machinery that reinforced their status at the bottom of the hierarchies. These elitist lenses when applied to the production of discourses made impossible for the subaltern to be fully represented. What was transversal in the varied groups that constituted the subaltern classes was the resistance to elitist domination.

The word served, and still serves as a tool for the African diasporic communities (in the United States as well as in the Caribbean) to display their humanity, intelligence, and well-roundedness, to be taken as full humans, full citizens, not as subalterns. Using the written word, these subjects are able to rework history, and through literature, they are able to remove themselves from the silence, enhancing empathy from the readership in the understanding of their conditions of oppression, as well as enhancing the possibility of the readership to engage in their behalf. Thus, these communities escape invisibilization when they get hold of the word, overcoming silences imposed by racial, sexual and class oppressions. The slave narratives are one example of this line of theorization, when the dominant language and the modes of representation of the colonizer were appropriated and many times subverted to become effective means to disseminate the abolitionist discourse.

Gayatri Spivak questioned the reality of subalternity with the widely read essay "Can the Subaltern Speak?", first published in 1983. Spivak points to the impossibility of the subaltern group to voice its claims, despite its mobilization, since there is not a place of enunciation where such claims can be heard. Moreover, Spivak critiques the essentialist nature of the classification "subaltern", which no methodology can escape, subjecting these groups once again to a reinstatement of their subalternity, as they once more are defined as subaltern (now by the researcher), instead of defining themselves in their own terms. Ashcroft, Tinning and Griffiths nonetheless clarify that Spivak's argument stems from the rigorous inquiry that the categorization of this identity must go through in order to affect change, stating:

> [...] Spivak's target is the concept of an unproblematically constituted subaltern identity, rather than the subaltern subject's ability to give voice to political concerns. Her point is that no act of dissent or resistance occurs on behalf of an essential subaltern subject entirely separate from the dominant discourse that provides the language and the conceptual categories with which the subaltern voice speaks. Clearly, the existence of post-colonial discourse itself is an example of such speaking, and in most cases the dominant language or mode of representation is appropriated so that the marginal voice can be heard. (219)

Literature may be seen as the site in which the concerns of the subaltern may be expressed, and heard. Their denounciations of oppression takes place in the written text that continuously serves the purpose of addressing their resistance to unfair experiences of reality created by different variables of violence, inscribing it in the macro-narrative of History.

2 From Intersectionality to an Ecology of Knowledges – The Knowledges of Literature

I am located in the margin. I make a definite distinction between that marginality which is imposed by oppressive structures and that marginality one chooses as site of resistance—as location of radical openness and possibility
bell hooks (*Yearning – Race, Gender, and Cultural Politics*).

Santos, Nunes and Meneses claim in the aptly titled *Another Knowledge is Possible – Beyond Northern Epistemologies* that, in the last decades, the development of the feminist epistemologies, cultural studies of science, and recent developments in the history and philosophy of science slowly expanded and made more complex the opposition between the humanities and the sciences. The authors affirm that this dichotomy has been challenged by "[...] a rather unstable plurality of scientific and epistemic cultures and configurations of knowledge" (XXIX). This change is attributed as the result of two different historical processes: one that delineates the limits between science and technology, placing the outcomes of scientific research in its applicability; and a second that makes a distinction between science and "[...] all other forms of relating to the world, taken to be non-scientific or irrational, including the arts, humanities, religion, and different versions of that relationship to the world [...]" (XXX).

This second historical process is responsible for the destabilizing of literary studies, which have come a long way in their crisis of legitimacy, as pointed by Ramalho and Ribeiro in "Dos estudos literários aos estudos culturais?". The field struggled from its early formalist claims, which were an attempt of creating limits to its own craft and consequently circumscribing literature and its knowledge in itself, evading any ideological contamination. The idea of a literature that was a form of high culture, fixed on strict canons, responsible for representing nationalist endeavors, relying on the figure of a self-sufficient author, was contested. As alternatives in this contestation through the late 1970s and 1980s, there are the incursions of literary studies into comparative literature and literary theory. Sociology however, known for its flexible standpoint between a scientific approach and a hermeneutic-literary stance, became the option in which literary studies came to find some legitimacy in the emergence of cultural studies. The authors state that the importance of the emergence of cultural studies for literary studies could not be overestimated, as it represented a challenge that produced more than beneficial destabilizing effects: it contributed to pushing to the last consequences the critique of an ontological concept of literature; it definitely made a narrow view of the canon falter, in particular by defending the inclusion in the field of analysis of a whole set of cultural practices associated with "mass

https://doi.org/10.1515/9783110752755-004

culture" by calling into question the "anxiety of contamination" inherited from modernism; forced to put on the agenda the urgency of a broad transdisciplinary opening, as well as demonstrated the limits of the philological-hermeneutic paradigm (72). Cultural studies have enlarged the scope of literary studies, without falling into the anxiety of becoming their end. Literary studies nowadays tend to be concerned with the critique of a series of different perspectives, with transnational questions, avoiding strictly formalist viewpoints, aiming at the overcoming of the opposition of the two cultures of knowledge production (the humanities and the sciences), with the opening to new areas of research, including the media, mass culture, gender studies, feminist studies, among many others. Which leads to a discussion regarding the role of literature in the production of knowledge in our times.

One possible question regarding this topic is how literature has an impact in the reality of readers. The findings of David Kidd and Emanuele Castano in "Reading Literary Fiction Improves Theory of Mind", published in 2013 in *Science,* are a possible answer. In addition, this research may be considered an interesting example of the overlapping of the two cultures of knowledge production, dealing with the intersection of literature and psychology. Such research endeavor is extremely relevant to the work developed in this study, since it demonstrates the effects of literature as a form of intervention in the realities of readers. Kidd and Castano are able to statistically quantify through experimentation the influence of literary fiction when compared with genre fiction in its capacity to identify and understand the subjective state of others, a set of abilities designated as theory of mind in psychological studies.

> Researchers have distinguished between affective ToM [Theory of Mind] (the ability to detect and understand others' emotions) and cognitive ToM (the inference and representation of others' beliefs and intentions) [...]. The affective component of ToM, in particular, is linked to empathy (positively) and antisocial behavior (negatively) [...] (1)

According to the authors, reading literary fiction boosts the empathy capacity of subjects, since it forces them to deal with the perspectives of others, expanding the knowledge about them, helping to recognize similarities among readers and the stories found in literature. "We submit that fiction affects ToM processes because it forces us to engage in mind-reading and character construction. Not any kind of fiction achieves that, though. Our proposal is that it is literary fiction that forces the reader to engage in ToM processes" (1). The authors present a discussion regarding the difference between literary fiction, taken to be fiction which is publicly recognized as so by the specialized critic, taking the shape of awards such as the Nobel Prize, the Pulitzer Prize, the National Book Award, among others; and genre fiction, which is understood as popular writing that falls in the categories of adventure, romance, science fiction, among others. This distinction is based on the capacity of literary fiction, with its stylistic devices and strategies, to defamiliarize the readers, breaking their expectations, while popular genre fiction tends to agree with the readers' expectations[1], in a more passive experience of entertainment. "Our contention is that literary fiction, which we consider to be both writerly [Barthes] and polyphonic [Bakhtin], uniquely engages the psychological processes needed to gain access to characters' subjective experiences" (1). In a second article, titled "Different Stories: How Levels of Familiarity With Literary and Genre Fiction Relate to Mentalizing", Kidd and Castano expand this analysis of literary fiction and theory of mind by incorporating the vision of Edward Morgan Forster about round and flat characters, establishing that round characters belong more often in literary fiction since they escape types and ready-made schemes in their construction, which challenges readers to constantly examine their mental states and cues, promoting theory of mind skills.

> Just as reading critically acclaimed fiction is theorized to promote ToM insofar as it draws attention to others' subjective experiences, it seems likely that other cultural practices may affect ToM in the same way. Biographies, memoirs, and narrative journalism are forms of nonfiction that could have the same effect [...], and recent research shows that playing a nonviolent narrative videogame [...] or watching acclaimed TV dramas [...] also improves ToM. ("Different Stories" 11)

Such views are corroborated by Valerie Smith in *Toni Morrison – Writing the Moral Imagination*, where the author asserts that Morrison's work is capable of creating a level of engagement with her readership that evades passivity. This defamiliarization, either by form or by content, is key for the creation of a space of action in the readers' mind, intended by the author herself: "I want to subvert [the reader's] traditional comfort so that he may experience an unorthodox one: that of being in the company of his own solitary imagination" (Morrison qtd in *Writing the Moral Imagination* 5). When deprived of the expectations provided by genres that conform to somewhat strict styles and patterns, the reader is forced to interact with his/her ideas to make sense of the narrative and the characters. Smith claims:

---

1 Additionally, they state: "[c]ontrary to literary fiction, popular fiction, which is more readerly, tends to portray the world and characters as internally consistent and predictable [...]. Therefore, it may reaffirm readers' expectations and so not promote ToM" ("Reading Literary Fiction" 1).

This quality of engagement is also important to her work because it is a means through which she dismantles the hierarchies that undergird systemic forms of oppression. For Morrison, language and discursive strategies are not ancillary to systems of domination. Rather, they are central means by which racism, sexism, classism, and other ideologies of oppression are maintained, reproduced, and transmitted. (4)

Commenting on the quality of the engagement between readers and text, Smith contends that this language work contributes to the undoing of inequalities in our social realities. Another possibility proposed by Kidd and Castano regarding the development of understanding and empathy concerns the opportunity literary fiction provides of offering the reader the experience of a life without its associated risks or dangers:

Just as in real life, the worlds of literary fiction are replete with complicated individuals whose inner lives are rarely easily discerned but warrant exploration. The worlds of fiction, though, pose fewer risks than the real world, and they present opportunities to consider the experiences of others without facing the potentially threatening consequences of that engagement. ("Reading Literary Fiction" 1)

The attempts to represent trauma and violence through literature provide glimpses into the experiences of subjects from a "safe" distance, promoting a broader understanding of such realities and contexts. This kind of knowledge production related to trauma and violence differs from those proposed by the regulatory discourses of politics, medicine, or even social science, since it deviates from a "scientific" mode of rigor, and therefore is "unqualified" as a valid form of explaining experience according to hegemonic modes of knowledge production. Literary texts provide a window into a representation of reality through the exercise of language that, because of its creative nature, makes the appropriation of the experience of trauma more flexible, giving it more possibilities of comprehension and representation than those related to hegemonically scientific forms of knowing. That appropriation is not normative or hierarchical, and more importantly, it does not claim to be complete or absolute. Its lacunar nature corroborates the above-mentioned perspectives, since it enriches the internal and external diversity of perspectives on the matter. In addition to serving as a means in which knowledge that otherwise would be discarded as mere aesthetic endeavor or entertainment, is socialized, absorbed, rethought, expanded, and so on.

2.1 Intersectionality

> The major source of difficulty in our political work is that we are not just trying to fight op-
> pression on one front or even two, but instead to address a whole range of oppressions. We
> do not have racial, sexual, heterosexual, or class privilege to rely upon, nor do we have
> even the minimal access to resources and power that groups who possess anyone of
> these types of privilege have (Combahee River Collective 1982: 18).

Understanding the position that black female subjects occupy in a society that is
blind to their experience begins with recognition. Recognition here undersood
with the promise of being transformative, as stated by Judith Butler and Athena
Athanasiou in *Dispossession – The Performative in the Political:*

> [...] recognition is not sufficient as the aim of politics, if we understand recognition as a
> static acknowledgement of what is. Recognition itself has to be a transformative category,
> or it has to work to make the potential for transformation into the aim of politics. (87)

The recognition of black female subjects and their plights is of utmost importance for the promotion of political and cultural change that envisions full equality to all citizens. The space occupied by black women as marginalized subjects
is the space where invizibilization occurs. Black women can neither set aside
their racialized selves when considering gender oppression, nor the opposite,
since both elements contribute to the construction of their social realities. Yet,
when these forms of oppression overlap, they create a form of obfuscation,
where neither claim for the adjudication of the oppressions is taken fully by
the justice system, which most often relies on perceptions that are external to
the lives of these subjects, leaving them in a precarious position. Also taking
into consideration the variable of class, bell hooks describes the reality of
black women in the following terms in *Feminist Theory From Margin to Center,*
published in 1984:

> As a group, black women are in an unusual position in this society, for not only are we col-
> lectively at the bottom of the occupational ladder, but our overall social status is lower than
> that of any other group. Occupying such a position, we bear the brunt of sexist, racist, and
> classist oppression. At the same time, we are the group that has not been socialized to as-
> sume the role of exploiter/oppressor in that we are allowed no institutionalized "other"
> that we can exploit or oppress. (14)

Intersectionality is a theory that evolved from these preoccupations and is based
on an analysis of the interaction between different categories, such as race,
class, gender, sexual identity, among others, as co-formative, and hence inextri-
cable from one another, with implications that go beyond the addition of a se-

quence of oppressions, a term coined by Kimberlé Crenshaw in 1989. The intersectional approach derived from the legal field; however, many different areas of social thinking have adopted the grammar of intersectional thinking. Crenshaw approaches three cases concerning the Title VII of the Civil Rights Act of 1964, which prohibits employment discrimination based on race, color, religion, sex, or national origin. She hoped to demonstrate how the judicial system failed at understanding the relevance of the intersection of sex and race. In one of them, a suit against General Motors, five plaintiffs accused the company of discrimination against black women, since all black women lost their jobs due to a seniority-based layoff in 1970. General Motors did not hire any black women before 1964, and claimed that no sex discrimination took place, since they hired women, all white, since before 1964. The court suggested a new suit should be constituted, based solely on racial discrimination, and "[t]he plaintiffs responded that such consolidation would defeat the purpose of their suit since theirs was not purely a race claim, but an action brought specifically on behalf of black women alleging race and sex discrimination" ("Demarginalizing the Intersection" 142). Crenshaw successfully demonstrates how the court ruling considered that these plaintiffs' claims were only valid inasmuch as they coincided with that of black men, regarding racism, or white women, regarding sexism. Intersectionality is the lens through which these claims could be made visible. "Through an awareness of intersectionality, we can better acknowledge and ground the differences among us and negotiate the means by which these differences will find expression in constructing group politics" ("Mapping the margins" 1299).

In celebration of the tenth anniversary of the Berkeley Women's Law Journal, in 1995, Trina Grillo delivered a speech called "Anti-Essentialism and Intersectionality: Tools to Dismantle the Master's House", pondering on the difficulty found in white mainstream feminism and the universality of the categorization "women" as the axis of un understanding of oppression:

> The perceived need to define what "women's" experience is and what oppression "as women" means has prompted some feminists to analyze the situation of women by stripping away race and class. To be able to separate out the oppressions of race and class (as well as sexual orientation and other bases of oppression), the theory goes, we must look at someone who is not experiencing those oppressions and then we will see what oppression on the basis of gender alone looks like. This approach, however, assumes that the strands of identity are separable, that the experience of a white woman dealing with a white man, or raising a white child, is the same experience that a Black woman has dealing with a Black man, or raising a Black child. But as the intersectionality critique has taught us, they are different and not just additively. (32)

White feminism began to show some awareness of its false universal premises, as shown by Adrianne Rich in "Notes Towards a Politics of Location", published in 1986. Rich's radical feminism questions the so-called centrality of the white experience, reexamining the ways in which white women oppressed all the others with such claims. She states: "[m]arginalized though we have been as women, as white and Western makers of theory, we also marginalize others because our lived experience is thoughtlessly white, because even our 'women's cultures' are rooted in some Western tradition" (219). The production of knowledge that departures from such false universalisms is bound to create more oppression than to liberate. She also questions:

> How does the white Western feminist define theory? Is it something made only by white women and only by women acknowledged as writers? How does the white Western feminist define "an idea"? How do we actively work to build a white Western feminist consciousness that is not simply centered on itself, that resists white circumscribing? (219)

Intersectionality became one of the answers to some of these questions. Black feminist thought interrogated these privileges, and proposed different approaches that expand and complicate the premises stated by white mainstream feminism since abolitionist times. In "The Master's Tools Will Never Dismantle the Master's House", a speech delivered during a conference in 1979 held at the New York University Institute for Humanities, Audre Lorde denounces the exclusionary practices she finds in mainstream white feminist theoreticians, which emulate the same kind of discrimination of the patriarchy as it dismisses the experiences of subjects who inhabit the margins. "It is a particular academic arrogance to assume any discussion of feminist theory without examining our many differences, and without significant input from poor women, Black and Third World women, and lesbians" (110). The author is demanding the acceptance of difference as a positive asset for a true inclusion in a community, rather than a cause for separation or mistrust. Only in making connection with others, and specifically with other women, can liberation be achieved. Specificity must be recognized as a transformative power which will enable all subjects under oppressive systems to resist and consequently explore all their possibilities. Discriminating and hierarchizing are tools from the master's house therefore, as Lorde would put it, they are not effective in a long-term process of emancipation:

> *For the master's tools will never dismantle the master's house.* They may allow us temporarily to beat him at his own game, but they will never enable us to bring about genuine change. And this fact is only threatening to those women who still define the master's house as their only source of support. (112)

The tools that are useful here are the ones that are inclusive of different realities, and that are able to recognize and incorporate different struggles in their efforts to construct a social reality that is more equitable. Intersectionality is a tool of analysis and intervention that is recognizant of the complexities of subjects who inhabit different parts of the specter of oppression, and require different layers of identification. Yet, the workings of intersectionality were present in black feminist thought from much before. Black women had to make their position in the intersections very clear since male sexism and white domination were concomitant sources of oppression. Barbara Smith, author of "Toward a Black Feminist Criticism", published in 1978, demonstrates how the lack of privilege experienced by female black subjects imposes an invisibility regarding their lives and their production, as in their accounts the marker of race obscures the marker of sex. The lack of privilege mentioned in the epigraph by the Combahee River Collective is present in the lives of black subjects, and consequently in the representations of their lives, and therefore there is a need for a black feminist perspective which would be able to account for these interlocking realities. Smith exposes the lack of recognition these subjects experience in life, and in literature, and the consequences of such lacking:

> When Black women's books are dealt with at all, it is usually in the context of Black literature, which largely ignores the implications of sexual politics. When white women look at Black women's works they are of course ill equipped to deal with the subtleties of racial politics. A Black feminist approach to literature that embodies the realization that the politics of sex as well as the politics of race and class are crucially interlocking factors in the works of Black women writers is an absolute necessity. Until a Black feminist criticism exists we will not even know what these writers mean. [...] without a Black feminist critical perspective not only are books by Black women misunderstood, they are destroyed in the process. (21)

Smith and The Combahee River Collective are examples of the effort in the recognition of the complexity of the identities of black women in the United States. The urgency for a black feminist critique of literature implies that to be able to fully comprehend these representations, one must be able to perceive the subtleties and nuances of sexual, racial, and class politics, meaning the complex set of influences that differentiate black lives. This requires trained and empathic eyes that will be capable of recognizing that their oppression is a set of intermingling factors.

2.1.1 A Critique of Intersectionality

Collins analyzes the work of Crenshaw and intersectionality in the foreword of *Emerging Intersections: Race, Class, And Gender In Theory, Policy, and Practice*, stating that Crenshaw made evident the ways in which programs that were developed via gender-only frameworks were limiting and deeply flawed, since they failed to realize how intersecting power relations conditioned the available possibilities for women. Crenshaw also made evident how knowledge and hierarchical power relations are co-created, as the same structures that designed understandings of violence against women concurrently influenced the violence itself in addition to the organizational responses to it, which would limit social institutions if they did not adopt intersectionality in their views (VIII). Lately another facet of intersectional thinking emerged, namely its application and recent interest in identity studies, constituting a shift from the social beginnings of the theory. Collins points to the practices of intersectional thinking and their aptitude to matters related to identity narratives and subjectivity construction:

> In recent years, intersectional analyses have far too often turned inward, to the level of personal identity narratives, in part, because intersectionality can be grasped far more easily when constructing one's own autobiography. This stress on identity narratives, especially individual identity narratives, does provide an important contribution to fleshing out our understandings of how people experience and construct identities within intersecting systems of power. (IX)

This predisposition goes hand in hand with the kind of investigation proposed in this study, where personal stories, autobiographical or fictional, expose the different systems of power, namely the kinds of structural violence and intersectional oppression experienced by women of the African diaspora in the United States and the Caribbean. hooks urges us to see the position of black women as a privileged point from which to construct a critique of racism, sexism, classism, and any other form of oppression that upholds the *status quo*. "It is essential for continued feminist struggle that black women recognize the special vantage point our marginality gives us and make use of this perspective to criticize the dominant racist, classist, sexist hegemony as well as to envision and create a counter-hegemony" (hooks 1984: 15). Through an intersectional analysis, the unveiling of their social problems can be achieved, denouncing how institutions and practices continue to perpetuate intersecting forms of oppression. The development of intersectionality and its future as a theory is uncertain, yet, Collins investigates if intersectionality as a theory could overcome its status of a theory of the oppressed and become a more universal tool for building knowledge in a transformative way. But should intersectionality aim at universality as one of its intented objectives? As long as intersectionality functions as a tool for the comprehension of co-formative tipes of oppression, at the same time it is able to recognize categorizations in their own terms, which means, avoiding a secondary oppression by the imposing of categories/identities, it might be able to be used as a transformative methodology in multiple contexts. According to Collins, the problem is twofold: since intersectionality is a theory created by minorities in the United States, it may be seen as a situated political tool for social justice, too specific to the needs of these minorities, not "serving" a wider audience; but its extreme opposite is also dangerous: the universalization of intersectionality, as it runs the risk of occurring through the application of a co-opted version of intersectionality that addresses the need to respond to a complex set of subjects and realities without really engaging with them, thus increasingly evading any political effort first intended by Crenshaw. This version accounts only for the pretense of equality and openness, without harnessing any effective change, or worse, steering the change toward a capitalist market-driven logic. This trend may be related to the description of the hegemonic kind of research carried out by the New Americanists, as described in Maria Irene Ramalho's perspective in the first chapter. Collins comments:

> Unlike the invisibility that plagued the field of intersectionality at its inception, it now faces an entirely new challenge of being hyper visible within equally novel conditions of global, commodity capitalism. In U.S. academic settings, a superficial version of intersectionality is routinely packaged, circulated, and sold to faculty and students alike, only to be prematurely discarded when the product's performance fails to match the promises on the package. (XIII)

Collins' premise rings true, since a universal perspective would obliterate the situational character of a theory that focuses in understanding the different oppressions and specific interrelations among them as co-constitutive determinants. Intersectionality cannot effectively provide an applicable universal knowledge, since it must generate understanding through an analysis that is sensible to context, considering the synchronic circumstances that create oppression and social inequalities. What might be universal, or at least seems to be the aim of current research in social issues is the recognition of such conditions, which are evermore undeniable, be it the struggles of the oppressed or the realities of the hegemonies. Thus, no singular vision is adequate in times when specificity, the recognition of difference, and the struggle against discrimination are imperative. However, as a commodity, intersectionality might undermine its original purpose, serving only as a veneer of engagement, which is bestowed upon theory and practice under the demands of the market.

2.1.2 Intersectionality, Decolonial Thinking, and the Ecology of Knowledges

Just as Collins is concerned with a hegemonic appropriation of the concept/theory, Julia Roth claims for the importance of an epistemological revision of intersectionality in "Entangled Inequalities as Intersectionalities: Toward an Epistemic Sensibilization" (2013). Colonialism and coloniality figure in her thoughts as forces that shape epistemological violence. The author states:

> [...] in Latin America intersectional/interlocking systems of oppression are not considered as theory, but they are rather experienced as everyday realities. In order to count as a critical tool, a "decolonization" of the methodological Occidentalism inherent to lots of theorizing on intersectionality thus requires a radical rethinking of what counts as knowledge. (30)

Intersectional theory might reinscribe the same hegemonic patterns it attempts to unveil when applied to contexts different from those in which it was first designed, for it could easily act as an imposition of European/North American/Western modes of knowledge production, failing to grasp reality due to some form of (un)conscious overlooking, and subsequently erasing experience/knowledge. This a perspective also found in the work of Gates, when he states: "[t]o attempt to appropriate our own discourses by using Western critical theory uncritically is to substitute one mode of neocolonialism for another" (15). To decolonize intersectionality is fundamental, since social inequalities spark from a state of limited citizenship, and the overcoming of such inequalities depends on the understanding of the matrix of oppressions that is particular to each context.

   Roth claims for a shift in categorization that is more attuned to social intervention, stating that categories such as "race", "class" and "gender" must be changed to "racism", "classism", "sexism", and "homophobia", since these categories denounce how Western knowledge production functions and maintains its privileged position in relation to other forms of knowing, and subsequently of being. Trying to understand the different modes of oppression as an interlocking system of co-formative realities will help to unveil the limits of knowledge production, Roth affirms: "An intersectional framing of interdependent inequalities might contribute to a critique and consequently the overcoming of methodological Occidentalism." (29). Occidentalism is a term clearly linked to Edward Said's *Orientalism*, however, it differs.[2] It functions similarly to Ameri-

---

2 Regarding the term Occidentalism, Roth adopts the perspectives of Fernando Coronil, author of *Beyond Occidentalism: Toward Nonimperial Geohistorical Categories* (1996). She clarifies: "Oc-

canity, as coined by Quijano and Wallerstein in 1992, since it makes clear the processes that characterize the formation of hierarchies[3] during modernity and the making of the New World. The creation of the Occident[4] as the sole producer of knowledge needed the creation of an alterity figure, which was utterly disqualified as inferior. This took place through the restructuring and strengthening colonialism and capitalism that already existed, such as racism and sexism, but that were recreated in much larger scale. The non-recognition of different cultures in the making of the New World was strategic to sustain the dominance of colonial power. The developments in anthropology, social sciences, and post-colonial theory show a tendency to a broader understanding of the multiplicity of cultures.

The discourses on multiculturalism have been an attempt to deal with the diversity within populations and the formation of hierarchies among them, but also with forms of knowing. Santos, Nunes, and Meneses state that such discourse functions, either simultaneously or alternatively, both as a description and a project.

> As a description it may refer to: 1) the existence of a multiplicity of cultures in the world; 2) the coexistence of diverse cultures within the same political space; 3) the existence of

---

cidentalism can be considered a process that turned difference into hierarchy and then naturalized these hierarchies, which provided the condition of possibility for Eurocentric concepts of modernity and related enlightenment discourses. Following Coronil, I understand Occidentalism as a regime of knowledge production and an epistemic standpoint that provides a hierarchical perspective of the world. While the Occident could thus construct itself as modern, all other locations were denied to be part of (European/Eurocentric) modernity" ("Entangled Inequalities" 6). Coronil states: "Occidentalism, as I define it here, is thus not the reverse of Orientalism but its condition of possibility, its dark side (as in a mirror). A simple reversal would be possible only in the context of symmetrical relations between "Self" and "Other" – but then who would be the "Other"? In the context of equal relations, difference would not be cast as Otherness. The study of how "Others" represent the "Occident" is an interesting enterprise in itself that may help counter the West's dominance of publicly circulating images of difference. Calling these representations "Occidentalist" serves to restore some balance and has relativizing effects. Given Western hegemony, however, opposing this notion of "Occidentalism" to "Orientalism" runs the risk of creating the illusion that the terms can be equalized and reversed, as if the complicity of power and knowledge entailed in Orientalism could be countered by an inversion" (7).
3 For an overview on the creation of hierarquies and difference see "Difference and Hierarchy Revisited by Feminism", written by Maria Irene Ramalho Santos, published in *Anglo Saxonica* Serie III, No. 6, pp 23–45, in 2013.
4 For more information regarding the creation of the West see "The West and The Rest: Discourse and Power" by Stuart Hall in *Formations of Modernity* (1992), and also Catarina Martins' "The West and the Women of the Rest" in *The Edge of Many Circles Volume II* (2017).

cultures that influence each other, both within and outside the geo-political space of the nation-state. (XXI – XXII)

The vision of multiculturalism as a description rather than a project has been adopted by most; yet, it has been challenged by either conservative or progressive fronts.[5] The authors point to emancipatory versions of multiculturalism, which are "[...] based on the recognition of difference, and of the right to difference and the coexistence or construction of a common way of life that extends beyond the various types of differences" (XXV). Such notions are found frequently in places where the overlapping of histories has taken place in the shape of colonialism/imperialism, and the dynamics of diaspora and hybridity have fostered new understandings of concepts such as identity, law, citizenship, and justice (XXV). Nevertheless, the inequality which is also produced by these dynamics through the capitalist/colonial/patriarchal world systems must be taken into account. Boaventura de Sousa Santos in *Para Descolonizar Occidente: Más Allá Del Pensamiento Abismal* uses the metaphor of the abysmal line to describe the formation of hierarchies in knowledge production:

> Modern Western thought is abysmal thinking. This consists of a system of visible and invisible distinctions, the invisible ones constitute the foundation of the visible ones. Invisible distinctions are established along radical lines that divide social reality into two universes, the universe on "this side of the line" and the universe on the "other side of the line." The division is such that "the other side of the line" disappears as reality, becomes non-existent, and is in fact produced as non-existent. Non-existent means not existing in any relevant or understandable way of being. What is produced as non-existent is radically excluded because it lies beyond the universe of what the accepted conception of inclusion considers to be its other. Fundamentally, what most characterizes abysmal thinking is the impossibility of the co-presence of both sides of the line. This side of the line prevails insofar as it narrows the field of relevant reality. Beyond this, there is only non-existence, invisibility, non-dialectical absence[6]. (11 – 12)

Such distinction imposed by the western forms of knowledge, through racism, sexism, classism, and so forth, have served to maintain the epistemological sovereignty of the West. The social experience in the world is much wider and more varied than what Western scientific or philosophical tradition knows and considers significant ("A Critique of Lazy Reason" 158), and an understanding of the

---

5 For a more thorough critique of multiculturalism, see *Another Knowledge is Possible* (Santos, Nunes and Meneses 2008).
6 Translation provided by the author.

external diversity of knowledge would lead to the realization that all forms of knowledge are incomplete on their own.

According to Santos, the epistemologies of the South exist in opposition to what can be understood as the epistemologies of the North (meaning the hegemonic knowledge produced by Europe during the colonial period), which started in the 17th century, peaked in the 19th century, and extends to our days. Scientific knowledge clashed with the discourses of philosophy and theology, taking their place as the most effective mode of understanding reality. It was able to do so creating the illusion of objectivity: separating the subject and the object of inquiry, as well as separating the experience from the world, through the usage of controlled and meticulous experimentation. The discourse of science, as the only rigorous form of acceptable knowledge legitimizes the domination of all the Others and their different forms of understanding and interpreting phenomena. The question of what counts as knowledge is crucial for the decolonization of knowledge, since it dictates how the humanity of subjects who produce knowledge is perceived in the construction of societies and their interactions. Santos' epistemologies of the South and the ecology of knowledges are strategies that promote intersectional research intended to recognize and validate the knowledge produced by those who are systematically oppressed/dominated/excluded/ because of the capitalist/patriarchal/colonialist world systems.

2.1.3 Epistemologies of the South

In this section I will explore in more depth the circumstances of knowledge production, as sociology helps to support the argument of recognition of difference in forms of knowledge production, as well as fostering an interdisciplinary understanding of the issues that are going to be analyzed in this study through the perspectives of the Ecology of Knowledges and the Epistemologies of the South. The South implied here is not simply geographical, but geopolitical, meaning the nations, countries or social groups that have lived under systemic forms of exclusion/oppression. Santos recognizes the need to rehabilitate forms of knowledge that were discredited during/after the colonial process, at the same time that he preconizes the epistemological interdependence among diverse ways of knowing. This approach rejects the hierarchical dimension of descriptions such as "third world" or "developing" countries, which assert the inferiority of those who do not belong to the geo-political North.

The objective of knowledge is essentially the obtaining of truth, which can be understood here as the representation of the real. Scientific knowledge is based on rigorous criteria, and is achieved through the observation of regularities and patterns. These are translated into laws/theories that explain/represent nature/society/reality. Thus knowledge is meant in this configuration to be neutral and objective. This model, however, exists inside what may be called a politics of knowledge, which is far from neutral, since it takes place inside the main modes of Western-centric domination, capitalism, colonialism and patriarchy. Modern science, as a paradigm of transformation of the world, may legitimize the models of domination of these systems. To hold the control over knowledge production is to be able to hold the control of the processes of appropriation of reality, and therefore to dominate those who do not factor in this equation. A critique of science sparks from two different stances: internal and external. The internal critique of science takes into account the plurality of forms of knowing inside the hegemonic model, looking into the concurrent conceptions of knowledge that are at play and compete as the closest to truth/reality. An example of such internal plurality is the emergence of feminist epistemologies, which refigured the ways in which science was thought within the hegemonic model, showing that truth was a social concession that depended on the subject of enquire and his/her conditions/location. These ways of knowing may be not only related to sex/gender, but to different factors which were used as suppressive devices in the same way. Thus, the feminist critique of science enlarged the experience of science, democratizing its practices not only for women, but for all, especially for the oppressed groups. Santos, Nunes, and Meneses make this more explicit:

> From the extensive body of literature on this subject, it is obvious that the consequences of feminist critique and of the debates over the science-gender link are, first, the denaturalization of the male dominance of modern science, sustained by a range of institutions, practices, and occupational ideologies; and second, the identification of the conditions associated with the constitution of knowledge subjects – not only gender, but ethnicity, class, nationality, or religion, to name only a few – and the consequent development of 'strong' forms of objectivity, linked to the idea of the 'positioned' or 'situated' subject. (XXXII)

The external critique of science regards all the other forms of knowing that were/are seen as not sufficiently rigorous ("scientific") to deserve being incorporated in the dialogue of modernity. Those forms of knowledge are frequently referred as indigenous, primitive, backwards, mystical, magical, obscurantist, and many other dismissive characterizations, which ultimately lead to an epistemicide as mentioned above. Gayatri Spivak was already aware of such questions related to epistemic violence, in "Scattered Speculations on the Question of Value" she stresses how the canon came to be as a result of the obliteration of the knowledge produced by those deemed as Others:

We cannot avoid a kind of historico-political standard that the "disinterested" academy dismisses as "pathos." That standard emerges, mired in overdeterminations, in answer to the kinds of counter-questions of which the following is an example: What subject-effects were systematically effaced and trained to efface themselves so that a canonic norm might emerge? Since, considered from this perspective, literary canon formation is seen to work within a much broader network of successful epistemic violence, questions of this kind are asked not only by feminist and Marxist critics, but also by anti-imperialist deconstructivists. (110)

What Spivak is making evident is that epistemic violence exists when there is a deliberate effort to erase forms of knowing and subjectivities that were not part of the colonial norm, an effort performed either by the colonizer or by the colonized who has absorbed this logic, relegating non-canonical perspectives to the realm of "pathos" and inadequacy, imposing the worldview of the colonizer as the only valid form to deserve interest. Similarly, Santos states in *A Crítica da Razão Indolente:*

The epistemological privilege that modern science grants itself is, therefore, the result of the destruction of all alternative knowledges that could come to jeopardize that privilege. In other words, the epistemological privilege of modern science is the product of an epistemicide. The destruction of knowledges is not an epistemological artifact without consequences, but rather implies the destruction of social practices and the disqualification of social agents who operate according to the knowledges in question[7]. (242)

This text was revisited by the author in English, two years later, "A Critique of Lazy Reason: Against the Waste of Experience", in which Santos examines the model of rationality proposed by Gottfried Leibniz in 1710. Such model is based on the inevitability of the future: since it will happen with or without our intervention, it is better to simply enjoy the present. This form of reasoning is deemed lazy for it gives up thinking in the face of necessity and fatalism. Santos critiques this model in four different instances, demonstrating how the "laziness" of such reasoning occurs:

- *impotent reason* does not exert itself because it thinks it cannot do anything against necessity conceived of as external to itself;
- *arrogant reason* feels no need to exert itself because it imagines itself as unconditionally free and, therefore, free from the need to prove its own freedom;
- *metonymic reason* claims to be the only form of rationality and, therefore, does not exert itself to discover other kinds of rationality; or, if it does, it only does so to turn them into raw material; and

---

7 Translation provided by the author.

– *proleptic reason* does not exert itself in thinking the future because it believes the future is already known – it conceives of the future as linear, automatic, and infinitely overcoming the present. (159–160)

Santos states that there are three initial points for the critique of these forms of reasoning: firstly, an understanding that the experience of the world exceeds the western understanding of the world; secondly, an understanding of the world and the ways in which it creates and maintains social power is heavily related to concepts of time and temporality; and finally, the most fundamental characteristic of western rationality (lazy reason), is the contraction of the present and the enlargement of the future. The contraction happens due to the selection of a totality, in detriment of a plurality, making room for an infinite expansion of the future, with an infinite number of expectations. This kind of reason is responsible for locking us in the fleeting present, between the past and the future. Santos states:

> Under its various forms, lazy reason underlies the hegemonic knowledge, whether philosophical or scientific, produced in the West in the past 200 years. The consolidation of the liberal state in Europe and North America, the industrial revolutions and capitalist development, colonialism, and imperialism constituted the social and political context in which lazy reason evolved. ("A Critique of Lazy Reason" 160)

As a reaction to this mode of knowledge production, Santos proposes an alternative form of reason, which he calls cosmopolitan reason. This kind of reasoning is aimed at expanding the present and contracting the future. This model is substantiated in three different procedures: the Sociology of Absences, the Sociology of Emergences and the work of translation. The stress in the contextual understanding of different struggles reiterates the ideas connected to intersectionality previously discussed, with one difference: the work of translation, aided by the Sociology of Absences and the Sociology of Emergences, is capable of providing a contextual analysis by comparative diatopical hermeneutics, while trying to avoid the possibility of imperialism and colonization, since its principle is based on the non-hierarchical interdependence of knowledges. Therefore, the decolonization of intersectionality is fostered by the epistemologies of the South and the ecology of knowledges, as they propose a situational and relational model of knowledge production that demonstrates the possibility of some form of cognitive and epistemological justice for the objects of analysis of the present work. Santos, Nunes, and Meneses summarize:

> The ecology of knowledges is an invitation to the promotion of non-relativistic dialogues among knowledges, granting "equality of opportunities" to the different kinds of knowl-

edge engaged in ever broader epistemological disputes aimed both at maximizing their respective contributions to build a more democratic and just society and at decolonizing knowledge and power. (XX)

2.1.4 Sociology of Emergences

The present work aims at identifying the possibilities and tendencies of possible futures in a corpus of literature produced by black diasporic female authors, trying to identify in the texts tendencies of emancipatory practices and knowledges. The Sociology of Emergences is concerned with "replacing the emptiness of the future according to linear time (an emptiness that may be all or nothing) by a future of plural and concrete possibilities, utopian and realist at one time, and constructed in the present by means of activities of care" ("Public Sphere" 54). It is focused on the endeavor of finding/creating alternatives that are encompassed in the present only as tendencies, yet not fully visible, or as Santos defines it: "[...] alternatives that are contained in the horizon of concrete possibilities" ("Public Sphere" 56). Santos cites the works of Ernst Bloch to help understand how modern philosophy has neglected this state of possibility, in detriment of reality and necessity ("Public Sphere" 55), stating that Block focuses on the concepts of *Alles* (All), *Nichts* (Not), and *Noch Nicht* (Not Yet), to investigate the nature of the possibility of change. To expand the possibilities of the future, one must perceive how the state of "possible" comes to exist in our conception, especially in time where over determinism is ever present in our speculations of the future. The Not yet is the most relevant theme since:

> The Not Yet is the more complex category because it expresses what exists as mere tendency, a movement that is latent in the very process of manifesting itself. The Not Yet is the way in which the future is inscribed in the present. It is not an indeterminate or infinite future, but rather a concrete possibility and a capacity that neither exists in a vacuum nor is completely predetermined. ("Public Sphere" 54)

Not yet is described as a form of anticipatory consciousness, as capacity and possibility. It exists in an element of uncertainty: though the conditions of such potency are only partially known, the outcome is still not yet defined. The possibility is not neutral, being a change to either the positive or the negative, but yet, a change of state. Santos states:

> The sociology of emergences consists in undertaking a symbolic enlargement of knowledges, practices and agents in order to identify therein the tendencies of the future (the Not Yet) upon which it is possible to intervene so as to maximise the probability of hope vis-à-vis the probability of frustration. ("Public Sphere" 56)

The intervention in the future, through the concrete possibilities of the present result in efforts to combat the despair and frustration of deterministic thought, which Santos aptly describes as the maximizing of hope. This symbolic expansion objective is twofold: to better understand the conditions of such possibility of hope, and to define the principles of action to promote such conditions. According to the author, the Sociology of Emergences replaces the idea of determination by the idea of care, caring for a future that is plural and more egalitarian, which tries to minimize the abysmal line between different subjects. In this sense, the ecology of knowledges is a form of epistemology that is inherently destabilizing, since it refuses conformity, and is focused in "action-with-clinamen[8]" ("Para Além do Pensamento Abissal" 32), meaning something that is inexplicable and alters the relation between cause and effect. This kind of action does not constitute a radical break with what exists, but functions with the cumulative effect of slight deviances, resulting in the possibility of new relations between people, social groups, beings, and so forth. Santos cites several fields in which the multiplicity and the diversity might come up, among which the most relevant for the present task are the experiences of knowledges, meaning the possible dialogues and conflicts among different forms of knowing; the experiences of development, work, and production, which are related to the possible dialogues and conflicts among different modes of production; and the experiences of recognition, which deal with the possible dialogues and conflicts among the classification systems ("Para uma sociologia das ausências" 259 – 260). The abundance of possible experiences comes with its own set of complexities and anxieties, as pointed by Santos. The gravest one is the atomization and extreme fragmentation of reality, which is responsible for the impossibility of ascribing meaning to social transformation. Such chasm has been responded for long with metonymic reason and proleptic reason, resulting in the massive waste of experience throughout modernity up to our days, producing a contracted present and an ever-expanding future. Cosmopolitan reason, on the other hand, states: "[...] the task before us is not so much to identify new totalities or to adopt other meanings for social transformation, but to propose new ways to think about such totalities and meanings" ("A Critique of Lazy Reason" 178). In this regard the Sociology of Absences is an interesting tool to consider what is there and cannot be seen, as it is going to be further explored.

---

8 Clinamen is a concept originally used by Epicurus to explain the inclination power of atoms so they cease to be perceived as inert.

2.1.5 Sociology of Absences

The Sociology of Absences represents a mode of understanding and producing knowledge that intends to bring to the foreground forms of knowing that have been made absent, or, as Santos states: "[...] research that aims to show that what does not exist is actually actively produced as non-existent, that is to say, as an unbelievable alternative to what exists. Its empirical object is impossible from the point of view of conventional social sciences" ("Public Sphere" 52). This strategy is of utmost importance in the context under scrutiny here, and that of the contribution of artistic/literary/creative discourses in the unmaking of oppression. The previous theorizations demonstrated how the discourses of black women writers have been made impossible/invisible/discredited/absent to the eyes of History, society, and subsequently, of the social sciences/scientific discourse. It also aimed at demonstrating how the overcoming of such condition is a constant struggle for these subjects. Morrison comments on such invisibility in "Unspeakable Things Unspoken – The Afro-American Presence in American Literature", stating that:

> We can agree, I think, that invisible things are not necessarily "not there;" that a void may be empty, but it is not a vacuum. In addition, certain absences are so stressed, so ornate, so planned, they call attention to themselves; arrest us with intentionality and purpose, like neighborhoods that are defined by the population held away from them. (136)

Santos explores how objects of inquire are made absent, in which a common denominator is the monocultural rationality, expressed through metonymic reason. The whole is described not as the sum of all its parts, but as a part turned into the reference to all others. Santos states:

> Metonymic reason is obsessed with the idea of totality in the form of order. There is no understanding or action without reference to the whole, the whole having absolute primacy over its parts. Therefore there is only one logic ruling both the behavior of the whole and each of its parts. ("A Critique of Lazy Reason" 161)

This totality must disregard all that it is unable to encompass, imposing itself over all the other parts. Thus, metonymic reasoning relies on the creation of dichotomies which create/reinforce hierarchies for its own benefit of dominance. This kind of reasoning, in conjunction with proleptic reason, is responsible for the contraction of present time, which "[...] conceals most of the inexhaustible richness of the social experiences in the world" ("A Critique of Lazy Reason" 163). This kind of reasoning is responsible for the creation of inadequacies of all that does not fit the reference model, resulting in violence, destruction and silencing of all the Others. Metonymic reason is accountable for the active crea-tion of non-existence or absence. The Sociology of Absences is an attempt at ex-panding the possibilities of the present by questioning how nonexistence is pro-duced. The author describes five different logics behind four modes of absence production: ignorant, backward, inferior, local or particular, and unproductive or sterile, he also describes the five ecologies that aim at combating such logics. Firstly, ignorant logic is related to the ideas or rigor and the monoculture of knowledge. It works by making modern scientific knowledge and high culture as the only possible criteria of truth and aesthetic value, where the production of a canon of knowledge and aesthetics produce exclusion and erasure. "All that is not recognised or legitimised by the canon is declared non-existent. Non-existence appears in this case in the form of ignorance or lack of culture" ("Public Sphere" 52). Once again, it is relevant to stress that literature produced by African-American and Afro-Caribbean writers has been long excluded from the canon, which does not mean that it did not fashion its own separate tradition in the United States or in the Caribbean in its Anglophone and Francophone in-stances. The ecology of knowledges is the response to such logic, trying to foster the interdependence among the various forms of knowing, indicating that a bet-ter comprehension of the "whole" depends on the different contributions of all the different parts that really comprise this entirety.

The second one, backward logic, states that everything that does not reso-nate with the monoculture of linear time, the notion that history has one mean-ing and direction (modern progress), is considered to be backward. Colonialism and coloniality are the main drives that created backward subjects, who seem not to inhabit the same time and history of those in power, and therefore are made absent from the process. "In this case, non-existence assumes the form of residuum, which in turn has assumed many designations for the past 200 years, the first being the primitive, closely followed by the traditional, the premo-dern, the simple, the obsolete, the underdeveloped" ("Public Sphere" 52). All these designations have been used to describe the African diaspora in the Amer-icas, during and after the colonial process. Conversely, the ecology of temporal-ities states that societies are formed through several temporalities, and that the disregard, suppression, or unintelligibility of practices is a result of such differ-ence, which escapes the canon of capitalist western modernity. To overcome such issue, one needs to relativize linear time, and recognize and value the different temporalities, so as to expand the present with a plurality of experiences of un-derstanding of time. The issue related to different temporalities is explored more evidently by Kincaid and Danticat in the following analysis, as they experience a different kind of perception of time when contrasting the island life in the Car-ibbean and the urban life in the United States.

The third logic identified by Santos is the logic of social classification, which is based on the principle of classifying populations into groups that create hierarchies, which are eventually naturalized. Santos states:

> Racial and sexual classifications are the most salient manifestations of this logic. Contrary to what happens in the relation between capital and labour, social classification is based on attributes that negate the intentionality of social hierarchy. The relation of domination is the consequence, rather than the cause, of this hierarchy, and it may even be considered as an obligation of whoever is classified as superior (for example, the white man's burden in his civilising mission). ("Public Sphere" 53)

These are the dynamics presented by Quijano and Wallerstein, as well as Walter Mignolo in their conjectures about the coloniality of power, and Maria Lugones about the coloniality of gender, which is concerned with determining difference as an inferior trait, and consequently as inequality, as it reiterates its entitlement to define what/who is different. Lugones states:

> I am interested in the intersection of race, class, gender and sexuality in a way that enables me to understand the indifference that men, but, more importantly to our struggles, men who have been racialized as inferior, exhibit to the systematic violences inflicted upon women of color. I want to understand the construction of this indifference so as to make it unavoidably recognizable by those claiming to be involved in liberatory struggles. This indifference is insidious since it places tremendous barriers in the path of the struggles of women of color for our own freedom, integrity, and wellbeing and in the path of the correlative struggles towards communal integrity. The latter is crucial for communal struggles towards liberation, since it is their backbone. The indifference is found both at the level of everyday living and at the level of theorizing of both oppression and liberation. The indifference seems to me not just one of not seeing the violence because of the categorial separation of race, gender, class, and sexuality. That is, it does not seem to be only a question of epistemological blinding through categorial separation. ("The Coloniality of Gender" 1)

The context hitherto presented is an example of this logic at play, where the social classifications of race and sex, and consequently of class, create an invizibilization of the production of knowledge of the African diaspora in the Americas. White supremacy presents the knowledge of the lesser Other as inexistent, or inferior, strengthening the idea of its own superiority in a naturalized fashion. The inferiority of the group cannot be overcome given its essentialist nature, and therefore, cannot be perceived as a legitimate alternative to the dominant discourse. The ecology of recognitions is based on the principle of the mutual recognition of equal differences, which is pursued through the questioning of hierarchies by a procedure denominated critical ethnography. Santos states:

> This [submitting hierarchy to critical ethnography] consists in deconstructing both difference (to what extent is difference a product of hierarchy?) and hierarchy (to what extent is hierarchy a product of difference?). The differences that remain when hierarchy vanishes become a powerful denunciation of the differences hierarchy reclaims in order not to vanish. The feminist and the indigenous movements have been in the forefront of the struggle for an ecology of recognitions. ("The World Social Forum" 21)

The fourth logic, the dominant scale, understands that the scale implemented as model determines the irrelevance of all other possibilities. Santos points to two forms of the dominant scale in the West, namely the universal and the global. They are superior to any reality that is contextual. "According to this logic, non-existence is produced under the form of the particular and the local. The entities or realities defined as particular or local are captured in scales that render them incapable of being credible alternatives to what exists globally and universally" ("Public Sphere" 53). This logic applies to the examples in this study because the particularity of the African diaspora experience in the Americas is produced as contextual, and therefore inferior and not representative face the white supremacist dominant narrative. The experience of Black women in particular is taken to be specific as opposed to the experience of white women, which is understood as universal. The ecology of trans-scales is the response to this issue provided by the Sociology of Absences. It works through the identification and retrieval of what has not been absorbed by hegemonic globalization. All that which has been absorbed in such order is designated as a "localized globalism" by Santos. The identification and retrieval of the local also allows for the possibility of counter-hegemonic globalization, since it broadens the scope of diversified social practices as it offers alternatives to localized globalisms. By focusing on the literature produced by diasporic African female subjects, the scope of what may be identified as American literature, or even African-American literature, is expanded and complexified, contributing to a more pluralistic understanding of society, as well as fostering the creation of a reality that is more accepting of difference.

The final logic is the logic of productivity, which dictates that economic growth is the sole objective that is possible,: "[...] non-existence is produced in the form of non-productiveness. Applied to nature, non-productiveness is sterility; applied to labour, sloth or professional disqualification" ("Public Sphere" 54). Though black labor in the Americas was extremely productive for the capitalist world order, it was never presented/perceived as the fruit of the effort of these subjects, but as the result of the white man's diligence and management of these forces, thus rendering black labor to a realm of symbolic invisibility. Later, after the abolishing of slavery, the paradigm shifted since the introduction of black subjects in the labor market was tainted by the stigma of disqualification. Another perverse facet of this logic can be seen in the relation of black writers to the white canon, whose production is seen as lesser in comparison, due to the lack of tradition in written literature in the white sense. The Sociology of Absences works against such logic by retrieving and valuing alternative production systems, popular economic organizations, labor cooperatives, among other forms of labor organization and production that have been made absent or inferior by the capitalist world order. It is possible to claim that by focusing on diasporic African female subjects in the United States, as well as presenting data that corroborates the importance of these authors in regard to their production, we contribute to the valorization of the work of these subjects, contradicting the logic of productivity imposed by colonialism/capitalism/sexism.

2.1.6 Intercultural Translation

Another important concept for the understanding of the problematic that is about to be approached is (inter)cultural translation. Homi Bhabha coined the concept of cultural translation in the context of a critique of multiculturalism, in the sense of trying to overcome essentialist readings of pure and unique cultural identities/communities, and consequently, overcoming the hierarchies imposed by these communites. Bhabha's "third space" functions as the site of hybridity and ambivalence in which negotiation of difference between cultures is possible. Bhabha states: "[c]ultural translation desacralizes the transparent assumptions of cultural supremacy, and in that very act, demands a contextual specificity, a historical differentiation *within* minority positions" (*The Location of Culture* 228). According to the author, the difference, which always imbued in the ambivalence inherent to the colonial discourse (*The Location of Culture* 85), is similar to the ambivalence present in the act of cultural interpretation, in the sense that the production of meaning derived from two or more cultural identities/systems/communities depends on the hybridity of a third space. The act of cultural translation happens during this negotiation process that can be seen in the light of radical subversiveness, as it offers the dialogical space for the creation of meaning and the possibility for change.

Judith Butler explores cultural translation in the context of feminist transnational endeavors of solidarity in *Precarious Life – The Powers of Mourning and Violence*, evidencing the ethical stance of cultural translation when dealing with different contexts, paying attention to the risk of inscribing oppressive language of Western politics on bodies that live in different contexts, stating that "[w]e have to consider the demands of cultural translation that we assume to be part of an ethical responsibility (over and above the explicit prohibitions against thinking the Other under the sign of the "human") as we try to think the global dilemmas that women face" (49). Cultural translation is seen by these authors as the possibility to deal with the diversity of experiences, knowleges, and demands in the world, trying to find ways to connect among several different and non-hierarchical forms of knowing. Translation may happen among hegemonic and nonhegemonic knowledges, as well as among different nonhegemonic knowledges. Counterhegemony is possible through the aggregation of nonhegemonic knowledges, fostered by the mutual intelligibility brought about by the work of translation. Such articulation also happens when dealing with social practices and their agents, dealing not only with the knowledges produced by them, but also with the organization and objectives of action of these applied knowledges, which are transformed into practices and materialities, resulting once again in the possibility of a stronger stance of counterhegemony. Santos, Nunes and Meneses, claim that:

> The theory of translation allows common ground to be identified in and indigenous struggle, a feminist struggle, an ecological struggle, etc., without erasing the autonomy and difference of each of them. Translation is also fundamental to articulation between the diverse and the specific intellectual and cognitive resources that are expressed through the various modes of producing knowledge about counter-hegemonic initiatives and experiences, aimed at redistribution and recognition and the construction of new configurations of knowledge anchored in local, situated forms of experience and struggle. (XXVI)

It is possible to claim that literature produced by diasporic African female subjects in the United States stands for a kind of cultural translation, in which the anti-racist struggle, the anti-classist struggle and the feminist struggle find a common ground of identification, as they denounce their local realities of oppression in their literary production.

Following the thought of Bloch, Santos proposes that if the world is an inexhaustible totality, maintaining many totalities, this reinstates the partiality of such totalities, as well as the totality of all parts. The result of such proposition is finally the rejection of the model of a single great theory. Which leads to the question of what is the other option to such model. Santos proposes the work of translation as an alternative:

> Translation is the procedure that allows for mutual intelligibility among the experiences of the world, both the available and the possible ones, as revealed by the sociology of absences and the sociology of emergences. This procedure does not ascribe the status of exclusive totality or homogenous part to any set of experiences. The experiences of the world are viewed at different moments of the work of translation as totalities or parts and as realities that are not exhausted in either totalities or parts. ("A Critique of Lazy Reason" 179)

This intelligibility is fostered by a form of interpretation that Santos calls diatopical hermeneutics, the interpretation of two different cultures, trying to identify similar concerns between them and the answers they provide. Diatopical hermeneutics assumes that all cultures are incomplete, and therefore, both could benefit from the perspective of the other, in their dialogues and conflicts. By exposing the different kinds of struggles experienced by diasporic African female subjects in the United States, the authors are capable of fostering an environment in which diatopical hermeneutics may take place, as the readership has the opportunity of achieving a better intelligibility of the struggles of others, through the identification of similarities/differences between their experiences and the ones being reported by the text. Translation in this sense focuses on knowledges and practices (and their agents), fostering as its final goal the ecology of knowledges, generating what Santos calls a form of negative universalism.

Santos states that the Sociology of Absences and the Sociology of Emergences enlarge the number of possible experiences of the world, but, since there is no single principle of social transformation, it is not possible to determine the articulations and hierarchies of these experiences in an abstract context. The work of translation of practices is capable of assessing these articulations in materialistic terms, and therefore, is capable of identifying the reciprocal instances in which the different knowledges are mutually understandable and beneficial, boosting their counterhegemonic and antisystemic potential because of these alliances. Santos states: "[...] The work of translation becomes crucial in defining, in each concrete and historical moment or context, which constellation of nonhegemonic practices carry more counterhegemonic potential" ("A Critique of Lazy Reason" 182). Although the translation of practices and their agents is more connected to the streak of knowledge produced by social movements in the above mentioned theorization, it is also possible to think about its presence in texts since the authors can be taken as the agents and their texts as their interventions in the social reality they inhabit. I argue that literature, especially when created by the margins, is a form of cultural translation of social struggles and the subsequent trauma that inevitably is present in the objects to be analyzed. The margins here are a reference to the theorization of bell hooks who states in *Yearning – Race, Gender, and Cultural Politics:*

> This site of resistance is continually formed in that segregated culture of opposition that is our critical response to domination. We come to this space through suffering and pain, through struggle. We know struggle to be that which is difficult, challenging, hard and we know struggle to be that which pleasures, delights, and fulfills desire. We are transformed, individually, collectively, as we make radical creative space which affirms and sustains our subjectivity, which gives us a new location from which to articulate our sense of the world. (153)

3 Danticat and Kincaid

"My past was my mother; I could hear her voice, and she spoke to me not in English or the French patois that she sometimes spoke, or in any language that needed help from the tongue; she spoke to me in language anyone female could understand. And I was undeniably that—female" (Kincaid, *Lucy*)

"According to Tante Atie, each finger had a purpose. It was the way she had been taught to prepare herself to become a woman. Mothering. Boiling. Loving. Baking. Nursing. Frying. Healing. Washing. Ironing. Scrubbing. It wasn't her fault, she said. Her ten fingers had been named for her even before she was born. Sometimes, she even wished she had six fingers on each hand so she could have two left for herself" (Danticat, *Breath, Eyes, Memory*)

Considering the line of argumentation previously presented, literature may be considered a rich source of knowledge, in which processes of identity construction are represented and are capable of informing the readership of the life experience of others, contributing to the construction of a more plural knowledge about such experiences and the world. Backing such perspective, Isabel Caldeira writes in "Toni Morrison and Edwidge Danticat: Writers as Citizens of the African Diaspora, or 'The Margin as a Space of Radical Openness'": "I submit that literature is an important mode of reappropriating history and offering a counterhegemonic perspective to create social awareness and promote a critical competence to resist the entanglements of wealth and power" (207). Through its political, ethical, and aesthetic features, literature is capable of disturbing the ways in which hegemony functions, demonstrating the radical possibilities of other points of view, as well as humanizing subjects who have long been produced as inferior and unequal by the *status quo*. Caldeira expands: "[l]iterature has indeed the power to give us the emotional access to experience and either humanize our world or reveal its/our (in)humanity, displaying the universe in a new light, and sharpening our senses and intellectual perspicuousness" (Caldeira 2017: 208). Toni Morrison calls this ability of literature the "sharpening of the moral imagination", especially in its capacity to translate the experience of trauma through the capacity of refiguring language in meaningful ways:

Certain kinds of trauma visited on peoples are so deep, so cruel, that unlike money, unlike vengeance, even unlike justice, or rights, or the goodwill of others, only writers can translate such trauma and turn sorrow into meaning, sharpening the moral imagination. A writer's life and work are not a gift to mankind; they are its necessity. (*Burn This Book* 4)

Both works about to be analyzed in this chapter are examples of these paradigms. They are fierce representations of how trauma and violence, taking differ-

https://doi.org/10.1515/9783110752755-005

ent shapes and being expressed in different dynamics, demonstrate that they are also formative forces in the identity construction of subjects.

Drawing from the work of Édouard Glissant, Jana Evans Braziel makes use of the concepts of rhizomes and relations to think about the diasporic reality of writers such as Kincaid and Danticat and their complex history as migrant subjects from the Caribbean:

> Thus, the Caribbean's histories, according to Glissant, are not those of roots, but of rhizomes: its entangled histories (of colonialism, violence, indigenous genocide, slavery, plantation economies, diaspora, racial and cultural hybridity) are points of rhizomic contact and proliferation – new creations, as Derek Walcott envisions it – not the singular and deeply-rooted history of Empire, wholly and intactly transplanted from Europe through conquest, as the colonialist may imagine. In contrast to rooted notions of history, rhizomic histories unfold multilaterally, since rhizomes spread underground through sending out shoots sideway. ("Daffodils, Rhizomes, Migrations" 112)

Both texts are comprised of the work and life of the authors, since Jamaica Kincaid and Edwidge Danticat draw the main themes of their fiction from their life experiences of entangled histories in different degrees. A common feature of these works is the refusal to subsume to a history of Empire and colonialism, as both authors defy the unitary account of the histories that are usually associated with the African diaspora.

Kincaid's work has been read in the light of autobiographical theories, as demonstrated later on, while Danticat's remains in the realm of fiction, though an appendix has been added to her later publication in which the author writes a letter to her protagonist addressing the complexities related to how the novel has been received. Another common feature in these authors is the migration to the United States, both to New York, and most especially, how growing up in this new context affects the construction of a sense of identity. These works exist within a context of hybridity, where the actually lived and the subsequent degree of fictionalization co-exist in their multiplicity, or in rhizomatic form, to use Glissant's terminology.

3.1 Jamaica Kincaid

Jamaica Kincaid was born Elaine Potter Richardson, renaming herself after migrating to New York at the age of seventeen. The desire to move to the United States was present in the writer's life from an early age. Her father, who had worked as a carpenter in the American base in Antigua during the Second World War, had met several Americans during the period. This exchange left him impressed with the American creativity and sense of possibility, which he transmitted to Kincaid. In a column in "Talk of the Town", published in *The New Yorker* on the 4th of July 1976, Kincaid writes of her wish to move to the United States:

> When I was nine years old, I added an extra plea to my prayers. Up to then, I would say the Twenty-third Psalm and the Lord's Prayer, and I would pray that God would bless my mother and father and make them live long enough to see me become a grown woman, and would bless me and help me to be a good girl. But when I was nine years old I started adding, "And please, God, let me go to America." I did this for six years straight. As I grew older, I got my own ideas about why I wanted to go to America. It had to do with pink refrigerators; shoes that fall apart if you get caught in the rain (because that way you can get a new and different pair); the flip in Sandra Dee's blond hair as she played a pregnant teenager in the movie A Summer Place; Doris Troy, the way she looked and the way she sang "Just One Look"; and, of course, Negroes, because any place that Negroes are is cool. (*Talk Stories* 58)

Lucy, the protagonist in the homonymous book, is inspired in the identity Kincaid tried to fashion during her early years in the United States. The form of the *Bildungsroman* is evocative here, but as previously mentioned, the authors selected for this study disturb the canonical form associated with this genre. Claudia Marquis, in her article "'Not At Home In Her Own Skin': Jamaica Kincaid, History and Selfhood" states that "[...] the novel looks like a modified *Bildungsroman*; modified in taking as its frame the period in Lucy's life that appears to matter most in forming her mature identity" ("Not at Home"). The main themes associated with the novel are migration, colonialism, metaphors of domination and ultimately the mother-daughter relationship, themes that have been continually explored throughout Kincaid's production as a whole. Marquis shares a small report of Kincaid in a conference, summarizing what her work, and herself, are all about:

> At a reading of her fiction in November, 1997, asked how her works might be taught, Kincaid described herself—backchat fashion—as "a woman deeply interested in power," "a woman who has reduced all the great issues of worldly politics to a quarrel between a mother and her daughter". ("Not at Home")

In *A Small Place*, Kincaid describes the inability that people from small places, such as Antigua, have to make order out of chaos, namely in the organization of events in logical sequence in their lives, as well as their role in this ordering, either as subjects or as spectators. Kincaid's production might be seen in this perspective, where the domestic is representative of the public, and repetition seems to be the modus operandi of this way of perceiving reality, exposing power structures through the particular. The issues that permeate Kincaid's life and work are all present in *Lucy*. The novel is by no means a definitive account of this period of the author's life, even though the life experience of Kincaid heavily informs her fiction. Marquis comments on the relation of history and identity construction in Kincaid's production, stressing how her personal history, in addition to the history of Antigua, come together in the text, assuming allegorical proportions:

> What we are constantly aware of, of course, is how the past, not such a small place indeed, haunts the story that she presents here, investing every move with significance beyond itself; in this it might be said that her history comes under the rule of allegory and becomes as well her island's story. ("Not at Home")

In the novel, published in 1990, Lucy is nineteen when she arrives in an unnamed city, which by its features resembles New York. The protagonist is sent to this place by her mother to work as an *au pair*, with the intent of sending money back home. At the same time, Lucy should be studying to be a nurse, one of the traditional tracks for girls from her small island. She describes it as a small place, without ever naming it: "I was born on an island, a very small island, twelve miles long and eight miles wide; yet when I left it at nineteen years of age I had never set foot on three-quarters of it" (134). This description fits perfectly Kincaid's original birthplace, Antigua. Lucy left her mother back in the island, and with this abandonment she hoped she could create a new identity that would be different from all the circumstances she was put through. When first arriving at her destination, the readership is given a glimpse of how this moment is relevant as a change of paradigm in the story of this character, though she is not completely at ease.

> In a day-dream I used to have, all these places were points if happiness to me; all the places were lifeboats to my small drowning soul, for I would imagine myself entering and leaving them, and just that – entering and leaving over and over again – would see me through a bad feeling I did not have a name for. I only knew I felt a little like sadness but heavier than that. (3)

Lucy demonstrates in its first pages the idealization the protagonist imbues into the idea of the United States, and how it seemed to be the opportunity to start anew, finally overcoming the sadness and depression she thought she had left back in Antigua. The expression "drowning soul" makes a clear allusion to the maritime universe the island invokes, and remaining in that situation would be fatal for the protagonist. The constant getting in and out of buildings indicates that the foreknowledge that the protagonist has of her future life was at most superficial, but enough to motivate her to leave. The landscape with its huge buildings and famous landmarks translate the awe of the immigrant before her new surroundings, where everything is the opposite she could find back home. The materialization of this reality is the first step towards the fashioning of a new self, starting from the unmaking of past idealizations.

> Now that I saw these places, they looked ordinary, dirty, worn down by so many people entering and leaving them in real life, and it occurred to me that I could not be the only person in the world for whom they were a fixture of fantasy. It was not my first bout with the disappointment of reality and it would not be my last. The undergarments that I wore were all new, bought for my journey, and as I sat in the car, twisting this way and that to get a good view of the sights before me, I was reminded of how uncomfortable the new can make you feel. (4)

The reality of the buildings is the first example of American fantasy being undone. When considering these buildings, what once was thought to belong to a realm of unattainable superiority, to see them for what they are, is the evidence that Lucy can no longer hold on to the childlike previous notions she had of the host country. The final analogy of the uncomfortable underwear gives the reader an example of how something new does not equate to something pleasing. By locating this discomfort in the body, and namely in an area related to her sexuality, Kincaid is also pointing to how much of the woman Lucy intends to be needs to adapt. Lucy is insecure of what may happen in this new chapter of her life, though it is one that she had long envisioned. Returning to the island comes as a first impulse at the first shock. The tension she feels is expressed in terms of the meta-analytical capacity of literature to name the unnamable, or in Morrison's terms, an unspeakable thing spoken at last.

> In books I had read—from time to time, when the plot called for it—someone would suffer from homesickness. A person would leave a not very nice situation and go somewhere else, somewhere a lot better, and then long to go back where it was not very nice. How impatient I would become with such a person, for I would feel that I was in a not very nice situation myself, and how I wanted to go somewhere else. But now I, too, felt that I wanted to be back where I came from. I understood it, I knew where I stood there. If I had had to draw a picture of my future then, it would have been a large gray patch surrounded by black, blacker, blackest. (8)

For Lucy, remaining in the island would be the confirmation of a previously determined life with relatives and acquaintances that produced in her extremely negative feelings. To imagine that homesickness would be the feeling that would dominate the first real contact with her idealized world demonstrates how much of a growing path the character would still have to go through to achieve a new sense of self, since she no longer fitted the reality she had come from.

Neither her bed (she had outgrown) nor the island (a small place) were capable of containing her anymore. The rage felt by the protagonist is mostly concentrated in her mother she had left in Antigua. The mother-daughter relationship is permeated by a feeling of resentment, which comes about for several reasons, such as the blame Lucy lays on her mother for never having given her the opportunities her younger brothers were offered, mainly due to her sex. She began her plans to leave this reality at an early age, in a clear defiance of the patriarchal economy that would limit her options profoundly, had she stayed. Interestingly, Lucy seems to be less bothered by the male part of the decision, her father, as she does not appear to pay him any respect, perceiving his flaws from the very beginning of her life and dismissing him as a prominent figure in her life. On the other hand, her mother, as a female, should understand better her longings and aspirations, so this misrecognition weighs much heavier in their relationship and ultimately in the identity formation of the protagonist.

> I was an only child until I was nine years old, and then in the space of five years my mother had three male children; each time a new child was born, my mother and father announced to each other with great seriousness that the new child would go to university in England and study to become a doctor or lawyer or someone who would occupy an important and influential position in society. I did not mind my father saying these things about his sons, his own kind, and leaving me out. My father did not know me at all; I did not expect him to imagine a life for me filled with excitement and triumph. But my mother knew me well, as well as she knew herself: I, at the time, even thought of us as identical; and whenever I saw her eyes fill up with tears at the thought of how proud she would be at some deed her sons had accomplished, I felt a sword go through my heart, for there was no accompanying scenario in which she saw me, her only identical offspring, in a remotely similar situation. To myself I then began to call her Mrs. Judas, and I began to plan a separation from her that even then I suspected would never be complete. (130–131)

The traumatic process of individuation from this maternal figure, a subject that will permeate all of Kincaid's production, is one of the main themes in this novel, even when the mother is not directly present in the plot. The feelings toward her mother that Lucy desperately tries to suppress are ultimately transferred to another maternal figure: Mariah, the host mother in the United States. Lucy claims at one point: "[m]y past was my mother" (90), reiterating what must be left behind, and making clear that to start anew she would need to reinvent herself as a completely independent being. The protagonist is able to do so through a relentless critique of (neo)colonialism, one that is perpetrated by most of the white people who surround her in her new life. Mariah is the focus of both Lucy's love and hatred, embodying the maternal figure Lucy left behind and the colonial logic of neoliberal points of view that she finds in American soil. The critique also looks backwards, as Lucy scrutinizes her mother's abidance to imperial England in Antigua, while Lucy sees the entangled histories of racism, sexism and privilege. In this representation of self-invention, Lucy realizes that most of the white characters seem to be unaware of their biases, which makes the protagonist angry and frustrated. These feelings are also frequently associated with Kincaid's writing at large.

3.1.1 Fiction, Autobiography and Truth Telling in Jamaica Kincaid

Moira Ferguson, in "A Lot of Memory: An Interview with Jamaica Kincaid", published in 1994, interrogates the author regarding her relation to the autobiographical discourse. Besides revealing the centrality of this mode of writing to her production, Kincaid points to the refiguring of her identity as a force for the creation of her characters.

> My writing has been very autobiographical. The events are true to me. They may not be true to other people. I think it is fair for my mother to say, "This is not me." It is only the mother in the books I've written. It is only the mother as the person I used to be perceived her. [...] There is no reason for me to be a writer without autobiography. There is none at all. I have no interest in writing as some sort of exercise of my class. I am not from a literary class. For me it was really an act of saving my life, so it had to be autobiographical. I am someone who had to make sense out of my past. (Ferguson and Kincaid 176)

Writing in autobiographical terms is what helps Kincaid develop an understanding of her personal history, as well as her collective history, as a peripheral black woman from the Caribbean. Leigh Gilmore, in *The Limits of Autobiography – Trauma and Testimony,* plays with the conception of Paul Valéry that every theory is the fragment of an autobiography, stating that theory is closer to the real and autobiography more related to subjectivity. By reversing Valéry's premise, Gilmore sets the stage for broader questions regarding the autobiographical discourse:

> [...] Every autobiography is the fragment of a theory. It is also an assembly of theories of self-representation; of personal identity and one's relation to family, a region, a nation; and of citizenship and a politics of representativeness (and exclusion). How to situate the self within these theories is the task of autobiography which entails the larger organizational question of how selves and milieus ought to be understood *in relation to each other.* (12)

Gilmore is questioning the centrality of the individual in Western tradition in detriment of the social. Her reasoning turns our attention to what are the factors that create the individual subjects, their relations, their access to power and its asymmetries. Autobiography is capable of making clearer the conditions in which the individual and the imagined collective arise in the popular imagination. Thus Gilmore states that the cultural work performed in the name of autobiography is relevant to representations of citizenship and the nation allying the representative person the national fantasy of belonging (12).

Making sense of the colonial history of her island and its consequences for present day Antiguans is also part of this task. In the same interview, when asked about the literary tradition she belongs to, Kincaid responds that in some way she belongs to an English tradition, as she comes from a former colony in the British Caribbean. Commenting on the education she received, the presence of the empire shaped the entirety of the history she was taught:

> We got kind of the height of empire. They were trying to erase any knowledge of another history, another possibility. So we learned Shakespeare, the King James version, Wordsworth, Keats. That's the tradition. I'm of the English-speaking-people tradition. British people, English people. (Ferguson and Kincaid 168)

This, however, does not compromise the anticolonial consciousness that Kincaid developed throughout her life and production. The collective history of the oppressed Antiguans, and colonized peoples at large, has fierce dialogues in her texts with the English colonialist narrative she was taught in school. The relational aspect of this collective history also comes at play when Kincaid juxtaposes her experience with the normative whiteness of the United States, and its collective history of erasure of black bodies and black experiences. On the matter, Marquis states that:

> [...] the novels trace with remorseless accuracy the gravitation of their central characters towards an identity inflected in innumerable ways by the presence of a master class of colonizers, and in doing so they seem to chart the struggles that variously constitute the national community. ("Not at Home")

Racism, sexism, and the colonial rationale are constantly under scrutiny is Kincaid's work, even when the writer chooses to display them in a personal stance, and as Marquis states, "[...] the personal inevitably exposes the scars where it has been touched by the political" ("Not at Home"). To be a writer in Kincaid's experience relates to what Morrison has described as the necessity of the author's life and work, and ultimately performs the shaping of the moral imagination in the transformation of trauma into meaning. Trauma has been present in Kincaid's personal and collective history, taking different shapes at different moments. Her serial autobiography, refracted and fragmentary as it is, is testament of the role of this mode of discourse in the shaping of the cultural performance of nation, belonging, and citizenship, as stressed by Gilmore. Another relevant matter in this case is that Kincaid writes in the United States but always with Antigua in her mind as a frame of reference, even if not directly, as in *Lucy*. Marquis also comments on the different aspects associated to Kincaid's production, stating that:

> The novels are written *against* paradise, but haunted by half-memories. They are written against romance, the work of an author who never willingly owns up to the softer feelings, and never flinches from getting things down as they undoubtedly are; observation seems almost sociological and the narrative like a case history. ("Not at Home")

Kincaid's production can be classified in several different genres, from fiction to essays, biography and travel writing, though none is capable of encompassing it exclusively, and all of them are connected to her lived experience. Kincaid started working as a writer for *The New Yorker*, writing columns for a section called "The Talk of the Town". Her editor and mentor was William Shawn, who also published many of her short stories in the magazine. Kincaid's serial autobiography can be traced to *At the Bottom of the River* (1983), *Annie John* (1986), *Lucy* (1990), *Autobiography of My Mother* (1996), *My Brother* (1997), *Mr. Potter* (2002), and *See, Now, Then* (2013). The non-fiction titles – *A Small Place* (1988), *My Garden Book* (2001), *Talk* Stories (2001) a collection of her writings for *The New Yorker*, and *Among Flowers – A Walk in the Himalayas* (2005) – can all, someway, also express this sense of the autobiographical continuum[1].

Revisiting her life story and its episodes in writing helps to (re)construct memory in a way that is more complete. Paradoxically, while memory is constantly altered at the time it is recollected, revisiting these episodes serves to compose a more complex paradigm of any given event. Kincaid tells Ferguson:

> MF: You said a few years back, "*To say exactly what happened was less than what I knew happened.*" That's a vital statement. I wonder if you would comment on it.

> JK: I would say that still. It is always less; whatever you say is always less. I think that is always true. [...] I go over my life all the time – events in my life – and suddenly I remember that I was with my mother somewhere. Over and over again, I remember my mother and I went somewhere and only now it has dawned on me what was happening. So when I say, "My mother and I go somewhere and this is the event of that day," all of that is true, but that would be less than what really happened. And at some point you really remember more of what happened. (Ferguson and Kincaid 183)

---

1 The dates presented here are reference of the first publication of each title.

In a 2002 interview with Kay Bonnetti, Kincaid makes clearer her relation to truth telling and to fiction, stating: "[e]verything I say is true, and everything I say is not true. You couldn't admit any of it to a court of law. It would not be good evidence" ("Interview with Jamaica Kincaid"). Kincaid explains her relation to the real and its representations, confiding that there is always more to be said than simply how a sequence of events happened. Yet, truth comes into question, especially when the label of autobiography is conferred upon writing. In *My Brother*, her only work labeled as memoir, Kincaid writes about the life and death of Devon Drew, her youngest brother, who died of AIDS in 1996. They were never close; yet, Kincaid investigates the family relations, and once again returns to the mother figure, one that is ever-present in her production. In a memorable passage, Kincaid's notion of truth telling comes at play again, as her brother enquires about a character in a book published some time earlier.

> He had read in a novel written by me about a mother who had tried and tried and failed and failed to abort the third and last of her three male children. And when he was dying he asked me if that mother was his mother and if that child was himself ("Ah me de trow'way pickney"); in reply, I laughed a great big Ha! Ha! and then said no, the book he read is a novel, a novel is a work of fiction; he did not tell me that he did not believe my reply and I did not tell him that he should not believe my reply. (*My Brother* 174)

Kincaid reveals in this passage that the reader should always be suspicious when dealing with her texts. The book in question, ironically, is called *Autobiography of My Mother*, this one, marketed as fiction. This instability subverts the contracts and pacts between reader and writer, creating an experience that is hybrid and rich as truth finds its way into fiction in Kincaid's texts, and her personal life story is refigured in ways that expand both fictional and autobiographical discourses. "Her answer seems to lie in a kind of hybridity, an act of bricolage or quilting that grants to the colonial subject locally differentiated subjectivities that she refuses to rank: narratives and voices that effectively patchwork an individual world" ("Not at Home"). These practices of hybridity are what, for Marquis, characterize Kincaid's production, understanding the role of (post) colonial histories in the creation of her identity through text, and the crafting of her own distinct voice as a formerly colonized subject, as Kincaid scrutinizes the large (post)colonial master narrative, which supports her as a writer, while reassessing its power over her.

Gilmore also addresses the matter of the tension between fiction and autobiography. In "There Will Always Be a Mother – Jamaica Kincaid's Serial Autobiography", Gilmore explores how Kincaid is able to build a sense of autobiography in her fiction through a network of recurring themes, repeatedly exploring how fracturing the mother-daughter (trauma) theme is present in many different works. Kincaid subverts the contracts of the traditional autobiographical pact, in Philippe Lejeune's terms, firstly by refusing the end of the text (and therefore of life) as a definite account through her texts' open-endness, in addition to escaping the convention of the coincidence of the trifold confluence of the author's name with the narrator's and the protagonist's as a proof of veracity and verifiability of the story. One example of that is referred by Gilmore: "[h]arkening back to Kincaid's given name, Lucy's full name is Lucy Josephine Potter [...]" (102). In the same interview with Bonetti Kincaid states: "[s]he [Lucy] had to have a birth-date so why not mine? She was going to have a name that would refer to the slave part of her history, so why not my own? I write about myself for the most part, and about things that have happened to me" ("Interview with Jamaica Kincaid"). This pattern is not unique to *Lucy,* since other examples may be brought to mind, such as in *The Autobiography of My Mother* in which the protagonist is called Xuela Claudette Richardson, sharing her last name with Kincaid's real mother, Annie Richardson. Annie is also the first name of the protagonist of the homonymous work *Annie John.* Marquis also points to such peculiarity, stating that:

> As in other novels, Kincaid puts aside her authorial name and returns in her characters to the family names she abandoned when she invented for herself names that fitted her hard-won independence. [...] As family signatures these names operate to authenticate the novels as precisely *her* history—as autobiographical. ("Not at Home")

Marquis refers to Kincaid's work as autobiographical fiction, and some times as narrative fiction, and claims that the relation between reality and fiction are only approximate in Kincaid's work, stating that "[...] Kincaid's narrative fiction has the relation of the asymptote to real history—a tangential relation where narrative and the line of actual events come very close to each other, but never exactly meet, where, necessarily, the narrative can never deliver its central character complete" ("Not at Home"). More importantly, Marquis states that to write this kind of hybrid text may be a response to the imposition of colonial demands, just as slave narratives had an effect in their narrators' fashioning of subjectivity and its inscription in text and in society.

Gilmore categorizes Kincaid's works in a genre called serial autobiography, a mode of finding unity in complex networks of texts by the inscription of autobiographical instances in which the readership is capable of identifying common features, even if the variants of name, context, or progression are in dissonance. What one can find in any of Kincaid's works is labeled by Gilmore as the autobiographical scene. The autobiographical scene and the subject-in-process are the locus in which seriality expands the modes of autobiography, making room for what the author refers to as "returning to the scene", a practice found in Kincaid's texts that infuses the reader with a sense of continuation, despite the *corpus* being comprised of different characters and protagonists. This figure, then, returns to the scene not as a recurrent protagonist in different texts bearing the same name and a linear teleology, neither as different stages of the author's life (such as in the autobiographies of Maya Angelou, for instance); instead, Gilmore clarifies: "[...] this figure may be recognizable less for the features it shares with the autobiographer, or her textual simulacra, than for the preoccupations represented in and through it" (98). The construction of the autobiographical happens then through something called "emergence-through-enactment" (97), in which the writer and protagonist are joined as a subject in process. Summing up, *Lucy* is not entirely Kincaid's account of her life, but some of *Lucy* is. In the same way *Annie John* is not entirely about Kincaid's life in Antigua, but some of it is. Gilmore states:

> This new self-representational figure (not new to self-representation, but newly visible as a feature of serial autobiography), then, does not suggest a one-to-one correspondence between real and represented life. Instead, this figure, as a representation of identity, is capable of crossing all kinds of boundaries of discrete texts, to extend the autobiographical into an intertextual system of meaning. (98)

The same artifice of returning to the scene is applied inside the story, such as in *Annie John*, as noted by Bonetti, and when Kincaid was questioned if her choice of repeating a scene from different points of view was meant to figure as a series of short stories or as a whole novel, the author responded:

> I didn't conceive of it as either one. I just write. I come to the end, I start again. I come to the end, I start again. And then sometimes I come to the end, and there is no starting again. In my mind there is no question of who will do what and when. Sometimes I've written the end of something before I've written the beginning. Whatever a novel is, I'm not it, and whatever a short story is, I'm not it. If I had to follow these forms, I couldn't write. I'm really interested in breaking the form. ("Interview with Jamaica Kincaid" 2002)

Kincaid's position in relation to her writing process seems also to mirror the cyclical nature of trauma and the working-through processes. The returning to the scene appears to be connected to the difficult reality of making sense of a traumatic situation, and in this case, such trauma can be pointed either to the familial relations, the colonial education received in Antigua, or even the reality encountered upon moving to the United States. Finally, there is no return, and the cycle seems to be broken. It seems that Kincaid is interested in breaking the form of canonical writing as much as she is dedicated to making order out of chaos of her traumatic past. Gilmore expands:

Survivors of trauma are urged to testify repeatedly to their trauma in an effort to create the language that will manifest and contain trauma as well as the witness who will recognize it. Thus the unconscious language of repetition through which trauma initially speaks (flashbacks, nightmares, emotional flooding) is replaced by a conscious language that can be repeated in structured setting. Language is asserted as that which can realize trauma even as it is theorized as that which fails in the face of trauma. This apparent contradiction in trauma studies represents a constitutive ambivalence. For the survivor of trauma such an ambivalence can amount to an impossible injunction to tell what cannot, in this view, be spoken. (7)

Regarding the text, and the urge to speak about one's life, applying the modality of a term such as "autobiographical" to text, comes with its own set of expectations and tensions, which must not limit the reader's interpretation of the text, but rather, enlarge its possibilities. "The autobiographical may, [...], function critically as an expansive, extendible system of meaning, one that enables readers to do much more than search out sources, proof, or evidence of a corresponding reality" (100). What Kincaid's work does, instead, through this autobiographical instance, is to question the limits of truth telling, or rather, expose the limits of the construction of reality as text in autobiography. This refusal to contain life into a single text, or rather, to contain experience into a single character as a definite account, demonstrates that the modes of canonical autobiography are under scrutiny and new forms of experimentation in self-representation are possible. Gilmore states: "Instead of respecting the sufficiency of each text unto itself (and why should she adopt that constraint without being able to rework it? And why should her readers?), Kincaid extends what appears to be the same character, with different names, into book after book" (101). What Kincaid makes clear in the construction of Lucy as a character is the possibility of eternal ambiguity, of not being on either side of a spectrum, but to live with the fracturing condition of both sides.

Lucy shares some associations with Annie John, though their names and stories seem to be disconnected at first glance. Kincaid interweaves some other aspects in their stories that are capable of creating a connection between both of them. In *Annie John,* published originally in 1986, the protagonist is made familiar to John Milton's *Paradise Lost* as a punishment for defacing a picture of Columbus in her schoolbook. This anecdotal account is also referred by Kincaid in some interviews when describing her childhood.[2] The identification with the devil is also present in a second moment, when looking at her reflection in a

---

2 Kincaid reports this story in an interview during the 25th Anniversary Chicago Humanities Festival, in 2014.

shop window, Annie recognizes how strange she had become after the first spurts of growth brought about by puberty. After enumerating the features that make her a stranger to herself in the reflection, she remembers a painting where a young Lucifer is depicted. This mirroring, however, does not serve as a force to strengthen her sense of self, but really captures how miserable these changes make her.

> [...] I didn't know that it was I, for I had got so strange. My whole head was so big, and my eyes, which were big, too, sat in my big head wide open, as if I had just had a sudden fright. My skin was black in a way I had not noticed before, as if someone had thrown a lot of soot out of a window just when I was passing by and it had all fallen on me. On my forehead, on my cheeks were little bumps, each with a perfect, round white point. My plaits stuck out in every direction from under my hat; my long, thin neck stuck out from the blouse of my uniform. Altogether, I looked old and miserable.
>
> Not long before, I had seen a picture of a painting entitled The Young Lucifer. It showed Satan just recently cast out of heaven for all his bad deeds, and he was standing on a black rock all alone and naked. Everything around him was charred and black, as if a great fire had just roared through. His skin was coarse, and so were all his features. His hair was made up of live snakes, and they were in a position to strike. [...] I was standing there surprised at this change in myself, when all this came to mind, and suddenly I felt so sorry for myself that I was about to sit down on the sidewalk and weep, already tasting the salty bitterness of my tears. (*Annie John* 94)

The mirroring of her image takes place first in the physical shop window, where adolescence starts to be noticed, and her self-image is different from what she constructed in her mind. The presence of acne, denoting a hormonal change, associated with the comment "My skin was black in a way I had not noticed before" takes the attention of the reader to questions regarding colorism and desirability. This scene takes place just before an encounter with older boys from her school, in which their sarcastic tone makes clear that Annie is not eligible as an object of desire in their standards. Her skin is compared to soot, mirroring the image of the young Lucifer in its charred background, her hair is likened to the snakes ready to charge. The young Lucifer serves then as a perfect icon of Annie's dissatisfaction with her body, and prefigures the depression that is about to set in. This episode might also be read in the light of the degradation of black and brown bodies under the Western colonial gaze, taking its toll over the young teenager.

When Lucy asks her mother about her name, she is informed that it derived directly from Lucifer, a comment meant to assert her mother's, even if temporary, resentment towards the burden of bringing a life into the world. She states: "I named you after Satan himself. Lucy, short for Lucifer. What a botheration from the moment you were conceived" (*Lucy* 192). This revelation, however, did not distress the protagonist for long, instead, she feels invigorated by such relationship with this other character, one she knew well.

> The stories of the fallen were well known to me, but I had not known that my own situation could even distantly be related to them. Lucy, a girl's name for Lucifer. That my mother would have found me devil-like did not surprise me, for I often thought of her as god-like, and are not the children of gods devils? (*Lucy* 192)

In her interview with Ferguson, Kincaid reveals that writing about this mother figure has been a task that will always figure in her production, a trend that has been this far maintained in the subsequent titles released after this statement. However, the mother figure also evolves, transfiguring from the maternal being into the political stance of the country, as the mother symbolically becomes Antigua. In *Annie John* we face the painful process of individuation experienced by the young protagonist, who one day had to figure out who she was without her mother's identity to mirror herself. In *Lucy*, the mother is always far removed physically, back in Antigua, but always present in the daughter's mind, while the young woman explores her identity in the United States. Kincaid states.

> [...] In my first two books, I used to think I was writing about my mother and me. Later I began to see that I was writing about the relationship between the powerful and the powerless. That's become an obsessive theme, and I think it will be a theme for as long as I write. And then it came clear to me when I was writing the essay that became "On Seeing England for the First Time" that I was writing about the mother – that the mother I was writing about was really Mother Country. It's like an egg; it's a perfect whole. It's all fused some way or other. (Ferguson and Kincaid 176–177)

3.1.2 Mother-Daughter, Mother-Island

"I wondered then, for the millionth time, how it came to be that of all the mothers in the world mine was not an ordinary human being but something from an ancient book" (*Lucy* 150). This is one way that Lucy describes her awe in relation to the maternal figure, this otherworldly creature. Laura Niesen de Abruna states: "Kincaid's greatest contribution to the full presentation of the female life is her exploration of the mother-daughter bond, and specifically, the effects of the loss of the maternal matrix on the relationship between mother and daughter" (Abruna 173). This perspective is backed by a series of studies focusing on mother-daughter relationships as frequent features of women's writing, and more specifically of Caribbean women writers[3]. Kincaid explores both how connecting with this maternal figure, as well as severing this connection, is key to her identity formation. Marquis states that in *Lucy* there are two competing stories related to the issue of coloniality and the mother figure:

> Firstly there is the story that centres on the mother who has collapsed herself into the colonial matrix and necessarily, therefore, figures the imposition of mother-country rule. This mother she must deny. Secondly there is the mother who is doubly subjugated, as colonial subject and woman, but who, paradoxically, possesses in reality a power that the imperial, colonialist, political structures cannot accommodate. ("Not at Home")

Beyond exploring the personal liaison and its effects in her identity formation the relationship with this complex mother figure serves as the allegorical groundwork in which Kincaid explores how colonialism takes place and is enforced. Seriality in autobiography is an important feature here, since the preoccupations expressed by Kincaid's fictional characters throughout the texts inform the reader of her developing criticism of colonialism. Similarly, Abruna states: "[i]n *Annie John*, as well as in *Lucy* and *The Autobiography of My Mother*, the alienation from the mother becomes a metaphor for the young woman's alienation from an island culture that has been completely dominated by the imperialist power of England" (Abruna 173). The mother figure is seen firstly as a positive image of empowerment, one that helps the characters construct their identity through mimicry during their infancy; later, during her teenage years, this figure becomes the object the characters build their identity in opposition to, since they perceive the maternal figure to be an instrument of coloniality. Commenting on the matter, Abruna states that:

> Much of Kincaid's distrust of the postcolonial environment went unnoticed by the reviewers of *Annie John* and *Lucy*. Like *Annie John*, *Lucy* was received in many academic circles as a book about mothers and daughters, a popular topic in feminist literary criticism, especially since the late seventies, when Nancy Chodorow and Carol Gilligan published their influential studies. (179)

These issues intersect in the mother figures of Lucy, in opposition to whom she forms her identity, while developing a postcolonial consciousness. Irline François, drawing from Adrianne Rich's theories on motherhood, writes in "The Daffodil Gap: Jamaica Kincaid's Lucy" about the complex relation of identification

---

3 She makes reference to the studies of Betty Wilson and Pamela Mordecai, *Her True-True Name*, as well as Rhonda Cobham and Merle Collins' *Watchers & Seekers: Creative Writing by Black Women*.

and the desperate need for individuation of daughters who perceive their mothers to be patriarchal instruments of domination: Adrienne Rich uses the term *Matrophobia*, or the fear of becoming one's mother, to describe daughters who see their mothers as having taught a compromise of self-hatred and as having transmitted the restrictions and degradations of a female existence, explaining that it is easier to direcly reject a mother instead of seeing beyond the forces acting upon her. This experience also comprises antithetical forces, as there may also be an underlying pull toward the mother, encompassing a fear that if the subject lowers their guard, they will completely identify with this figure. "Rich concludes that matrophobia may be seen as the splitting of the self in the desire to become purged once and for all of our mother's bondage, to become individuated and free" (80).

Abruna claims that the separation from the mother-island creates a higher level of anxiety and is manifested through cultural and psychic alienation in the narrative. In *Annie John*, the protagonist experiences the trauma of separation from the mother from the onset of adolescence, while in *Lucy* the separation is already set in two levels: physically, since the novel takes place in New York, and emotionally, as the protagonist refuses to contact her family back in Antigua.

Annie John deals with a younger protagonist, one that enjoys very much the company and treatment of her mother, and experiences this identification very intensely. In a chapter entitled "The Circling Hand", there are long descriptions of mother-daughter daily life, when both share many hours together while Annie receives an education. There are accounts of their family dynamics, the role of Annie's father, and how she was frequently in awe of her mother's presence and command in the world, her social skills and more importantly, how much the mother cared for the rearing of her daughter. Kincaid describes Annie and her mother's relationship as very close, and allegorically translates this closeness into dresses made out of the same cloth: "[...] my mother and I had many dresses made out of the same cloth, though hers had a different, more grownup style, a boat neck or a sweetheart neckline, and a pleated or gored skirt, while my dresses had high necks with collars, a deep hemline, and, of course, a sash that tied in the back" (*Annie John* 25). The choice of dresses is interesting here namely because they reinforce the idea of identity construction around symbols that are centrally feminine. In a deeper reading, this characterization alludes to a pre-oedipal state, in which mother and daughter are one, and Annie is a continuation of her mother, deriving complete pleasure from this state of symbiosis, though in clear different stages of development. This state is indicative of a connection to the island culture imbued in the mother figure. As it is perceived as the origin and continuation of the protagonist, it also celebrates their blackness in several aspects. In the same chapter, there are also descriptions of rituals that both have shared, namely the storytelling associated with the items stored in a wooden trunk the mother had brought from her island of origin. The trunk will later belong to Annie, evincing the social aspect of bonding among female generations which has been mentioned above, and the importance of stories in the process of identity and memory construction. Kincaid writes:

> From time to time, my mother would fix on a certain place in our house and give it a good cleaning. If I was at home when she happened to do this, I was at her side, as usual. When she did this with the trunk, it was a tremendous pleasure, for after she had removed all the things from the trunk, and aired them out, and changed the camphor balls, and then re-folded the things and put them back in their places in the trunk, as she held each thing in her hand she would tell me a story about myself. (*Annie John* 21)

Annie accompanies her mother during many house chores, including washing, grocery shopping, cleaning and cooking. The voice of this kind of mother also figures in *At The Bottom of the River*, Kincaid's seminal work as a fiction writer, in the short story "Girl", in which a mother lectures a daughter in a series of commands meant to shape her identity as the desirable prototype of femininity in the Caribbean. Commenting on the piece, Kincaid states: "[...] [w]ell, it is my mother's voice exactly over many years. There are two times that I talked in my life as a child, as a powerless person. Now I talk all the time" (Ferguson and Kincaid 171). The text is formatted as a continuous set of instructions, with only two interjections coming from the recipient girl, who never disputes what she is being told. This may be the reference to the powerlessness that Kincaid referred earlier. The instructions range from cooking and housekeeping to how to behave in front of men, how to love and how to boil herbs to cause an abortion. Although it seems that the intention of the mother is teaching her daughter the skills a woman needed to succeed in the island, sharing the knowledge associated with heteronormative gender roles and performance, as well as emancipatory information regarding contraception, the mother seems to be unforgiving, focusing repeatedly on controlling the girl's blossoming sexuality, emphasizing the Victorian mores of her own upbringing:

> [...] on Sundays try to walk like a lady and not like the slut you are so bent on becoming; [...] this is how to hem a dress when you see the hem coming down and so to prevent yourself from looking like the slut I know you are so bent on becoming; [...] this is how to behave in the presence of men who don't know you very well, and this way they won't recognize immediately the slut I have warned you against becoming. (*At the Bottom of the River* 3–4)

Fearing the devaluation of the daughter's marriage prospects, the mother seems to have adopted the preconception of the oversexualized black female body, and tries to avert this situation by controlling any demonstration that could be perceived as sexual, especially by the male gaze, that ultimately means protection for for the young woman in this social economy. This voice surfaces again in *Annie John* with the same claim, when the mother, after seeing her daughter talking to some boys in the street after being repeatedly admonished against doing so, scolds the daughter for her apparent wrong behavior. Kincaid writes:

> She went on to say that, after all the years she had spent drumming into me the proper way to conduct myself when speaking to young men, it had pained her to see me behave in the manner of a slut (only she used the French-patois word for it) in the street and that just to see me had caused her to feel shame. (*Annie John* 102)

Puberty seems to be the fact that separates Annie from her mother, as noted by Abruna (174), when the daughter, who once believed she was a continuation of her mother, in pre-oedipal fashion, feels isolated when the severance of the bond occurs. As a young woman, Annie must create an identity that differs from her mother's expectation, and rejection of that same expactation seems to be the only way out. The falling out happens shortly before the "long rain", a period when Annie becomes ill, physically and psychologically, which seems to be a consequence and a trope of their symbolic separation. At this time, Ma Chess, Annie's grandmother, is summoned from Dominica to care for the child. The argument with Annie's mother, the final straw in their already fraught relation, consists of a personal attack of Annie, who uses the same weapon her mother used to attack her, a shaming of her sexual behavior, with the aggravating layer of inheritance:

> The word "slut" (in patois) was repeated over and over, until suddenly I felt as if I were drowning in a well but instead of the well being filled with water it was filled with the word "slut," and it was pouring in through my eyes, my ears, my nostrils, my mouth. As if to save myself, I turned to her and said, "Well, like father like son, like mother like daughter". (*Annie John* 102)

Lucy continues to make reference to this problematics when, in the later part of the story, after ignoring nineteen letters from her mother, she decides to respond. The slut-shaming rhetoric is re-appropriated by the protagonist and now she has the opportunity to tell her mother how she has been exercising her liberty and her sexuality, which involves a platonic lesbian relationship, as well as other heterosexual relationships.

I reminded her that my whole upbringing had been devoted to preventing me from becoming a slut; I then gave a brief description of my personal life, offering each detail as evidence that my upbringing had been a failure and that, in fact, life as a slut was quite enjoyable, thank you very much. (*Lucy* 117–128)

Another element that helps the reader to understand the mother-daughter bond throughout the texts is food. Food particularly seems to be a way through which the mother communicates her love for her child, at the same time it illustrates her local identity and affiliation as a Caribbean woman through ingredients that are found in the island. The motifs of food and food culture in immigrant literature are investigated by Erick Greyson in "The Most Important Meal: Food and Meaning in Jamaica Kincaid's *Lucy*". The author writes: "[a]s a marker of class, a factor in cultural alienation, or a means for establishing bonds with other people, food draws its power and symbolic plasticity from its association with primal human drives" (15). These drives, namely nourishment, pleasure and protection, are also associated with motherhood. In the following excerpt, the presence of antroba, a dish made out of crushed eggplants common in the Caribbean, in addition to salt fish and pepper pot (a dish originated originally in Guyana but popular throughout the Caribbean) denotes the specificity of the cuisine of the islands (*Annie John* 17).The abundance of tropical produce (even in an island as draught-ridden as Antigua) and fish, is made present in many instances of the novels, when the setting of the plot takes place in the island, or even in memory when it happens in the United States. The short remark about the market is also indicative of the social aspect of food, as well as the availability of ingredients, highlighting a different model of consumption and production *vis-a-vis* the model later found in the United States or the imperial English counterpart with its industrialized goods. In *Lucy*, when the protagonist first arrives in the United States, the presence of a refrigerator and its convenience, seemed to be something that Lucy would enjoy, but turns out to be overwhelming, along with all the new things she would have to adapt to (6). Lucy imagined that leaving her life and family in Antigua would be a comfort, and that the access to things different and new would grant her the opportunity she needed to develop a new sense of selfhood, apart from the colonial upbringing she had finally escaped. However, the adaptation process proves to be harsh, and the ostracized adolescent dreams of the small comforts she had left back in Antigua. Food appears once again as a vehicle for nostalgia, and the figure of her grandmother as caregiver represents something Lucy realizes she needs. Tropical fruit and fish are here a metonym for the Caribbean. Ma Chess, a Carib woman, is fundamental here to understand the alliances that Kincaid is interested in keeping when thinking about her island life, and food seems to be the instrument that she finds to fulfill this longing for the past.

> Oh, I had imagined that with my one swift act—leaving home and coming to this new place —I could leave behind me, as if it were an old garment never to be worn again, my sad thoughts, my sad feelings, and my discontent with life in general as it presented itself to me. In the past, the thought of being in my present situation had been a comfort, but now I did not even have this to look forward to, and so I lay down on my bed and dreamt I was eating a bowl of pink mullet and green figs cooked in coconut milk, and it had been cooked by my grandmother, which was why the taste of it pleased me so, for she was the person I liked best in all the world and those were the things I liked best to eat also. (7)

Breakfast in Antigua is also a showcase of ingredients in *Annie John*, but this time, the colonizing presence of England is undeniable. Although the poverty experienced in the island transpires in many aspects of the characters' lives, Annie's family seems to enjoy a satisfactory level of income, represented by the plentiful breakfasts enjoyed by the protagonist. The presence of porridge, cheese, and grapefruit, as well as bread and butter is telling:

> My mother would then give me my breakfast, but since, during my holidays, I was not going to school, I wasn't forced to eat an enormous breakfast of porridge, eggs, an orange or half a grapefruit, bread and butter, and cheese. I could get away with just some bread and butter and cheese and porridge and cocoa. I spent the day following my mother around and observing the way she did everything. (*Annie John* 15)

In "On Seeing England for the First Time", an essay published originally in 1991, Kincaid explores the issue of being surrounded by a foreign culture and having to abide to its order, even in matters as simple as food. The breakfast items consumed in Kincaid's table as a child are imported from the "mother country", eliciting the economic and commercial dependence Antigua experienced even in post-colonial times. Also present in this text is the critique of the adoption of England's culture and way of living in detriment of more sensible local practices. The amount and type of food consumed by the colonies due to the rule of England did not suit their daily life and climate, as noted by Kincaid. Nonetheless, the symbolic and material authority of colonial ruling hegemonically coerced the colonial population into practices unfit for their wellbeing:

> At the time I was a child sitting at my desk seeing England for the first time, I was already very familiar with the greatness of it. Each morning before I left for school, I ate breakfast with half a grapefruit, an egg, bread and butter and a slice of cheese, and a cup of cocoa; or half a grapefruit, a bowl of oat porridge, bread and butter and a slice of cheese, and a cup of cocoa. The can of cocoa was often left on the table in front of me. It had written on it the name of the company, the year of the company was established, and the words, "Made in

England." Those words, "Made in England", were written on the box the oats came in too. [...] we somehow knew that in England they began the day with this meal called breakfast and a proper breakfast was a big breakfast. No one I knew liked eating so much food so early in the day; it made us feel sleepy, tired. But this breakfast business was made in England like almost everything else that surrounded us, the exception being the sea, the sky, and the air we breathed. (32–33)

Kincaid shows the conflation of the mother and English domination later in the essay, by exposing the appreciation the maternal figure has when the child successfully emulates the modes of colonial education, while the child would transgress these teachings anytime she had an unsupervised moment, enjoying her meals with her bare hands (33). Kincaid is educated in a way that mimics the values, modes, and aesthetics of the colonizer, but ultimately her refusal to partake in this order is evident in her choices for her protagonists. The mother here serves as a mouthpiece for England, along with breakfast, her school, the authors she had to read, the history she was taught, the people she was made to admire, her uniform, every item of clothing that was tagged "made in England". Ultimately, England becomes a symbol of negativity in her existence, and rebellion seems to be the way to resist the continuos process of acculturation that so pervasively plagues her coming of age. Later in *Lucy*, the protagonist attends a party in which guests seem to be all from an artistic background. Interestingly enough, in *A Small Place*, Kincaid refers to the Antiguan population as artists, referring broadly to their sensibilities:

I look at this place (Antigua), I look at these people (Antiguans), and I cannot tell whether I was brought up by, and so come from, children, eternal innocents, or artists who have not yet found eminence in a world too stupid to understand, or lunatics who have made their own lunatic asylum, or an exquisite combination of all three. (57)

She remarks that the artists in the party spoke freely about the world, they drank too much, and often died paupers. Lucy wonders then that she is not part of such a group, for she is not an artist, but she would always like to be associated with such kind, since they stood apart from the norms. Conversely, she identifies the other side of the population with the manners she was taught as the correct ones, hence English, and attributes to them the benefits of conformity.

I had just begun to notice that people who knew the correct way to do things such as hold a teacup, put food on a fork and bring it to their mouth without making a mess on the front of their dress—they were the people responsible for the most misery, the people least likely to end up insane or paupers. (*Lucy* 98–99)

Lucy's relationship with her mother is explored less explicitly in the homonymous novel than in *Annie John,* since the readership is only granted access to what the protagonist remembers or thinks about their connection. When there is finally some communication between Antigua and New York, as an acquaintance of the family, whom Lucy clearly dislikes, physically goes to Lucy's apartment to deliver the news of her father's death, there are explicit references to the comparison between mother and daughter. After receiving the news, the messenger seems to be pleased to see Lucy at the brink of a breakdown, and with a spiteful smirk declares that Lucy reminds her very much of her mother. Lucy sees this claim as a lifeline, since her efforts have all been directed at creating a new sense of identity, and while still shocked by her father's untimely death, she is able to muster the strength to respond. Defiantly, she states everything that makes them, mother and daughter, different. The identity construction process, which was made in opposition to her mother's own identity, seems to be continuously taking place, as the protagonist refuses to define her identity in anyone's terms but her own. Lucy makes clear in this confrontation what caused them to be apart.

> I am not like my mother. She and I are not alike. She should not have married my father. She should not have had children. She should not have thrown away her intelligence. She should not have paid so little attention to mine. She should have ignored someone like you. I am not like her at all. (*Lucy* 123)

Lucy believes that her choices are what separate her from her mother, attributing much of her mother's situation, financial and otherwise, to decisions she had made in the past, suspending any belief connected to the structural reality of these choices. The remarks concerning her marriage and children can be related to the abidance to gender roles and expectations projected upon women by the Victorian model, things that Lucy tries to avoid by experiencing her sexuality in a free and fluid manner. When it comes to the intelligence of both mother and daughter, Lucy criticizes that her mother's was wasted in the island, living such a traditional life, when she should have sought to free herself from the education she had been given, instead of reproducing it with her own children. In regards to Lucy's intelligence, she feels that her mother should have recognized in her all the potential she was often told she had, either in school or otherwise. Kincaid states "[a]s a child I had always been told what a good mind I had, and though I never believed it myself, it allowed me to cut quite a figure of authority among my peers" (*Lucy* 92). She comes to question the ideology that determines the position which a woman should occupy in society, the one that forced her mother to assume this position, and possibly would have made her follow the same path had she not succeeded in moving to the United States. "Why did someone not think that I would make a good doctor or a good magistrate or a good someone who runs things?" (*Lucy* 92). Instead, Lucy is sent to the United states to be a nurse, a common expectation for girls and a stereotypical position of female subordination to a male doctor:[4] "Whatever my future held, nursing would not be a part of it. I had to wonder what made anyone think a nurse could be made of me. I was not good at taking orders from anyone, not good at waiting on other people" (*Lucy* 92). When Lucy reaches out to her mother, she stresses, in teenage angst, what had driven her apart from the maternal figure.

> I wrote my mother a letter; it was a cold letter. It matched my heart. It amazed even me, but I sent it all the same. In the letter I asked my mother how she could have married a man who would die and leave her in debt even for his own burial. I pointed out the ways she had betrayed herself. I said I believed she had betrayed me also, and that I knew it to be true even if I couldn't find a concrete example right then. I said that she had acted like a saint, but that since I was living in this real world I had really wanted just a mother. (*Lucy* 127)

Mariah, the host mother in New York, becomes the figure onto which Lucy projects her longings and aggravations related to the mother-daughter bond in her new life. The relationship with Mariah serves also as the groundwork for the critique of colonialism but this time it develops in the nuances of pervasive racism and privilege. Lucy is amazed at the capacity Mariah and her peers have of ignoring the material inequalities of life in the United States, and sees the liberal ideology professed by Mariah as naïve at best and cruel at worst.

3.1.3 How do you get to be this way?

How do you get to be this way? This is the question that Lucy asks repeatedly when dealing with Mariah, the host mother. Though the woman is mainly caring and affable towards Lucy, she still inspires moments of rage in the protagonist. Mariah expresses her affection for Lucy in many different ways, from sharing her personal history and memories, to paying her more money than the agreed, buy-

---

4 "A nurse, as far as I could see, was a badly paid person, a person who was forced to be in awe of someone above her (a doctor), a person with cold and rough hands, a person who lived alone and ate badly boiled food because she could not afford a cook, a person who, in the process of easing suffering, caused more suffering (the badly administered injection)" (*Lucy:* 92).

ing Lucy things that she thought she would like, as well as giving Lucy a large sum of money to send back home when she finds her father is deceased.

> [...] [I]t is difficult for Lucy not to be seduced by (and thereby give in to) Mariah's disarming disposition, warmth, humanity and extraordinary good will towards her. Paradoxically, Lucy cannot help being enraged (but also intrigued) by her employer's profound naïveté, simplistic world view, complacent ethnocentrism, and lastly (perhaps unforgivably for a young woman in the full throes of rebellion and egotism), Mariah's insistence on placing Lucy's discourse within an intellectual and homogenous cultural paradigm—a discourse which the protagonist categorically refuses to accept. (François 81)

The disagreement experienced by Lucy is based on the frequent display of racial thoughtlessness performed by the host mother, her family and friends. This type of prejudice, mostly resulting from a cultural insensitivity and a lack of situational thinking, seems to condensate in the unselfconscious whiteness that Lucy meets in the United States. Their clash comes from the problematic identification Lucy develops with this mother figure, conflating her to the Motherland, as pointed by François: "[...] therein lies Lucy's conflict with Mariah, who unwittingly recalls both the mother and (by her very appearance, yellow hair and blue eyes), the totalizing values of the "Motherland" whose values Lucy must evade" (80). This reaction is not limited to Mariah, but extends to other white characters in the narrative.

One example of the impact of this insensitivity performed by a white subject is found in Dinah, who is one of Mariah's friends. When first meeting Lucy, she causes the protagonist to become enraged by the way she addresses the matter of her national origin (Kincaid 1990: 56). The antipathy for Dinah continues, as she fails to recognize in Lucy a full human being, relegating her to the role of caretaker, and nothing beyond that. The stark contrast in treatment between Lucy and Mariah's children, for instance, denounces that for Dinah Lucy is inferior, or even a case of reification, crystalized in the emptiness of "the girl", destitute of history, desire or specificity. For Marquis, Dinah represents the embodiment of coloniality, and commenting on Lucy's lack of reply, she states: "Lucy's response is silent, for all its being more backchat, but underlines the fact and force of historically conditioned first world superiority" ("Not at Home"). In a contrasting instance, there is a clear difference between the treatment between characters. Kincaid writes: "Dinah now showered the children with affection—ruffling hair, pinching cheeks, picking Miriam up out of my lap, and ignoring me. To a person like Dinah, someone in my position is 'the girl'—as in 'the girl who takes care of the children'" (*Lucy* 58). The readership is also introduced to Hugh, Dinah's brother and briefly Lucy's lover, who is able to introduce the question without causing a negative impression on the protagonist, simply by recognizing her specificity:

> Dinah came with her husband and her brother, and it was her brother that Mariah had really wanted me to meet. She had said that he was three years older than I was, that he had just returned from a year of traveling in Africa and Asia, and that he was awfully worldly and smart. She did not say he was handsome, and when I first saw him I couldn't tell, either; but when we were introduced, the first thing he said to me was "Where in the West Indies are you from?" and that is how I came to like him in an important way. (*Lucy* 65)

The recognition of her specificity grants Lucy's sympathy, since it helps her positioning herself more clearly in her own narrative of this new reality. The characterization of Hugh as someone who was well traveled and cultured, transpiring an aura of open-mindedness contrasts with the other people whom Lucy meets, who despite the means and access to culture, continue to categorize difference as inferiority. In the first chapter, called "Poor Visitor", Lucy is greeted by the reality she had long dreamed of, far from the Island. Lewis, Mariah's husband, during dinner in the first week after Lucy's arrival, gives the protagonist the epithet that names the chapter. This makes evident the difficult adaptation of the protagonist, who is seen as an outsider by the family that has received her. The American family seems to interpret her reserved manner as detachment, and her astonishment as mockery:

> It was at dinner one night not long after I began to live with them that they began to call me the Visitor. They said I seemed not to be a part of things, as if I didn't live in their house with them, as if they weren't like a family to me, as if I were just passing through, just saying one long Hallo!, and soon would be saying a quick Goodbye! So long! It was very nice! For look at the way I stared at them as they ate, Lewis said. Had I never seen anyone put a forkful of French-cut green beans in his mouth before? [...] He said, "Poor Visitor, poor Visitor," over and over, a sympathetic tone to his voice, and then he told me a story about an uncle he had who had gone to Canada and raised monkeys, and of how after a while the uncle loved monkeys so much and was so used to being around them that he found actual human beings hard to take. (*Lucy* 13–14)

The story told by Lewis seems to exacerbate the difference between them, at least from the perspective of the white interlocutor. In Lewis anecdote, Lucy and her family might not figure as the uncle who is unused to humans, but as the primates, devoid of a full humanity. The white family at this point does not yet see her black body, a marker of difference, as "actually human". The veiled racial difference between them ostracizes the protagonist, even if she does not explicitly perceive the situation in this sense.

Following this dialogue, Lucy shares with the family a dream she had in which they all took part, intending to demonstrate that while she seems detached from them, they now are special to her. In this dream, Lewis chases a naked Lucy, while Mariah commands him to catch her. They all fall to the ground on a bed of snakes. The chapter's final words are in Lucy's voice: "I had meant by telling them my dream that I had taken them in, because only people who were very important to me had ever shown up in my dreams. I did not know if they understood that" (*Lucy* 14). The contrast between Lucy's willingness to accept these new figures in her life is made evident in the dream, while her outside performance of sociability seems to misguide the American family's perception. The acceptance of difference seems to come mostly from Lucy's perspective, while on the other side, there still seems to be the idea that she does not belong, hence the epithet "poor visitor". After the retelling of the dream the couple falls into awkward silence, which is only broken by remarks of ironic reproach, as Mariah says that she should meet Dr. Freud. Lucy is left disconcerted by the comment, since she does not know the referent and the related legacy of sexuality studies, feeling further alienated from the family.

On the matter of familiarity, Lucy describes the welcoming reception displayed by the American family as something positive. However, Lucy's own ideas related to family and its cumbersome weight in her life tinge this reception with a color of discomfort.

> How nice everyone was to me, though, saying that I should regard them as my family and make myself at home. I believed them to be sincere, for I knew that such a thing would not be said to a member of their real family. After all, aren't family the people who become the millstone around your life's neck? (*Lucy* 7–8)

Lucy's position in the family is unclear from the beginning, since she does constitute a part of the help, but she is not a maid (who promptly declares that she does not like Lucy simply for the way she speaks, denoting the difficult and tortuous lack of sisterhood among women in situations of vulnerability, migrant or not). As an *au pair*, she enjoys a level of independence from the family that hosts her, but maintains a closer personal relation to the family members, especially the children and the mother. The spaces in the house serve as a metaphor for the commodification of servants, hinting the similarities with slave ships, when black bodies were treated as cargo, or even the triangulation of goods in the Atlantic slave trade[5]:

---

5 In *See Now Then*, published in 2013, Kincaid makes another reference to ships and cargo, stressing the idea of humanity/inhumanity once again. "Mr. Sweet could retreat from the dis-

The room in which I lay was a small room just off the kitchen—the maid's room. I was used to a small room, but this was a different sort of small room. The ceiling was very high and the walls went all the way up to the ceiling, enclosing the room like a box—a box in which cargo traveling a long way should be shipped. But I was not cargo. I was only an unhappy young woman living in a maid's room, and I was not even the maid. I was the young girl who watches over the children and goes to school at night. (*Lucy* 7)

The following chapter, named "Mariah", deals mainly with how Lucy develops her relationship with this mother figure. Later in the novel, Lucy describes their relation in the following terms: "[t]he times that I loved Mariah it was because she reminded me of my mother. The times that I did not love Mariah it was because she reminded me of my mother" (*Lucy* 58). Abruna comments on this relationship, stressing the different dynamics between Lucy and Mariah:

With Mariah she [Lucy] has the closeness of conversation and intimacy that she could not have experienced with her 'saint-like' mother. It is Mariah who points out to Lucy that she is filled with anger and later suggests, even as her own marriage is falling apart, that Lucy must forgive her mother in order to thaw her cold heart [...] (176)

It is through the mother-daughter relation that Kincaid continues to explore questions of dominance and power, even if the mother is not the same. Similar reasons that maintained Lucy apart from her mother are found in the relationship she develops with Mariah, who embodies the privileges associated with white supremacy. This character informs much of the post-colonial issues in the novel, making explicit the matters related to race, gender, history and power. She is the mother of four children, in a fragile marriage, living a very comfortable life. The family's upper class status is shown throughout the narrative, in their surroundings, their trips, and their relations. Kincaid writes:

Mariah wanted to rescue me. She spoke of women in society, women in history, women in culture, women everywhere. But I couldn't speak, so I couldn't tell her that my mother was my mother and that society and history and culture and other women in general were something else altogether. (*Lucy* 131–132)

---

turbance of those children and the presence of that woman [Mrs. Sweet, Kincaid's protagonist and ultimately, one facet of herself] who had absolutely arrived on a banana boat or some vessel like that, for nobody knew exactly how she arrived; she had a story that began with her mother hating her and sending her away to make money to support her family and she had no father, there was no claim made on her, she was just sent away on a vessel that went back and forth, carrying cargo, human sometimes, of a nonhuman but commercial nature sometimes [...]" (97–98).

Race and class are not perspectives through which the Mariah character is used to seeing reality, resulting in clear conflicts with Lucy, who is overly (self)conscious of her circumstances. Mariah is heavily informed by a liberal feminist point of view, a point of view that has not assessed its privileges or questioned the influence of other factors other then gender in its analysis, one that is not intersectional. She speaks of women in a kind of homogeneous fashion, in which their differences and histories are all categorically erased in the quest for the establishment of a feminine experience. The critique of this kind of feminism is made explicit in Ferguson's interview, in which Kincaid denounces the impossibility of white women to give up their privileges in order to create a new and transformed reality, since they would also lose something in this process.

> Even as they [white women] are oppressed within their group, they are still of the privileged. I think that change for them would be very threatening to their status because when we rebel we [black women] want the whole thing washed away, turned upside down. But they can't do that because they would lose something too. Let's face it – a white woman earns more money in this society than a black man. She earns less than a white man, but she earns more than a black man. (Ferguson and Kincaid 171)

In the summer, during the trip to the Great Lakes to visit Mariah's childhood home, the family goes on a long train journey across the country. Mariah is very excited to show Lucy this side of the United States, as well as share her life experience. In the train ride, Lucy notices similarities between the passengers and Mariah's family, and herself and the servants. The color line seemed not to be something that the white party of characters noticed. Lucy is quick to grasp that the black people in the train were similar to her folk simply in color, but not in manner. Mariah seems to be unaware of these similarities and differences, exercising her entitlement as a white subject as she sees her experience as the norm:

> The other people sitting down to eat dinner all looked like Mariah's relatives; the people waiting on them all looked like mine. The people who looked like my relatives were all older men and very dignified, as if they were just emerging from a church after Sunday service. On closer observation, they were not at all like my relatives; they only looked like them. My relatives always gave backchat. Mariah did not seem to notice what she had in common with the other diners, or what I had in common with the waiters. She acted in her usual way, which was that the world was round and we all agreed on that, when I knew that the world was flat and if I went to the edge I would fall off. (*Lucy* 32)

By mentioning that their worlds, and consequently their worldviews were utterly different, Kincaid is stressing the abyssal gap, in Santos' terms, between their experiences. In an exchange regarding her own family, Mariah expresses a kind of confidence that Lucy is yet to experience, or even witness, in her racialized reality. The relation with Mariah develops into some kind of conflict that is still difficult to be named with precision. The mother-daughter bond that is developed between the characters comes under scrutiny, similarly to the relation Lucy had developed with her mother in Antigua. This time, however, the mother figure treats Lucy in a seemingly condescending manner, differently from the imposing Caribbean mother, who often left the protagonist in awe. On the same note, when the readership is introduced to Gus, a groundskeeper in Mariah's childhood home, the question of the bond between characters comes to the foreground, as Lucy perceives that their relation of belonging/possession is different from hers, and race seems to be one factor that would determine the nature of this difference.

> When we got to our destination, a man Mariah had known all her life, a man who had always done things for her family, a man who came from Sweden, was waiting for us. His name was Gus, and the way Mariah spoke his name it was as if he belonged to her deeply, like a memory. And, of course, he was a part of her past, her childhood: he was there, apparently, when she took her first steps; she had caught her first fish in a boat with him; they had been in a storm on the lake and their survival was a miracle, and so on. (*Lucy* 33–34)

Lucy realizes that Mariah "owns" Gus as a fixture of her identity, as he seems to be a formative part of her past and present. Their shared history is telling of the importance of maintaining this memory alive, even though Gus is not granted agency. Lucy, as a colonized subject, sees this relation as perverse, siding with her bias towards historical domination. François agrees with this perspective:

> Lucy interprets Gus's position as a subaltern as he is subjected to the whims of a complacent, privileged class among which a servant is often patronized, depersonalized and reduced into an object whose functional value is attached to the employer's needs and emotions. He is not a person in his own right. (Françoise 88)

The difference between a person from Sweden and a person from the Caribbean might also be seen through a perspective not primarily associated with race, but with their stereotypical temperaments, Lucy's anger as opposed to Gus's mild and rational Nordic behavior. Regarding Mariah's personal history, she expects Lucy to recognize in her specificity that she does not fully grant to the protagonist. One of these specificities, however, comes completely as a shock to lucy. Mariah's mother claims that she has got indigenous ancestry, which supposedly grants her the skills associated with the wilderness, skills she tried to showcase during their trip to the Midwest. This claim perplexes Lucy, since there is nothing apparent in Mariah that loosely resembles her alleged indigenous ancestry. Lucy questions herself on what would be the right way to take this declaration, making evident that Mariah is performing some kind of cultural appropriation, his is not justifiable in the eyes of someone who can legitimately feel connected to this ancestry and who has lived a side of history that has been fully affected by this reality, questioning the essentializing process of cultural appropriation.

> I myself had Indian blood in me. My grandmother is a Carib Indian. That makes me one-quarter Carib Indian. But I don't go around saying that I have some Indian blood in me. The Carib Indians were good sailors, but I don't like to be on the sea; I only like to look at it. To me my grandmother is my grandmother, not an Indian. My grandmother is alive; the Indians she came from are all dead. If someone could get away with it, I am sure they would put my grandmother in a museum, as an example of something now extinct in nature, one of a handful still alive. In fact, one of the museums to which Mariah had taken me devoted a whole section to people, all dead, who were more or less related to my grandmother. (*Lucy* 40)

Lucy's critique of the reification of minorities and the historicization processes that are involved in the making of the cultural supremacy of whiteness are all present in this little episode of cultural appropriation. This facet of Mariah might be seen as the continuous effort of whiteness to create consensus by erasing difference, even if it involves absorbing the "other". In the interview with Ferguson, when asked about a trip to Kenya and the racial perceptions between whites and blacks in the country, Kincaid comments on the issue of claiming a culture to oneself, making reference to Lucy and Mariah:

> It is a very peculiar thing. Again, you see how easy it is for winners to claim – it is very easy for Europeans to claim – they are Africans. But I have never met an African who was born in Europe who said, "I am European." It is so much regarded as a distinct racial thing European. You hear all the time about white Africans. How can that be? When you win, you can do anything. You can adopt another people's identity very easily. That is the point of Mariah in the chapter in Lucy where Mariah adopts Indian ancestors. (Ferguson and Kincaid 171)

Even though it seems that Mariah is simply trying to narrow the gap between their experiences, she senses that Lucy would not be able to read this statement in this light. Lucy is indeed enraged once again: "Mariah says, "I have Indian blood in me," and underneath everything I could swear she says it as if she were announcing her possession of a trophy. How do you get to be the sort of victor who can claim to be the vanquished also?" (*Lucy* 40 – 41).

When talking about her love for her children, Mariah expresses something Lucy misses in her own mother-daughter relationship, one that has been disturbed by the arrival or her brothers. At the same time, Mariah's confidence puzzles the protagonist, hinting at the privileges that must have forged this kind of worldview.

> She said, "I have always wanted four children, four girl children. I love my children." She said this clearly and sincerely. She said this without doubt on the one hand or confidence on the other. Mariah was beyond doubt or confidence. I thought, Things must have always gone her way, and not just for her but for everybody she has ever known from eternity; she has never had to doubt, and so she has never had to grow confident; the right thing always happens to her; the thing she wants to happen happens. Again I thought, How does a person get to be that way? (*Lucy* 26)

Even though Mariah belongs to a minority as a woman in the power hierarchy of contemporary American society, she ranks much above Lucy, who has never been able to experience the feeling of not having to struggle to achieve what she desires, or even to be perceived in her full humanity. Clearly, Kincaid depicts Mariah as a less complex subject than Lucy, especially in regard to her own development as a human being. Mariah is the archetype of white supremacy that has had history on her side. She is not capable of fully empathizing with Lucy's intersectional experience as a poor woman of color, focusing mainly on the "woman" axis of interpretation.

When spring comes, Mariah poses a question to Lucy: "You have never seen spring, have you?" (*Lucy* 17). This is a question that is not intended to do any harm, and serves the purpose of introducing Lucy to a new reality, one that Mariah is eager to share with the protagonist. However, Lucy perceives Mariah's fascination with spring as naïve, and is perplexed on how much something like the weather is capable of affecting her. Seasons are only a fixture of the imagination in Lucy's view, since the Caribbean weather remains mostly unaltered during the year. Mariah's plans for the trip to the Midwest are delayed when the first day of spring is marked by a snowstorm, one event Mariah describes in a scornful retort as "typical". Lucy interprets that Mariah believes it to be some sort of personal betrayal to her plans, and questions: "[...] How do you get to be a person who is made miserable because the weather changed its mind, because the weather doesn't live up to your expectations? How do you get to be that way? (Kincaid 1990: 20). François also comments on the issue:

> Lucy's reactions demonstrate that the people's lives in her society are marked primarily by hardship or that they are so shackled by the harsh conditions of their lives that they are unable to accept the luxury to indulge in the weather. Her attitude towards the weather mirrors the peasant mentality of the agricultural Antiguan setting in which she grew up. In brief, to Lucy, Mariah's leisured, fanciful preoccupation with the vagaries of the weather seems trivial and, in Moira Ferguson's words, "perilously self-indulgent". (François 84)

Reversing the gaze, one must ask how Lucy got to be the way she is. The impact of the mother-daughter relationship and its subsequent dismantling are clear factors, as well as colonialism and its violent presence in her life. Some other factors are also present in the forging of her identity, such as the weather, even as Lucy is shocked by its impact in Mariah's life and personality. Later in the novel, Lucy refers to how the Caribbean climate has had an impact, though uncertain, in her personality. While the sunny landscape is dominant in her life in the island, her mood does not reflect joy, the idea primarily associated with it.

> I was born and grew up in a place that did not seem to be influenced by the tilt of the earth at all; it had only one season—sunny, drought-ridden. And what was the effect on me of growing up in such a place? I did not have a sunny disposition, and, as for actual happiness, I had been experiencing a long drought. (*Lucy* 86)

The lack of happiness implied here may be translated in the rage and sulkiness that the character frequently displays in the narrative. The contrast between the beautiful weather and Lucy's cold temperament also comments on the ways in which this "visitor" contradicts the expectations set upon her, shunning stereotypes of an easy life in a paradise. The experience of the Caribbean as a playground for white Americans and white tourists in general is also relevant here and widely explored in *A Small Place* (77). Marquis aptly points that "[t]ourists have replaced the English plantation owners, and blacks remain the victims of a racist system, objectified, even in their poverty seductively open to the tourist gaze" ("Not at Home"). The monotonous weather, even if beautiful, poses the question of a different temporality for Lucy, who sees her life as a perpetual repetition of the same day, and possibly the same lack of opportunities. This permanence also hints at the continuous and unaltered lives the peoples in the islands live, as structural factors impede them from overcoming the oppressions and difficulties faced since the inception of colonialism. They seem to be simply locked in time.

The question of temporality and its perception is explored at length in *A Small Place*, in which Kincaid investigates the ways that this different understanding of time affects the lives of the people in places like Antigua.

> To the people in a small place, the division of Time into the Past, the Present, and the Future does not exist. An event that occurred one hundred years ago might be as vivid to them as if it were happening at this very moment. And then, an event that is occurring at this very moment might pass before them with such dimness that it is as if it had happened one hundred years ago. No action in the present is an action planned with a view of its effect on the future. When the future, bearing its own events, arrives, its ancestry is then traced in a trancelike retrospect, at the end of which, their mouths and eyes wide with

their astonishment, the people in a small place reveal themselves to be like children being shown the secrets of a magic trick. (54)

This alternative sense of time helps understand the claims Lucy makes in relation to her heritage, and consequently to how colonialism is still a very much present force in her life. Lucy sees the oppression and exploitation of people who look like her as having a heavy influence in her identity formation, in addition to the lives that her people are allowed to live under these systems of structural oppression and violence. This might be one way of seeing the permanence of these issues in Kincaid's writing, and post-colonial literature at large. When thinking about the exploitation of Africans in the New World, Kincaid points out that this reality is remembered as a distant experience and as very close at the same time, shaping everyday life; ironically, the narrative gains almost a mythological aura.

> In Antigua, people speak of slavery as if it had been a pageant full of large ships sailing on blue water, the large ships filled up with human cargo—their ancestors; they got off, they were forced to work under conditions that were cruel and inhuman, they were beaten, they were murdered, they were sold, their children were taken from them and these separations lasted forever, there were many other bad things, and then suddenly the whole thing came to an end in something called emancipation. Then they speak of emancipation itself as if it happened just the other day, not over one hundred and fifty years ago. The word "emancipation" is used so frequently, it is as if it, emancipation, were a contemporary occurrence, something everybody is familiar with. (*A Small Place* 55)

The author's recognition of a different set of temporalities helps to foster a broader understanding of how colonialism is still a very much present reality in the lives of many subjects, even those who inhabit the same geopolitical space, but who got there in different circumstances and with very different backgrounds. This recognition also helps create a basis from which the dominant might better comprehend the long and gripping reach of their influence outside the spheres of the lived, and how this understanding continually shapes history, as linear temporalities are challenged and scrutinized.

Halfway through the trip, Mariah calls Lucy's attention to one of her favorite views in the route, freshly plowed fields with the soil turned over and ready for seeding. Lucy however, sees a different landscape in front of her, denouncing how past and present are still intermingled in the colonized experience:

> Early that morning, Mariah left her own compartment to come and tell me that we were passing through some of those freshly plowed fields she loved so much. She drew up my blind, and when I saw mile after mile of turned-up earth, I said, a cruel tone to my

> voice, "Well, thank God I didn't have to do that." I don't know if she understood what I meant, for in that one statement I meant many different things. (*Lucy* 33)

Lucy is referring to the work slaves would have performed in those fields, meaning that not so long ago such work, the plowing of land, would have been executed by someone who looked like her. Backchat is one of Lucy's tools to respond to the pervasive racism that she encounters in the United States. This type of response is a constant feature of Lucy's personality, one that, according to George Lamming, is a common characteristic of the Caribbean mode of speaking ("Not at Home"). Marquis comments on this specific kind of rhetorical strategy as a way for the writer to find a voice, as her resistance to the discouses and practices that surround her, forge her own sense of self. This kind of responsive interlocution is often translated as anger in many cultures, and certainly in the American context; in the Caribbean, however, it is the cultural norm and does not necessarily indicate negative feelings. Kincaid herself has been described frequently as an angry writer, one that does not spare words to backchat against everything she finds unfair in her life. Backchat becomes a form of resistence for Lucy, and for Kincaid in a larger sense.

> Anger pervades the novels. It can be detected in the very body language of Kincaid's characters, their reflexive dismissals of others, quite as much as it finds a voice in their protests and judgments. If those protests have a larger, social target, register a more expansive, community history, everything from the line of the novels' actions to the family names of characters and the pattern of their relationships insists on our recognizing that they are not merely personal, but connect to Kincaid's own life. Anger strikes home as utterly personal, no matter how much it appears to speak for others. Yet it must be said that it also unsteadies the personal story, somehow leaving it estranged from itself, as voice is to echo, or as voice catches up the echoes of other voices. ("Not at Home")

Lucy's rage, and to an extent Kincaid's, is not simply connected to race, or, gender, or even geography and its political implications, but comes from the desire of recognition that she demands as legitimate, and the former affiliations are part of a hybrid and intersectional composite that helps channel this desire of full citizenship, either as a woman, as a black subject, or as a peripheral subject. The subject of anger in Kincaid's production is seen as a pervasive characteristic, reflecting the hybrid character of this feeling, both personal and collective. This anger is elicited in one of the central episodes in the narrative, when the power of symbolic violence is demonstrated in the simplest things, such as in a grove full of flowers, or even in a poem. The violence imbued in these symbols is analyzed and challenged by Kincaid in the novel, as well as in many other titles by the author.

3.1.4 Daffodils

In *Lucy*, Kincaid explores daffodils as a metaphor to depict the grave consequences of colonialism and the acculturation processes experienced in Antigua, and that are carried out through her adult life. Braziel characterizes Kincaid's model of depicting the daffodils, a motif previously named "flowers of evil" by Kincaid herself, as "[...] exemplary of racial privilege and colonial expansionism [...]" (113). The first mentioning of the flowers in *Lucy* happens in the Mariah chapter, when Mariah asks:

> Have you ever seen daffodils pushing their way up out of the ground? And when they're in bloom and all massed together, a breeze comes along and makes them do a curtsy to the lawn stretching out in front of them. Have you ever seen that? When I see that, I feel so glad to be alive". (*Lucy* 17)

Braziel comments on the performance of the daffodils as symbols of normative femininity, stating: "[t]he image of the daffodil curtsying *en masse* suggests the very conformity to femininity that Lucy scorns and resists" (115). Lucy is mystified by Mariah's reaction to these flowers, which prompt a "how do you get to be this way" response, and in mild irritation mentally chastises the host mother. It seems that Lucy disapproves of her reaction based not only on its frivolous character, but also because it showcases the limits of Mariah's sensibility. Following the exchange, Lucy delves into her personal history of the daffodil, one that is profoundly connected to the English imperial project, William Wordsworth's poem "I Wandered Lonely as a Cloud", also know as "Daffodils" [6], published in 1807. It describes the experience of unbridled joy the author feels when seeing these flowers, one type of experience that is unfamiliar to Kincaid, as she points out in *My Garden Book*, in 1999: "I am not in nature. I do not find the world furnished like a room, with cushioned seats and rich-colored rugs. To me, the world is cracked, unwhole, not pure, accidental; and moments of joy

---

[6] The poem inferred by Kincaid is the following: "I wandered lonely as a cloud / That floats on high o'er vales and hills, / When all at once I saw a crowd, / A host, of golden daffodils; / Beside the lake, beneath the trees, / Fluttering and dancing in the breeze. / Continuous as the stars that shine / And twinkle on the milky way, / They stretched in never-ending line / Along the margin of a bay: / Ten thousand saw I at a glance, / Tossing their heads in sprightly dance. / The waves beside them danced; but they / Out-did the sparkling waves in glee: / A poet could not but be gay, / In such a jocund company: / I gazed—and gazed—but little thought / What wealth the show to me had brought: / For oft, when on my couch I lie / In vacant or in pensive mood, / They flash upon that inward eye / Which is the bliss of solitude; / And then my heart with pleasure fills, / And dances with the daffodils" (Wordsworth 1807).

for no reason is very strange" (124). Kincaid also mentions the need for the English to tame their surroundings wherever they are, and even more so in their colonies, creating a landscape that is organized and orderly:

> Whatever it is in the character of the English people that leads them to obsessively order and shape their landscape to such a degree that it, the English landscape, looks like a painting (tamed, framed, captured, kind, decent, good, pretty) – and a painting never looks like it, the English landscape, unless it is a bad painting – this quality of character that leads to the obsessive order and shape of the landscape is blissfully lacking in the Antiguan people. (132)

Daffodils are not native to the Caribbean and are not suitable to thrive in this landscape as easily. As a reflex of the ordering will of the English empire, there was the cultural imposition of gardening as an activity of leisure, one that is completely absent in tropical cultures. The economic implications of dedicating time and effort to the achievement of simply aesthetic beauty in the ordering of plants and their care is completely alien to a landscape that is abundant with examples of beauty in its chaotic and wild reality. To Lucy, as well as to Kincaid, daffodils are a natural symbol of the English imperial project, as equally negative as the school and its syllabus, the uniforms, the English language itself, among many other forms of cultural imperialism. Its presence, even only theoretical in the educational apparatus of English Antigua, is telling of the force of the "Englishness" of the plant as a symbol. Marquis remarks on this, stating:

> Kincaid's *Lucy*, in fact, permits us to see through the "mystery" of those daffodils: for daffodils we must read "Daffodils," and for Wordsworth's iconic poem we must read the assault of the colonial education system. Kincaid herself has more recently enlarged the explanation, pointing out that what she calls "gardening cultures" typically have empires; furthermore, "You can't have the luxury of pleasure without somebody paying for it". Those daffodils came at a price, however unobserved the payment. In such a case, as in her fierce complaint about the loss of the Caribbean, wrath seems most righteously to target one or another generation of colonizers. Yet there is nothing remotely polemical in this rage; it is utterly personal in its fusion of grief, despair and fury. ("Not at Home" 2007)

Abruna also points to the traumatic imposition of English culture experienced by Lucy stating that "[...] the flowers are symbolic of the many ways British culture had been forced on the young women in Antigua" (178). Observing the issue of acculturation and the English presence, Kincaid writes in *A Small Place* a commentary on the colonizing will of England and its perverse consequences, employing the justified wrath and grief above mentioned by Marquis:

[...] they should never have left their home, their precious England, a place they loved so much, a place they had to leave but could never forget. And so, everywhere they went they turned into England; and everybody they met they turned into English. But no place could ever really be England and nobody who did not look exactly like them would ever be English, so you can imagine the destruction of people and land that came from that. (24)

The symbolic violence perpetrated by the colonial education is showcased here, as the Caribbean flora in its abundance is ignored in detriment of the English counterpart. The referents associated to the empire are naturalized in these small exercises of cultural identification, which may seem harmless at first, but work towards the subordination of the colonized subjects to the culture and symbols of the ruler, which are frequently alien to the daily material reality of the colony. Lucy's defiant nature surfaces here as she is conflicted by the praise she receives from performing the role of domesticated colonized subject after reciting the poem at school, one that contradicts her rebellious attitude. At the age of ten, the protagonist is already aware of the domination of the English and the exercise of Englishness as forms of oppression. The poem figures here as the arena in which the symbolic hierarchy of referents is at play, since it celebrates Englishness in its epitome, and the mimicry of such Englishness, stressed by the approval of its perfect presentation and delivery, is celebrated by the colonized audience. Lucy, however, has a difficult time coming to terms with such approval and resolves the issue by a matter of disassociation:

> I was then at the height of my two-facedness: that is, outside I seemed one way, inside I was another; outside false, inside true. And so I made pleasant little noises that showed both modesty and appreciation, but inside I was making a vow to erase from my mind, line by line, every word of that poem. The night after I had recited the poem, I dreamt, continuously it seemed, that I was being chased down a narrow cobbled street by bunches and bunches of those same daffodils that I had vowed to forget, and when finally I fell down from exhaustion they all piled on top of me, until I was buried deep underneath them and was never seen again. (*Lucy* 18)

The two-facedness is also evocative of Dubois's double consciousness, making explicit the difficult task of reconciling the multiple demands of the racialized experience. The internal desire to erase the verses of the poem is indicative of the refusal to subsume to the imperial order, even if at the time this effort could only happen in the most private sphere. The outside performance might be related to values associated to a regulated kind of femininity, where the demure acquiescence of approval is seen as the appropriate response a young girl should provide, a kind of response that is in its essence English. The dream Lucy experiences helps shape the idea that the presence of the English culture in her life is oppressive and threatens to eliminate her own identity, never to be seen again under the symbols of the colonizer culture.

After telling Mariah about her experience with the daffodils, Lucy senses that the gap between them widens, as it seems that there is no real understanding in Mariah's side regarding the colonial violence suffered by the protagonist and perpetrated by Mariah's ancestors. The anger expressed by Lucy seems to be recognized as not expected and therefore less legitimate by the white counterpart:

> I had forgotten all of this until Mariah mentioned daffodils, and now I told it to her with such an amount of anger I surprised both of us. We were standing quite close to each other, but as soon as I had finished speaking, without a second of deliberation we both stepped back. It was only one step that was made, but to me it felt as if something that I had not been aware of had been checked.
>
> Mariah reached out to me and, rubbing her hand against my cheek, said, "What a history you have." I thought there was a little bit of envy in her voice, and so I said, "You are welcome to it if you like". (*Lucy* 18 – 19)

To sense Mariah's reaction as envy is an indicative of how much Lucy relates the whiteness associated to Mariah to a desire of a continuous process of colonization, when once again Lucy's experience is an object to be absorbed by the colonial/imperial endeavor. When Lucy offers her history to Mariah in ironic retort, the readership is compelled to see that what she is really offering is a wish for a deeper comprehension of her circumstances of coloniality, one would no doubt trade for the "easier" experience she considers Mariah to have had.

The daffodil is to Lucy simply a mental image, an idea that exists in the indoctrination syllabus of the foreign English culture. One that specificallt stresses how much these subjects are forced to understand themselves as foreign within in their own environment and culture. When Lucy is finally presented to the flowers, her first desire is to destroy them, even if not knowing specifically that the flowers in front of her are daffodils. This desire for violence against something seemingly innocuous makes evident the refusal to identify with them and their significance.

> Along the paths and underneath the trees were many, many yellow flowers the size and shape of play teacups, or fairy skirts. They looked like something to eat and something to wear at the same time; they looked beautiful; they looked simple, as if made to erase a complicated and unnecessary idea. I did not know what these flowers were, and so it was a mystery to me why I wanted to kill them. Just like that. I wanted to kill them. I wished that I had an enormous scythe; I would just walk down the path, dragging it alongside me, and I would cut these flowers down at the place where they emerged from the ground. (*Lucy* 29)

Lucy's powerful response is understandable as long as it is considered under the scope of coloniality and Lucy's identity formation, as pointed out by François: "Lucy's violent reaction at her first sight of daffodils might seem disproportionate to Mariah's intentions if we do not place it firmly within the dialectic of colonial discourse and Lucy's personal trajectory" (François 2008: 85). Mariah, armed with the knowledge of Lucy's relation to daffodils is incapable of seeing beyond her experience and the affective memories of the flowers, ignoring the violence suffered by Lucy. The characterization of the flowers as objects of consumerism, made to be worn or eaten, resembling play teacups, or fairy skirts, allow for readings connected to ideas of stereotypical femininity and the desire to fulfill the expectations imbued into these flowers. The simplicity stands for the disguised history of violence and genocide, of slavery and colonialism, as pointed out by Braziel (117). When Lucy describes them as beautiful and simple, fashioned in a way that removes the possibility of a deeper reading, she could also be describing Mariah's position in relation to colonialism, in which there seems to be only a veneer of historical understanding, but no further inquire into about its consequences, one idea previously explored by Adrienne Rich in *The Politics of Location*. For Mariah the complicated and unnecessary idea might be the understanding that she, and her class, benefit from the privileges obtained through a history of colonialism and coloniality that is embodied by Lucy. All privileges build upon a legacy of domination, exploitation, prejudice, and poverty. Mariah, who is glad to share her love for the flowers with Lucy, acknowledges Lucy's pain, only to dismiss it once again:

> Mariah said, "These are daffodils. I'm sorry about the poem, but I'm hoping you'll find them lovely all the same."
> There was such joy in her voice as she said this, such a music, how could I explain to her the feeling I had about daffodils—that it wasn't exactly daffodils, but that they would do as well as anything else? Where should I start? Over here or over there? Anywhere would be good enough, but my heart and my thoughts were racing so that every time I tried to talk I stammered and by accident bit my own tongue. (*Lucy* 29)

Lucy is unable to tell Mariah the extent to which her life is permeated by symbols similar to the daffodils, and once again miscommunication seems to characterize the relation of this mother-daughter pair. François comments on the issue, stating that "[...] Mariah's class background and position make her assume that aesthetics and politics are separated, in a similar way she had been taught that literature was separate from politics; not so for Lucy" (85). Lucy is unable to find the words to explain the pervasiveness of coloniality in her experience, and the unintentional biting of the tongue is symbolically telling[7]. The misunderstanding of Mariah mirrors the difficulty white America continues to have when dealing with race related issues and colonialism. Kincaid exposes through Lucy and Mariah how issues that are central to the lives of those who do not belong to a white supremacy are (miss)understood by those who most benefit from them.

Finally, Lucy is capable of expressing the central problem with the daffodils story. Lucy's revelation is an attempt to make clear her position and her history, as opposed to the narrative that Mariah is accustomed to perceive as the normalized and unique perspective.

> Mariah, mistaking what was happening to me for joy at seeing daffodils for the first time, reached out to hug me, but I moved away, and in doing that I seemed to get my voice back. I said, "Mariah, do you realize that at ten years of age I had to learn by heart a long poem about some flowers I would not see in real life until I was nineteen?"
>
> As soon as I said this, I felt sorry that I had cast her beloved daffodils in a scene she had never considered, a scene of conquered and conquests; a scene of brutes masquerading as angels and angels portrayed as brutes. (*Lucy* 29 – 30)

The misinterpretation performed by Mariah is expressed in the unwanted hug Lucy is able to avoid, stressing once again how far apart the two parties are from un understanding of their differences, the affection Mariah feels for Lucy is put into question as a force that obliterates Lucy's attempts to communicate her side of history. Lucy's own conflicting feelings regarding this mother-daughter relationship are evident in the subsequent discomfort she feels when considering the difficult process of re-signifying that Mariah is going through, as Lucy reverts the expectations related to who are the real perpetrators of violence and who are the victims, something that Mariah never needed to consider in her experience, or did so from a safe distance. According to Braziel: "[...] the daffodils and their cultural signification shifts, migrates: Mariah, who tried to impose her view of the daffodils onto Lucy, is now forced to see the daffodils 'cast' into a new scene, a diabolic one, in which she becomes a brutalizing agent rather than the 'civilizing' one" (118). Lucy's assertiveness and rage are part of the voice she finds to state her position and inscribe her perspective as one of the sides in the complex dialogue of colonialism.

> This woman who hardly knew me loved me, and she wanted me to love this thing—a grove brimming over with daffodils in bloom—that she loved also. Her eyes sank back in her head

---

7 See, for instance, the work of Marlene Nourbese Philip *She Tries Her Tongue, Her Silence Softly Breaks*, originally published in 1988.

as if they were protecting themselves, as if they were taking a rest after some unexpected hard work. It wasn't her fault. It wasn't my fault. But nothing could change the fact that where she saw beautiful flowers I saw sorrow and bitterness. The same thing could cause us to shed tears, but those tears would not taste the same. We walked home in silence. I was glad to have at last seen what a wretched daffodil looked like. (*Lucy* 30)

Finally, the revelation of the daffodils to Lucy seems to be one way her colonial history comes full circle, as she is able to reorder the significance of the memory she had associated with it, in addition to refiguring it in a social context with Mariah.

The relationship between Mariah and Lucy is complex, since even when Lucy is able to pinpoint the flaws in Mariah's character, she still nurtures positive feelings for the host mother. She finds in Mariah the tenderness she had long lost from her mother in Antigua. The lack of recognition of her own privileges is what creates a distance between this mother and daughter, as Lucy's anger is surpassed by the love she develops for Mariah. In other words, the situational thinking that Mariah lacks is compensated by her acts of altruism in the eyes of the protagonist. At the end of the novel, Lucy looks back at these first years in the United States and reconsiders her position in relation to history, in a pessimistic fashion:

History is full of great events; when the great events are said and done, there will always be someone, a little person, unhappy, dissatisfied, discontented, not at home in her own skin, ready to stir up a whole new set of great events again. I was not such a person, able to put in motion a set of great events, but I understood the phenomenon all the same. (*Lucy* 147)

Lucy might not have been able to put in motion a set of great events in her narrative, but certainly Kincaid was.

3.2 Edwidge Danticat

Edwidge Danticat is a Haitian-American author born in Port-au-Prince in 1969, moving to New York at the age of twelve. She is the author of several titles, such as *Breath, Eyes, Memory* (1994), *Krik? Krak!* (1996), *The Farming of Bones* (1998), *The Dew Breaker* (2004), *Brother, I'm Dying* (2007), *Create Dangerously: The Immigrant Artist at Work* (2010), *Claire of Sea Light* (2013), *Untwine* (2015), *The Art of Death* (2017), among others. Her production encompasses fiction, short story, young adult literature, biography and essay, besides working as editor for anthologies and collections. The themes that are most often found in her writings concern the immigrant experience, Haitian life and culture, history, violence and trauma. *The Farming of Bones*, for instance, deals with the "Parsley Massacre", which took place in the Dominican Republic border with Haiti, and killed more than twelve thousand Haitians in 1937. *The Dew* Breaker is a short story collection that deals with a group of torturers from a paramilitary police directed by the Haitian dictators Jean-Claude and François Duvalier. In *Brother, I'm Dying* Danticat delves into a more personal account of immigration and crisis, as she reports her memories, including memories regarding her dying father and the death of an uncle at the hands of the Department of Homeland Security in Florida in 2004, while in *The Art of Death* she explores her mother's final days. She was the winner of the National Book Award in 1999, for *The Farming of Bones*, and the National Book Critics Circle Award in the memoir/biography category in 2007 for *Brother, I'm Dying*.

In an essay titled "We Are Ugly, But We Are Here", Danticat writes about her background and her parent's migration, stressing how the dictatorship as a marker of oppression, either economic or political, is one of the fundamental matters in her trajectory, becoming a theme she would explore in many different ways in her production:

> I was born under Haiti's dictatorial Duvalier regime. When I was four, my parents left Haiti to seek a better life in the United States. I must admit that their motives were more economic than political. But as anyone who knows Haiti will tell you, economics and politics are very intrinsically related in Haiti. Who is in power determines to a great extent whether or not people will eat. (Danticat 1996)

She was left in Haiti at the care of an aunt before being summoned to join her parents in the United States, similarly to her main character in *Breath, Eyes, Memory*. Some more details are given in *The Art of Death*, in which the author describes: "[m]y mother in her late thirties was an undocumented immigrant living in Brooklyn, away from her two small children. She was a factory worker who made handbags for pennies on the dollar" (22).

Danticat's life and fiction bear some resemblance, but, differently from Kincaid's project, Danticat's production sides better with fiction than with autobiography. Braziel characterizes *Breath, Eyes, Memory* as "semiautobiographical novel" (119), a definition that tries to translate how this complex narrative encompasses elements from different genres in order to deal with matters of trauma and violence. In other words, though some elements of her life story are present in her fictional writing, they do not constitute a network of recurring meaning that pieces together her trajectory, as we can find in Kincaid's novels.

In an interview with Deborah Treisman for *The New Yorker* on May 7th 2018, Danticat tells of her parent's experience as undocumented migrants in the United States, and the expectation of a better life, even with all the uncertainty that the migration experience creates, especially in the migration policies in the United States, during the Obama's administration, and then Trump's:

> I think that everyone who makes this type of journey—including my parents, who did not come by boat but were undocumented during their early years in the U.S.—hopes that, however difficult, it will have been worth it, if not directly for them, then for their children. Of course, there are no guarantees, especially in the current climate, in which immigrants are being scapegoated and even the few protections that have existed for some years—such as T.P.S. (Temporary Protected Status) and *daca* (Deferred Action for Childhood Arrivals)— are in jeopardy or are being taken away. (*The New Yorker*)

Borrowing the words from the protagonist of *Breath, Eyes, Memory*, Sophie's self description could easily be applied to Danticat: "I come from a place where breath, eyes, and memory are one, a place from which you carry your past like the hair on your head" (234). For Danticat, personal history, and history in general, are features that have shaped her life and her production in a more acute manner than usual, as she makes explicit in her writing, as well as in her positioning, the extent to which these discourses of past experience and legacy touch present day living and thinking, expressing similar views to Kincaid's observation in *A Small Place*, as previously mentioned. The common and historical amnesia, in addition to the repeating cycles of memory are present in Danticat's production in many forms, be it as she delves in the cyclical movement between the island and the United States, or even in the depiction of trauma in her characters. Memory, history, and representation are co-formative factors in her creation. In a sentence that remits to the novel's title, and that unites all these ideas, Danticat claims: "In our family, we had come to expect that people can disappear into thin air. All traces lost except in the vivid eyes of one's memory" (180). In this passage, the protagonist infers about the violent reality of the dictatorial regime existent in Haiti, as government militias would kidnap, torture and assassinate any dissident citizen, relegating their existence solely to memory.

Some other examples of the violent reality of Haiti are explored in "We Are Ugly, But We Are Here", in which the author stresses the invisibility of Haitian women in the media, claiming that:

> Watching the news reports, it is often hard to tell whether there are real living and breathing women in conflict-stricken places like Haiti. The evening news broadcasts only allow us a brief glimpse of presidential coups, rejected boat people, and sabotaged elections. The women's stories never manage to make the front page. However they do exist. I know women who, when the soldiers came to their homes in Haiti, would tell their daughters to lie still and play dead. I once met a woman whose sister was shot in her pregnant stomach because she was wearing a t-shirt with an "anti-military image." I know a mother who was arrested and beaten for working with a pro-democracy group. Her body remains laced

with scars where the soldiers put out their cigarettes on her flesh. At night, this woman still smells the ashes of the cigarette butts that were stuffed lit inside her nostrils. In the same jail cell, she watched as paramilitary "attaches" raped her fourteen-year-old daughter at gun point. ("We Are Ugly")

Telling stories about Haiti becomes only a possibility once assuming the hyphenated reality of migration, as pointed by Sharrón Eve Sarthou in "Unsilencing Défilés Daughters: Overcoming Silence in Edwidge Danticat's *Breath, Eyes, Memory* and *Krik? Krak!*", in which the author explores the ways in which Danticat is able to break the silence of Haitian violence and trauma, stating that for Danticat literature is a space where she voices Haiti's estimable and complex history, as she writes of a world that seems hopelessly broken, her people silenced by more than two centuries of violence and poverty in the island, and outside it by the dislocation and disruption resulting from immigration (101). *Breath, Eyes, Memory* deals with a particular story of violence and of trauma regarding a young migrant Haitian female character. The story's main theme is the development of Sophie Caco, a second generation Haitian-American, and her sexual awakening, paired with her mother's own personal history of sexual violence and trauma. In this novel, one is confronted with a mother who violates her daughter in the name of tradition through the traditional practice of virginity testing. Nonetheless, this mother is concomitantly dealing with her own unconfronted history of sexual violence and trauma. Danticat explores these complex realities in many of her titles, in which the lines between the perpetrator of violence and the victim of violence are blurred, as pointed by Airtor Ibarrola-Armendariz in "The Language of Wounds and Scars in Edwidge Danticat's *The Dew Breaker*", who states: "[u]nlike most trauma theorists, Danticat is a bit more reluctant to mark a clear division between victimizer and victimized, since all of them seem to be burdened by a history in which they have been pawns of forces they could not control" (54). In *The Dew Breaker,* for instance, the readership is confronted with the reality of an ex-torturer from the regime, but from the perspective of his daughter and wife.

Victoria Pinkston, author of "'Our Voices Will Not Be Silenced': Edwidge Danticat, Haiti, and the Silences of History", discusses the centrality o violence in Haiti's history, and even in its national formation myths, demonstrating the ways in which violence, and more specifically sexual violence, is an ever present force that shapes the social and the personal. The author points the ways in which Haitian history and folklore play a significant role in *Breath, Eyes, Memory*, as the myth of Sister Rose, a slave woman who gave birth to the entire nation as a result of her rape pinpoints the context of inherited violence in the identity of the nation. Pinkston explores how this narrative demonstrates the place that sexual violence holds from the inception of Haitian history, and the Cacos' experience then is framed as a continuation of a longstanding cycle of violence toward women; an issue that has permeated Haitian history for centuries (30).

In fact, regarding the reception of *Breath, Eyes, Memory*, the readership has frequently applied a metonymic reasoning process on cultural object, extending its particularity over the universal of Haitian experience, thus equating Sophie's particular narrative of trauma and violence to all Haitian women and girls throughout history. As a peripheral woman from the Caribbean, it seems that her experience assumes the position of the norm for all subjects beyond the abysmal line, forcing upon this individual representation the burden of speaking for all instead of being understood as a specific occurrence. In an afterword, Danticat writes a letter addressed to her protagonist explaining the impact of the work and its developments, clarifying the particularity of the story in the face of the reductionism of a metonymic reading:

> I write this to you now, Sophie, because your secrets, like you, like me, have traveled far from this place. Your experiences in the night, your grandmother's obsessions, your mother's "tests" have taken on a larger meaning, and your body is now being asked to represent a larger space than your flesh. You are being asked to represent every girl child, every woman from this land that you and I love so much. Tired of protesting, I feel I must explain. Of course, not all Haitian mothers are like your mother. Not all Haitian daughters are tested, as you have been [...] I write this to you now, Sophie, as I write it to myself, praying that the singularity of your experience be allowed to exist, along with your own particularities, inconsistencies, your own voice. And I write this note to you, thanking you for the journey of healing – from here and back – that you and I have been through together [...] (*Breath, Eyes, Memory* 236)

The criticism, especially from the Haitian community, seems to concentrate on the fact that the fictional depiction created by Danticat would represent a community, already struggling for acceptance in American society, in a bad light. These preoccupations reflected that the choice of a subject, even if fictional, may impact on the overall perception, usually negative, that mainstream United Stated has of immigrants. Danticat discusses the reception of the book later on in *Create Dangerously:*

> The virginity testing element of the book led to a backlash in some Haitian American circles. "You are a liar," a woman wrote to me [...]. "You dishonor us, making us sexual and psychological misfits". [...] Maligned as we were in the media at the time, as disasterprone refugees and boat people and AIDS carriers, many of us had become overly sensitive and were eager to censor anyone who did not project a "positive image" of Haiti and Haitians. (32)

She also asserts: "[...] though I was not saying that 'testing' happened in every Haitian household, to every Haitian girl, I knew many women and girls who had been 'tested' in that way" (32). Lying, as her critic has put it, is the job of a fiction writer in its essence, and though the reality presented in her representations might not have a direct (auto)biographical stance, the featured theme is present in the culture. Fictionalizing the question, making it a public matter up for scrutiny is the mission of a writer with social responsibility, fostering a broader sense of humanity, and ultimately delivering a character that is complex in its particularities. The African diaspora in Haiti, due to its unique circumstance of having overturned the colonial system in the eighteenth century, lived a particular reality of hardship in the (re)making of the new-world order during modernity, a hardship that has continued until today. Through the narrative in *Breath, Eyes, Memory* the readership is capable of getting to know a more complex reality than the one which is usually portrayed in Eurocentric historiography, or even Eurocentric literature.

3.2.1 Daffodils and Adaptations

Just like in Kincaid's writings, daffodils are present in Danticat's *Breath, Eyes, Memory*. However, in opposition to Kincaid's daffodils, to the protagonist Sophie Caco they are a familiar memory positively associated to the island and her infancy, and are used as a trope to symbolize her mother's love and attachment to Haiti, as her mother wore yellow frequently and expressed a fondness for the flower in many instances throughout the novel. There is also the distancing from the symbol, as a metaphor for the distancing from the island and the traumatic events that have changed their lives, when Martine, Sophie's mother, chooses hibiscus as her flower of choice, leaving daffodils behind. Unlike Kincaid's story, the daffodils were present in the Island of Haiti, as the transplantation of the species was made possible by a more favorable climate than Antigua, though their adaptation implied a change of color. In a dialogue with Atie, the origin of the flowers in the Island is revealed:

> "That is right," she said, "your mother, she loved daffodils."[8] Tante Atie told me that my mother loved daffodils because they grew in a place that they were not supposed to. They were really European flowers, French buds and stems, meant for colder climates.

---

8 The quotation marks are very frequent in Danticat's structuring of dialogue in the novel. Thus, they will frequently appear in the indented quotations, as they facilitate the identification of different types of discouse, direct and indirect, as presented by the author.

A long time ago, a French woman had brought them to Croix-des-Rosets and planted them there. A strain of daffodils had grown that could withstand the heat, but they were the color of pumpkins and golden summer squash, as though they had acquired a bronze tinge from the skin of the natives who had adopted them. (21)

The adaptation of the daffodils to the Haitian context may be seen as a process of creolization, in which the flowers absorbed the characteristics of the landscape to which they were transplanted to. The change of color may also be seen as a metaphor for the creation of a new identity in foreign lands due to the process of migration, which are ironically symbolized here in the figure of a European plant. Braziel comments on the transplantation of Sophie, the daughter of an absent mother raised by her aunt, as mirrored by the creolization of daffodils:

Like Sophie, who has been adopted and raised by her Tante Atie, the daffodils are adopted by the natives, and the flowers, like Sophie, are transformed by this experience- more resilient, more colorful, sturdier, better able to "withstand the heat." Daffodils, in Danticat's novel, are diasporic; the flowers form lateral, rhizomic roots; they migrate. Yet Sophie, like the daffodils, will also be transplanted elsewhere, even if she feels that she was not "meant for colder climates." This transformation requires Sophie to transfer – at least in part – her love from Tante Atie to Martine, from Haiti to the U.S. (121)

Sophie migrates to the United States at the age of twelve, and must also adapt to this new reality, developing new traits that are the result of this process of migration. Her mother, Martine, abandoned the island in her teenage years, moving to the United States, and now Sophie must follow her steps, leaving behind her "adoptive" mother and her grandmother. Atie raised Sophie in Croix-de-Rosets, a small village in the countryside of the island, and occasionally both would travel to the mountains to visit the Caco matriarch, Grandmè Ifé. Atie fulfills the role of mothering Sophie, and therefore is the main model of femininity for the protagonist. After being set aside by her love interest, she resigns to the position of caregiver in the family, namely by raising Sophie, and later on by caring for Grandmè Ifé. Interestingly, Atie develops feelings for another woman later in the novel, being left behind once again as her lover is able to migrate to the United States. This veiled interest is not fully explored by the author, yet it serves to demonstrate the sexual diversity among the generations in this Haitian family. Atie never explicitly deals with this love interest in the narrative, demonstrating one of the ways in which women are silenced in the traditional Haitian culture. Sophie, the protagonist will acquire the silences that are imposed on her own life, something she was instructed directly, or indirectly, by the Caco women. Sarthou comments that "[...] by word and by example, Atie teaches Sophie that Haitian women suffer in silence, and Sophie begins practic-

ing silence the morning after she learns she is leaving Croix-des-Rosets [...]"
(*Breath, Eyes, Memory* 105).

Sophie is temerous, just like Atie, about what she is going to face in her new
life, as she will need to get acquainted with her own mother, who was only indirectly present in her life, either as a photograph in her room, or through the
cassette tapes she would send every month telling the family about her life in
the United States. When describing her mother and her absence, Sophie likens
Martine to Erzulie, a vaudoo[9] female goddess, both virgin and sexualized. Sophie
receives advice about dealing with Martine from both her grandmother and her
aunt, indicating the close-knit relationship that the Caco women try to maintain: "'You must never forget this,' said my grandmother. 'Your mother is your
first friend.'" (*Breath, Eyes, Memory* 24). Atie, however, encourages Sophie to
love and accept her mother, even if she does not completely understand why
she was left behind. Atie states: "[...] you were never abandoned. You were
with me. Your mother and I, when we were children we had no control over anything. Not even this body." She pounded her fist over her chest and stomach"
(*Breath, Eyes, Memory* 20). The lack of control, even over their bodies, is also understood in the economic life of these women, who were peasant workers in the
sugarcane plantations for generations, a backbreaking occupation that continues to be underpaid, echoing times of slavery. The cane fields are also the
place in which much of the violence, colonial and dictatorial, took place, holding a special significance in the trauma that runs in the family.

Martine is the one who brings change to this cycle in their family after im-
migrating to the United States, as she is able to send money to her family
back home, drastically changing their living conditions. The other connotation
in Atie's comment is a veiled allusion to the circumstances that propelled this
migration, Martine's violation and the subsequent birth of Sophie. The mother
could only migrate to the United States as a refugee, seeking asylum from the
violent conditions of her life back in the island. When describing their living
conditions, Atie makes clear how much the migration to the United States is a
factor of change in the standards of the family, who could now be compared
to those who are seen as more prestigious and educated, such as Monsieur Au-
gustin, the school teacher and Atie's first love interest. In comparison, when the
family history is brought to the forefront, their peasant status is evidenced in different generations. Martine's migration breaks the cycle of extreme poverty,
showcasing how much the structural life of a family might change when the

---

9 The spelling of the tradional folk religion of Haiti adopted here, similar to the one used by
Danticat, follows the francophone tradition.

basic needs are guaranteed. Due to the American dollars that Martine is able to send her family back in the island, the "daughters of the hills" are elevated in social class. Atie narrates to Sophie their family history of ascension:

> When my father died, my mother had to dig a hole and just drop him in it. We are a family with dirt under our fingernails. Do you know what that means? [...] That means we've worked the land. We're not educated. My father would have never dreamt that we would live in the same kind of house that people like Monsieur and Madame Augustin live in. He, a school teacher, and we, daughters of the hills, old peasant stock, *pitit soyèt*, ragamuffins. If we can live here, if you have this door open to you, it is because of your mother. (*Breath, Eyes, Memory* 20)

Sophie's grandfather died while working in the cane fields, in working conditions similar to a chain gang. Life in Haiti is depicted as violent and full of tumultuous events, and poverty permeates many aspects of the characterization. When leaving Haiti, for example, on her way to the airport in Port-au-Prince, Sophie witnesses a student protest against the corruption of the Haitian government, in which the truculence of the army is characterized as commonplace. The trip was organized by Martine, who arranged with a fellow stewardess to meet Sophie at the airport and safely take her to the United States. Sophie's trip is made beside another unaccompanied child, the son of a government official, simply characterized as "très corrupt", who had been assassinated moments before during the protest that was taking place outside. Atie sees in the migration of Sophie to the United States the possibility of a future that is safer, and probably more prosperous. Atie's destiny, however, is remaining in the island, looking after Grandmè Ifé, the expected duty of an unmarried second daughter. The dream of moving to the New York metropolis is the dream of many of the characters that still inhabit the island. The crossing either by boat or plane, is a much-coveted desire, as the migration is seen mostly in the positive light of economic empowerment, for themselves as well as their families, glossing over the hardships associated to the immigrant life.

When on American soil, Sophie is met by her mother whom she knows so little about, her expectations regarding her are challenged. Sophie meets a woman that looks much more exhausted and worn out than the picture she used to see on her night stand. Later on Martine will disclose her past with cancer, going through a double mastectomy, in addition to a history of mental illness. However, upon meeting her child in New York Martine excitedly interrogates her in Haitian Creole about all the relatives back home and instructs her to carry the legacy of the Caco women, to which Sophie responds very little, enacting the silence learned from Atie. Education was in fact one of the limitations that hindered the Caco women from achieving their goals, especially in the case of Atie, who by the beginning of the novel is not capable of reading and writing, though Sophie urges her to attend the night school, and offers to teach Atie herself. However, when they were young, Martine and Atie had always dreamed of having careers, and they expect Sophie to achieve their goal (*Breath, Eyes, Memory* 44).

Language becomes a very important issue in Sophie's new life. In her first day in the United States, the protagonist is instructed to learn English as soon as possible, so as to complete her transition to this new reality: "My mother said it was important that I learn English quickly. Otherwise, the American students would make fun of me or, even worse, beat me" (*Breath, Eyes, Memory* 51). Creole, however, is the language the mother uses with her daughter most frequently. However, Danticat chooses to write the dialogues in English, only keeping some expressions in Creole, most often with an immediate translation after them. While Creole becomes the language of familiar relations, something that connects Sophie both to the Caco women and to the Haitian homeland, English is the language that is crucial for her development in the host country in which her subjectivity is being constructed anew. This split in language is telling of a hybrid reality, in which mastering both the Creole and American English is imperative to keep a sense of self that encompasses this complex reality of migration.[10] Martine's worries emphasize the fact that the migrant experience would be easier and more profitable to all if the child is able to communicate in the host country language, thus maximizing any opportunity for better conditions of living and realizing her mother's and her aunt's dream in a collective experience of female success and overcoming of adverse conditions. The threat of becoming an outcast is paired with the threat of physical violence, an indicative of the close connection between language and safety in Sophie's life. Her training, though, takes place in a bilingual school in which immigrants had most of their subjects taught in French. Sophie resents this choice, from the language, to the uniform; the education she receives singles her out, instead of creating a sense of integration:

> I never said this to my mother, but I hated the Maranatha Bilingual Institution. It was as if I had never left Haiti. All the lessons were in French, except for English composition and literature classes. Outside the school, we were "the Frenchies," cringing in our mock-Cath-

---

10 For a more in-depth analysis of language and hibridity, please see Sharrón Eve Sarthou's "Unsilencing Défilés Daughters: Overcoming Silence in Edwidge Danticat's *Breath, Eyes, Memory* and *Krik? Krak!*" in *The Global South – Special Issue: The Caribbean and Globalization*, edited by Adetayo Alabi, 99–123.

olic-school uniforms as the students from the public school across the street called us "boat people" and "stinking Haitians." (*Breath, Eyes, Memory* 66)

Prejudice is very much present in different institutions, which are predominantly occupied by white children, who see the children of immigrants as otherized subjects. Martine's co-workers at the nursing home reports that their children are also hostilized by the Americans due to their national origin, which is implicitly coupled with their race. Other prejudices are associated with Haiti, such as AIDS, making explicit the link between poverty and infection in the eyes of the dominant white society:

> A lot of other mothers from the nursing home where she worked had told her that their children were getting into fights in school because they were accused of having HBO—Haitian Body Odor. Many of the American kids even accused Haitians of having AIDS because they had heard on television that only the "Four Hs" got AIDS—Heroin addicts, Hemophiliacs, Homosexuals, and Haitians. (*Breath, Eyes, Memory* 51)

Her schooling, however, is not circumscribed to the institution, and her mother plays an important role in the acquisition and command of English during the first years. Martine's desire of better opportunities for Sophie is made clear in her dedication to the improvement of her daughter's language in addition to her enormous workload.

Martine works two different jobs as a caretaker in a nursing home for the elderly to be able to support her family both in the States and in the island. The routines are draining and often she would take the shifts from other workers to make ends meet, making explicit the difficult path of underpaid jobs an unskilled migrant must withstand in order to survive in the host country. It is also during her first years in the United States that Sophie is introduced to Marc, another Haitian immigrant, who is a lawyer and her mother's boyfriend. Marc is also a marker of coloniality in the novel, as he is portrayed in a manner that is different from the Caco women. When describing Marc's background to Sophie, Martine is clear that his status is different from hers, due to his ancestry, which is frequently deployed through a series of names and family names, all French, which he uses when introducing himself. Class differences, however, are made less sharp in the United States. Martine claims: "'He helped me a lot in getting you here', she said, 'even though he did not like the way I went about it. In Haiti, it would not be possible for someone like Marc to love someone like me. He is from a very upstanding family. His grandfather was a French man'" (*Breath, Eyes, Memory* 59). Before helping Martine bring Sophie to the United States, he helped Martine obtain her asylum status as a refugee, and soon became her partner. He is the only man both women are in contact with during Sophie's teenage years. Both Martine and Marc form a union that shows different facets of the Haitian immigration. When first meeting Sophie, Marc takes the family to a Haitian restaurant in New Jersey, where fellow immigrants gather to eat traditional Haitian food and discuss politics. Here Danticat gives the readership a slight glimpse of the Haitian community in the United States, providing a brief characterization of the political positioning they might take, as well as their opinion regarding the American presence in the island. The preoccupation of Haitians and Haitian-Americans with their original country is demonstrated, regarding the occupation of their land by American troops and the subsequent mistreatment of its people by the imperialistic government that was implemented, a poor infrastructure, countless deaths, refugees, discrimination and prejudice experienced by Haitians in the United States:

> "Never the Americans in Haiti again," shouted one man. "Remember what they did in the twenties. They treated our people like animals. They abused the konbit system and they made us work like slaves". "Roads, we need roads," said another man. "At least they gave us roads. My mother was killed in a ferry accident. If we had roads, we would not need to put crowded boats into the sea, just to go from one small village to another. A lot of you, when you go home, you have to walk from the village to your house, because there are no roads for cars". "What about the boat people?" added a man from a table near the door. "Because of them, people can't respect us in this country. They lump us all with them". (*Breath, Eyes, Memory* 54)

During the same first dinner the newly formed family had, Sophie is enquired about her plans for the future, which are challenged by Martine once again, who dismisses her daughter's perspective. Martine's dream is that Sophie becomes a doctor, though Sophie's childhood dream profession is becoming a *dactylo*, a secretary (Danticat 1998: 56). Ultimately, Sophie becomes an English speaker, and the narrator acknowledges the words her mother has imprinted in her psyche: "'There is great responsibility that comes with knowledge,' my mother would say. My great responsibility was to study hard. I spent six years doing nothing but that. School, home, and prayer" (*Breath, Eyes, Memory* 66). This was the first step towards a larger goal: obtaining a university education. The plans for Sophie's education seem to be going the right direction at the end of the first part of the novel, as the protagonist finished her basic education and is admitted to college at the age of eighteen. Sophie and Martine move into a new neighborhood, the same where Marc lives, experiencing a more suburban kind of life, and thus attaining a piece of the American dream that Martine so much strived for.

3.2.2 *Ou byen?* Are you all right? PTSD and representation

Sophie was a child born from rape, which was possibly perpetrated by a Tonton Macoute (an officer of a paramilitary militia under the command of the dictator "Papa Doc Duvallier"). Since then Martine is haunted by PTSD episodes, involving night terrors, nightmares, and flashbacks, which now are made more severe in the presence of Sophie. During the fist night in her new home in the United States, still in their Brooklyn house, Sophie is awakened by one of Martine's night terrors. Kincaid illustrates the episode, exploring it from the perspective of the protagonist, who must deal with her mother's unresolved trauma:

> Later that night, I heard that same voice screaming as though someone was trying to kill her. I rushed over, but my mother was alone thrashing against the sheets. I shook her and finally woke her up. When she saw me, she quickly covered her face with her hands and turned away. "*Ou byen?* Are you all right?" I asked her. She shook her head yes. "It is the night," she said. "Sometimes, I see horrible visions in my sleep."
>
> "Do you have any tea you can boil?" I asked. Tante Atie would have known all the right herbs. "Don't worry, it will pass," she said, avoiding my eyes. "I will be fine. I always am. The nightmares, they come and go." (*Breath, Eyes, Memory* 48)

Another instance that stresses the PTSD episodes in Martine's life happens later on, when Sophie is already a teenager, showcasing the extent to which these reactions were common in the Caco household. This second mentioning is relevant since it suggests questions related to the motives for Martine's night terrors, indicating that the rape incident, and he terror caused by the memory of the faceless figure of the rapist, are still present in her psyche, and materialized in Sophie's appearance. The repeated and invasive traumatic memories are represented here, as Martine grapples with flashbacks of the painful event that changed her life:

> Whenever my mother was home, I would stay up all night just waiting for her to have a nightmare. Shortly after she fell asleep, I would hear her screaming for someone to leave her alone. I would run over and shake her as she thrashed about. Her reaction was always the same. When she saw my face, she looked even more frightened. "Jesus Marie Joseph." She would cover her eyes with her hands. "Sophie, you've saved my life." (*Breath, Eyes, Memory* 81)

In this description the readership is introduced to a dual perspective of Sophie in Martine's eyes, as both the savior who rescues her from the PTSD episodes, and as a trigger of the same episodes, since her resemblance to her father terrifies Martine even more. Sophie's identity is strongly connected to the women in her family, and the history of her own birth has long been concealed from her, yet the missing paternal figure would create a set of questions that needed to be answered. As a child, Atie would create stories to distract the girl, as the curiosity about her own history would grow.

This kind of strategy, used to spare the child from traumatic information, does help to satisfy her curiosity, but not for long. The paternal/masculine figures in Sophie's infancy are few, and they do not perform important roles in her life. Professor Augustain, Atie's frustrated love interest, may be seen as one with some influence over the child, mainly due to his educational role in her life; other than this, the paternal figure is absent in the Caco family. The absent paternal figure also comes to mind when Sophie is confronted with a picture of herself as a baby, and there is no familiarity in features between her and the family that surrounds her, making clear that her physical characteristics were inherited mainly from the father she has never met.

Later, Martine informs Sophie about the gruesome circumstances in which she was conceived, and the physical resemblance between Sophie and the rapist deeply affects her psyche. Martine claims: "I did not know this man. I never saw his face. He had it covered when he did this to me. But now when I look at your face I think it is true what they say. A child out of wedlock always looks like its father" (*Breath, Eyes, Memory* 61). Consequently, Martine sees in Sophie, even if unconsciously, a reminder of the traumatic event that changed her life. Martine shares with Sophie the memory of her rape, imagining that Atie had already given the child some idea of what had happened:

> "Did Atie tell you how you were born?" From the sadness in her voice, I knew that her story was sadder than the chunk of the sky and flower petals story that Tante Atie liked to tell. "The details are too much," she said. "But it happened like this. A man grabbed me from the side of the road, pulled me into a cane field, and put you in my body. I was still a young girl then, just barely older than you." I did not press to find out more. Part of me did not understand. Most of me did not want to. (*Breath, Eyes, Memory* 61)

Even though the telling of the traumatic memory is very lacunar, Martine is able to convey the story's violence, even without describing it in detail. The young listener is forced to make sense of this snippet of information, though it is sufficient to communicate the negative impact that it had on the life of the mother, becoming now Sophie's own trauma. Sophie comments on Martine's telling, referring to the matter-of-fact tone she used, as if the mother had already consciously made sense of the event: "She did not sound hurt or angry, just like someone who was stating a fact. Like naming a color or calling a name. Something that already existed and could not be changed. It took me twelve years to piece together my mother's entire story. By then, it was already too late" (*Breath, Eyes, Memory* 61). Martine's PTSD, conversely proves that this calmness was not a consequence of a resolved issue, but instead, of a form of dissociation. What is clear is that Martine does not possess the language to access this traumatic moment. According to Sarthou, following the studies of Carole Sweeney, Martine's silences are a result of her being trapped in her own trauma, as she is raped of her language, and therefore is unvoiced (106). Her silences, that have been socially constructed, as in the case of Atie, and later on reinforced by the trauma of rape, stretch over her adult life since the memory of the rapist continues to exert power over her, hindering her capacity of addressing the trauma. Sarthou comments, in broader terms, on the colonial rape of Haiti:

> Through Martine's story, Danticat explores the unnatural silence that is the consequence of pernicious violence in the lives of Haitians. Rape, murder, torture, and other forms of violence are very effective tools for the suppression of any opposition, and were used extensively by the Duvaliers. Political campaigns instigate chaos and insecurity, and impose silence on victims, as well. In addition, as Danticat's stories show, too often these Haitian families internalize and regularize that silence, perpetuating it as tradition and shutting off the possibility for individuals to mediate the traumatic consequences. (107)

Later on, as an adult, Sophie is the one who is capable of describing her mother's rape in more detail (after overcoming her own sexual trauma), stressing that the identity of her father is the direct result of the violence perpetrated upon Martine. Sophie's characterization points to the possibility of a political identity for the perpetrator, in contrast to Martine's general description (a man), and the process of silencing is made literally physical, as the physical violence of being pounded in the head impedes her from speaking:

> My father might have been a Macoute. He was a stranger who, when my mother was sixteen years old, grabbed her on her way back from school. He dragged her into the cane fields, and pinned her down on the ground. He had a black bandanna over his face so she never saw anything but his hair, which was the color of eggplants. He kept pounding her until she was too stunned to make a sound. When he was done, he made her keep her face in the dirt, threatening to shoot her if she looked up. (*Breath, Eyes, Memory* 139)

Though Sophie is certainly loved by her mother, and her family as a whole, she is the living consequence of a memory of violence, which causes the new and fragile relationship with her mother to be even more challenging. The night terrors that Martine experiences are indicative of the permanence of this trauma in her life, as Martine continues to battle the memory in her sleep.

Looking back, Martine's mental condition starts to deteriorate right after the rape incident, and the fear of a new attack now targeting her child causes her to lose any mental stability she might have had. The nightmares start at this point in her life as intrusive flashbacks of the violence inflicted upon her body. After this her mother, Grandmè Ifé, relocated Martine to another town, the one in which Sophie grew up, where she was taken in as help by a rich family (*Breath, Eyes, Memory* 139). Her mental health did not improve completely during this period, as Sophie reports that her mother tried to commit suicide many times during her childhood. Sarthou points out that Danticat's text addresses "[...] the ways in which maintaining silence about trauma and family history can result in personal and cultural insanity" (101), an issue explored mainly in Martine's character, which would surface in Sophie in a minor degree as well. This detail is relevant as it shows the extent to which trauma affects the life of people, disrupting the possibility of normalcy, especially when this mental wound is not treated adequately.

> My mother came back to Dame Marie after I was born. She tried to kill herself several times when I was a baby. The nightmares were just too real. Tante Atie took care of me. The rich mulatto family helped my mother apply for papers to get out of Haiti. It took four years before she got her visa, but by the time she began to recover her sanity, she left. (*Breath, Eyes, Memory* 139–140)

The network of women in Martine's family is responsible for saving her during this time, as well as the help from the family that took her in. Her visa as a refugee in the United States would allow her to distance herself from the history of violence that characterized her homeland, giving her the headspace to recover and develop a sense of security.

3.2.3 You're a good girl, aren't you? – Control and the Female Body

Grandmè Ifé's perspective on the birth of a girl, and the practices that are associated with it, are very much telling of the space women inhabit in Haitian society. She reports that when seeing a lantern in the distance that would come and go, she could tell that a baby was being born. The movement of the light indicated that a midwife was running from the shack to the yard where the clean sheets and the rags would be boiling over a pot, and based on what would happen next, Grandmè Ifé would be able to tell the sex of the baby, since the lantern would be put outside in the case of a male, as a man would be outside all night long with the newborn. In the case of a girl, there would be no light, as the midwife would simply perform the action related to birthing the child and would soon leave the mother and the baby (*Breath, Eyes, Memory* 146). This distinction is revealing, since it is possible to apprehend that the birth of a girl was no reason to be celebrated in this patriarchal society, in addition to suggesting that the fate of the mother and the daughter were literally associated with darkness. The traditions and values associated to girlhood/womanhood in Haitian culture are exposed and discussed in Danticat's novel, allowing the readership to gain access to a reality that must be scrutinized and refigured. Later, Grandmè Ifé would verify if her daughters were still virgins, thus guaranteeing that they would be fit for marriage later on, however, Martine's personal history of trauma and violence made this practice unnecessary in her case, something that did not prevent Martine from recuperating the traditional ritual with her own daughter. This form of control over the female body and sexuality is a form of traditional rearing of girls followed in the Caco family, and Atie was also a victim of this practice. She comments on the issue, revealing that the obsession with a girl's purity was overwhelmingly pervasive in her life, as well as the preoccupation with marriage. There is also an education of femininity, especially in matters concerning the ideals of beauty and behavior that are expected from a good Haitian girl. Some cues are indirectly given, such as the desire for fairer and lighter skin, demands that exist in the intersectional space of race, sex and even class. Instances of this education often figure in the second part of the narrative, as Sophie watches her mother go through her routines, as in "[m]y mother brought some face cream that promised to make her skin lighter" (*Breath, Eyes, Memory* 51), or in her mother's choice for a doll: "She was standing there with a tall well-dressed doll at her side. The doll was caramel-colored with a fine pointy noise" (*Breath, Eyes, Memory* 44). Another question regarding beauty standards is related to Sophie's weight, as Marc scrutinizes her eating habits during their meal in the Haitian restaurant (*Breath, Eyes, Memory* 55–56). Atie also reveals the societal expectations imposed on women's bodies, stating: "Haitian men, they insist that their women are virgins and have their ten fingers" (*Breath, Eyes, Memory* 151). Furthermore:

> They train you to find a husband, she said. They poke at your panties in the middle of the night, to see if you are still whole. They listen when you pee, to find out if you're peeing too loud. If you pee loud, it means you've got big spaces between your legs. They make you burn your fingers learning to cook. Then still you have nothing. (*Breath, Eyes, Memory* 136–137)

Atie's impression evidences the centrality of such heteronormative praxis of education of women in Haiti, as all the efforts of the females verge to the patriarchal order. The physical evidence of the effort, either being the virginity test or the stove-burnt fingers, demonstrate the extent to which these subjects are rendered alien to their own needs and subjectivities in detriment of the prospect of a good husband. Comparing Atie's experience to her own, Sophie reflects upon this vigilance, showcasing the extent to which Grandmè Ifé's practices continued to be executed by Martine in the United States. The protagonist adds some other precautions to the list given by Atie, stressing that the burden of purity and virginity would reach out of the realm of sexuality, being understood as a full performance of modesty that encompassed different aspects of a girl's life:

> I have heard it compared to a virginity cult, our mothers' obsession with keeping us pure and chaste. My mother always listened to the echo of my urine in the toilet, for if it was too loud it meant that I had been deflowered. I learned very early in life that virgins always took small steps when they walked. They never did acrobatic splits, never rode horses or bicycles. They always covered themselves well and, even if their lives depended on it, never parted with their panties". (*Breath, Eyes, Memory* 154)

In addition to the reports by Sophie, Martine, and Atie, once again the narration turns to fable, to deal with the traumatic and violent events related to the protagonist. After the explanation of the testing given by Grandmè Ifé, there is also a fable regarding a rich man that had chosen a poor black girl for marriage because of her virtue and purity. He arranged the whitest sheets for their wedding night, so as to have the blood stain paraded after they had intercourse for the first time, this way claiming her purity and eliciting that her honor, and consequently his own, were untouched. The girl, however did not bleed, and the man decided what he should do to keep his reputation intact, making evident the frequency in which tradition and ignorance are intermingled:

> He did the best he could to make her bleed, but no matter how hard he tried, the girl did not bleed. So he took a knife and cut her between her legs to get some blood to show. He got enough blood for her wedding gown and sheets, an unusual amount to impress the neighbors. The blood kept flowing like water out of the girl. It flowed so much it wouldn't stop. Finally, drained of all her blood, the girl died. (*Breath, Eyes, Memory* 154–155)

The vigilance and discipline of the female body turn out, as Atie states, to leave these women with nothing of their own, or at least, only that which is related to the masculine figure. On the matter, the narrator reports on Atie's impressions, which denounce the oppressive traditions of womanhood in Haiti. The duties expected of a woman (named in the epigraph of this chapter) are mostly domestic, eliciting the sphere where they are supposed to have some form of agency. All the duties that are placed under a woman's responsibility create a demand that makes it impossible for them to pursue their own desires, especially if they fall out of the patriarchal/heteronormative order. Atie's wish to have her own purposes named is testament that there are centers of resistance, or at least a wish to change the traditions that are imposed upon these subjects. Martine also comments on her own testing by Grandmè Ifé, making clear once again that and it was a mother's duty to watch over her daughter's honor and purity, since this was the way women were raised in their tradition:

When I was a girl, my mother used to test us to see if we were virgins. She would put her finger in our very private parts and see if it would go inside. Your Tante Atie hated it. She used to scream like a pig in a slaughterhouse. The way my mother was raised, a mother is supposed to do that to her daughter until the daughter is married. It is her responsibility to keep her pure. (*Breath, Eyes, Memory* 60–61)

Martine is socialized to see it is a woman's responsibility to uphold her daughter's purity, as it ensures the daughter's future, meaning that virginity is key to find a good marriage prospect, a husband who will provide her a comfortable future. The mother, who after her own personal history of sexual violence sees her opportunities in this social economy dwindle, imposes the patriarchal ideology over her daughter's body, as she also choses to test Sophie. One of the conversations that sets the tone of keeping the tradition takes place in the first day after Sophie's arrival to the United States. After admonishing the girl that she must learn the language and that her studies are her priority, Sophie is questioned if she is a "good girl[11]", a veiled question related to any history of sexual activity. At twelve, Sophie is read by Martine as a possible sexual being, and the discourse of purity is conflated with her role as a mother and protector:

"You need to concentrate when school starts, you have to give that all your attention. You're a good girl, aren't you?" By that she meant if I had ever been touched, if I had ever held hands, or kissed a boy. "Yes," I said. "I have been good. You understand my right to ask as your mother, don't you?" I nodded. (*Breath, Eyes, Memory* 60)

The patriarchal duty of preserving the virginity of a young woman as a sign of character, when paired with the traumatic act of testing, is a stressor that generates a new layer of trauma in this family. Sarthou points out that in Danticat's stories it is women who support and perpetuate the destructive and demeaning conventions of patriarchy (Sarthou 2010: 108). As a second-generation immigrant growing up in the United States, there is a tension between the traditions and culture of her family and the reality of her present community. Though at this time this is just a theoretical conversation that leads to her mother's confession about having been tested herself and why it stopped, the theme is retrieved in

---

11 The stereotype of a studious, quiet and obedient daughter is also explored by Haitian-American author Roxane Gay in *Hunger – A Memoir of (My) Body*, defining Haitian-American girlhood in the following terms: "The only way I know of moving through the world is as a Haitian American, a Haitian daughter. A Haitian daughter is a good girl. She is respectful, studious, hardworking. She never forgets the importance of her heritage. We are part of the first free black nation in the Western Hemisphere, my brothers and I were often told. No matter how far we have fallen, when it matters most, we rise" (49).

the second part of the narrative, in which Sophie, at the time a nineteen-year-old woman, starts a relationship with a neighbor.

Knowing about her mother's expectations, this relation is concealed from Martine. Joseph, the neighbor, is fifteen years older and an African American, working as a musician, certainly not a suitor that would fulfill Martine's hopes for a future son-in-law. Upon meeting Sophie, Joseph establishes that she still has an accent, allowing the protagonist to display her desire to assimilate to American culture, foreshadowing her identification with the host country: "'I detect an accent,' he said. Oh please, say a small one, I thought. After seven years in this country, I was tired of having people detect my accent. I wanted to sound completely American, especially for him" (*Breath, Eyes, Memory* 69). Joseph is fast establishing a connection, declaring that he too speaks a different language:

> "We have something in common. *Mwin aussi.* I speak a form of Creole, too. I am from Louisiana. My parents considered themselves what we call Creoles. Is it a small world or what?" I shook my head yes. It was a very small world. "You live alone?" he asked. My mother's constant suspicion prodded me and I quickly said, "No." Just in case he was thinking of coming over tonight to kill me. This was New York, after all. You could not trust anybody. (*Breath, Eyes, Memory* 70)

Men had been in Sophie's life a somewhat unknown reality, as her family was very much female centered, after the death of her grandfather. Thus, even if interested in becoming closer with this man, Sophie's first response is fear. Later on, she stresses this lack of familiarity with men by making a comparison to her lack of acquaintance with whiteness in Haiti: "Aside from Marc, we knew no other men. Men were as mysterious to me as white people, who in Haiti we had only known as missionaries. I tried to imagine my mother's reaction to Joseph. I could already hear her: 'Not if he were the last unmarried man on earth.'" (*Breath, Eyes, Memory* 68).

Concerning the awareness of her transgression in Martine's point of view, when going over Joseph's house once to listen to him play, Sophie would consider her mother's perspective on this, stating: "I knew what my mother would think of my going over there during the day. A good girl would never be alone with a man, an older one at that. I wasn't thinking straight. It was nice waking up in the morning knowing I had someone to talk to" (*Breath, Eyes, Memory* 72). Nevertheless, Sophie had already developed feelings towards Joseph, and the sexual exploration of her own body and pleasure is connected to the idea of him. However, she perceives this pleasure to be some form of wrongdoing, making clear the impact of Martine's sense of morality and control in her development:

He waited for me to go inside. I locked the door behind me. I heard him playing his key-board as I lay awake in bed. The notes and scales were like raindrops, teardrops, torrents. I felt the music rise and surge, tightening every muscle in my body. Then I relaxed, letting it go, feeling a rush that I knew I wasn't supposed to feel. (*Breath, Eyes, Memory* 76)

Martine's multiple jobs allow for some easy concealment of the relationship, as Sophie is able to meet Joseph during the time her mother is not at home. Joseph is frequently absent too, as his job as a musician takes him wherever there is an opportunity, so any time he is allowed to share with Sophie is cherished.

The first time Sophie is tested happens when she arrives late one evening, after being out with Joseph, and Martine is back home earlier than usual from work. Noticing her daughter's absence, she waits for Sophie to arrive, and predicting that she must have been out with a man, she submits Sophie to a virginity test.

I tried to tell her that I had not done anything wrong, but it was three in the morning. I wished that I had not asked Joseph to let me go in alone. Perhaps if he had been there. Who knows? "Where were you?" She tapped the belt against her palm, her lifelines becoming more and more red. She took my hand with surprised gentleness, and led me upstairs to my bedroom. There, she made me lie on my bed and she tested me. (*Breath, Eyes, Memory* 84)

Martine's concern with Sophie's virginity is paired with violence primarily in an indirect manner, made present in this passage through the belt and the palm, belying the threat of physical harm in the face of transgression. The gentleness of Martine's touch contrasts with the harsh reaction she displayed at the suspicion of Sophie's sexual activity, and here the depiction of violence is summarized in the passivity of Sophie's abidance to Martine's command. The language chosen to depict the violent act is direct and clear, as the verbs portray a sequence of actions that culminate in the actual testing.

Sophie faces the testing trying to resist the humiliation and shame that the circumstances caused, focusing firstly on religion, and then on her family. The purity question seems to be a subject that, though highly implicated in the masculine figure, belongs to the realm of women in Danticat's depiction. Sophie appeals to the Virgin Mary, a reflex of her catholic upbringing, demonstrating the extent to which her faith is the refuge she is able to find when facing the violence perpetrated by the one who was supposed to protect her. The response denounces, in a secondary symbolic level, the prominent role of the female in her life, as she resorts to God's mother instead of God himself in a moment of need (later on, Sophie will also resort to Erzulie to deal with questions related to herself and her mother). In addition, Sophie also tried to remember positive moments of her life with the figures that were most important to her, including even her mother, and finally Joseph, as their memory would serve to counteract the traumatic event she was going through, using this effort as a tool for self-preservation.

> I mouthed the words to the Virgin Mother's Prayer: Hail Mary... so full of grace. The Lord is with You ... You are blessed among women ... Holy Mary. Mother of God. Pray for us poor sinners. In my mind, I tried to relive all the pleasant memories I remembered from my life. My special moments with Tante Atie and with Joseph and even with my mother. (*Breath, Eyes, Memory* 84)

Martine, during the testing moment, narrates a story to Sophie: a fable about two lovers that were one and the same. In the words of the narrator, this embedded narrative serves the purpose of primarily distracting the young woman during the procedure. Nevertheless, through this story, Martine is providing an explanation regarding her own fear of rejection. Dreading being left alone, she describes a kind of connection between mother and daughter that is the opposite of the one depicted in Kincaid's *Lucy*. Here, there is a mother who takes extreme measures not to sever the bond between mother and daughter, delaying the process of individuation started by puberty.

> As she tested me, to distract me, she told me, "The *Marassas* were two inseparable lovers. They were the same person, duplicated in two. They looked the same, talked the same, walked the same. When they laughed, they even laughed the same and when they cried, their tears were identical. When one went to the stream, the other rushed under the water to get a better look. When one looked in the mirror, the other walked behind the glass to mimic her. What vain lovers they were, those *Marassas*. Admiring one another for being so much alike, for being copies. When you love someone, you want him to be closer to you than your *Marassa*. Closer than your shadow. You want him to be your soul. The more you are alike, the easier this becomes. [...] You would leave me for an old man who you didn't know the year before. You and I we could be like *Marassas*. You are giving up a lifetime with me. Do you understand? (*Breath, Eyes, Memory* 84)

The strategy of distraction is employed by the mother without success, as Sophie cannot escape the brutality of the moment, not even by focusing on the tale. What Martine achieves instead is demonstrating how perverted her understanding of a healthy relationship has become. By suggesting that both mother and daughter become *Marassas* in these terms, she is close to suggesting a form of incest. More gravely, this suggestion is made at that precise moment, being paired with the forceful penetration of her daughter's body. Mireille Rosello comments on the use of the fable in "Marassa With a Difference – Danticat's *Breath, Eyes, Memory*", stating that:

As a trick, it did not work, but now it is also another textual performance embedded into the narrative that the daughter controls: ultimately, she chooses to reinsert Martine's little parable into her own account of this life. Its function and meaning are both transformed in this process. (119)

Martine's desire of their becoming *Marassas* depends on Sophie's consent, in the sense that she must choose to identify only with her mother figure in order to achieve the sameness that characterizes the characters in the fable. The violation that is being performed though is exactly the opposite of consensual agreement, as the mother imposes her will over her daughter's body and psyche. This problematic *Marassas* tale, which serves in the end not as advice but as a warning, illustrates the fear of abandonment experienced by Martine, who wishes Sophie would never leave her, thus proposing a complete identification with her. This proposal reveals the extent of the pathological consequences of sexual violence in Martine's experience, as she becomes the rapist as well as the raped in the renovation of the traumatic cycle. In this sense, Rosello also agues that "[t]he mother's yearning for a Marassa lover is presented as a replication of the violence she continues to both endure and inflict: the mother story and her testing are inseparable." (120). The pair of identical beings function as a representation of the purest form of love in Martine's perspective, consolidated here in the bonding of mother and daughter. The threat of abandonment is combined with the obsession with virginity, as Martine fears Sophie will leave her behind for Joseph. This narrative serves the purpose of chastising Sophie for trying to find identification with another entity, such as a male partner. The fear of abandonment might be testimony of the unresolved trauma that continues to haunt Martine's experience, as she equates Sophie's sexual awakening to a negative experience, and furthermore, attaches to it the destiny of leaving her life and family behind. This reality of rejection is exemplary of the consequences of Martine's own rape, yet, she is unable to see that her actions are also violent towards her daughter's body and psyche. Martine words after the end of the testing assert that, from that moment on, her virginity, or its lack thereof, would be constantly verified, as well as display her own sense of shame and disappointment at her daughter. Denise Shaw also writes about this episode in "Textual Healing: Giving Voice to Historical and Personal Experiences in the Collective Works of Edwidge Danticat", adding one more layer to this interpretation:

This scene is confusing to readers as the maternal figure is both protective (assuring a daughter's virginity) but also violent (testing constructed as sexual violation). Martine places a psychological burden on Sophie by "testing" not only her virginity but also her loyalty as a daughter. [...] This convoluted enmeshment between Martine and Sophie represents much more than a mother-daughter relationship rife with complications, it is an extension of the metaphor of mother as motherland, both person and place that incites contradictory memories and emotions. (5–6)

Shaw relates this mother(land)-daughter relationship to the myths of the creation of Haiti that have been previously discussed, in their forceful and violent similarity, stating that Martine and her body/place continue to represent for Sophie a violent experience that her mother had not yet been able to work through and overcome. Sophie extends the act of doubling present in vaudou to the political life of Haiti, stressing that what made the state violence possible was the dissociation that the perpetrators could establish between their acts and their own psyche. She combines this line of reasoning with the vaudou traditions found in Haiti, stating that this doubling is composed by flesh, meaning the human side, and by shadow, characterizing the evil side:

> There were many cases in our history where our ancestors had doubled. Following in the vaudou tradition, most of our presidents were actually one body split in two: part flesh and part shadow. That was the only way they could murder and rape so many people and still go home to play with their children and make love to their wives. (*Breath, Eyes, Memory* 155)

Martine's shadow part accounts for the violence that had been inflicted upon her body and psyche and the subsequent inability to address this trauma; as for her light, her acts are intended as a form of loving and care for her daughter, even if misguided, and ultimately for the preservation of the family honor. The episode of the first testing and the narration of the story will be remembered in another occasion, as Sophie confronts Grandmè Ifé about the tradition, in which the protagonist stresses the physical description of the testing act, combining it with its psychological consequences:

> I closed my eyes upon the images of my mother slipping her hand under the sheets and poking her pinky at a void, hoping that it would go no further than the length of her fingernail. Like Tante Atie, she had told me stories while she was doing it, weaving elaborate tales to keep my mind off the finger, which I knew one day would slip into me and condemn me. I had learned to double while being tested. I would close my eyes and imagine all the pleasant things that I had known. The lukewarm noon breeze through our bougainvillea. Tante Atie's gentle voice blowing over a field of daffodils. (*Breath, Eyes, Memory* 155)

Sophie reports that, like the *Marassas*, she had also learned to double, meaning that she had learned to dissociate from the experience that was taking place. This experience of detachment will extend to some other moments in Sophie's life, especially in aspects related to her body. "After my marriage, whenever Joseph and I were together, I doubled" (*Breath, Eyes, Memory* 155), Sophie would confess, stressing that the traumatic experience of violation perpetrated by her mother through the testing would alter her relation to her body and her sexuality. The matrilineal character of this violence does not obfuscate the sexist ideology that upholds it, even as it connects past and present generations. Sophie thinks of Atie and her reaction to the practice, also feeling invaded and mortified:

> "There are secrets you cannot keep," my mother said after the test. She pulled a sheet up over my body and walked out of the room with her face buried in her hands. I closed my legs and tried to see Tante Atie's face. I could understand why she had screamed while her mother had tested her. There are secrets you cannot keep. (*Breath, Eyes, Memory* 85)

Danticat recently revealed in *The Art of Death* that she also had experiences of sexual abuse at a young age in Haiti, and that a dissociative response out of fear was the way she survived it:

> When I was ten years old, an older boy – Tante Denise's godson and nephew, Joël – moved into their house after his grandmother died. Many nights, over several weeks, he would walk into the room where three other girls and I slept and would slip his hands under our nightgowns and touch our private parts. Our bunk beds were lined up near the armoire from which he needed to get a set of sheets, and before he'd pick up the sheets he would touch us. Sometimes it was one or two of us, sometimes it was all four of us, all terrified even to discuss among ourselves what was going on, all of us too afraid that he might kill us if we screamed or told anyone else.

> During those moments, I would pretend that my body was no longer mine and that I had merged with the bed sheet. In the daytime, I would find certain objects to keep me from thinking about the night: self-made amulets in the form of beautiful black and brown women on toothpaste boxes. I would cut out these faces and their gleaming with teeth and I would think how lucky these bodiless women were because no one could touch them in terrible ways. These women could also shield me, I thought, by drawing me into an imaginary world where people laughed all the time and had no vulnerable flesh. (44)

Danticat's own reaction is extended to her characters, as they frequently move to a world of imagination and fable when dealing with traumatic moments. The author would comment on how vividly these images of bodiless toothpaste women would appear in her dreams during this period of her life, and during the moments in which sexual violence took place: "[t]hese cut-out images would come fully alive while my body was being violated. And even though these faces were as mute as I was – at times this boy would put his hand over my mouth and I would think I was finally going to die – at least they would witness my death" (*The Art of Death* 43). Another aspect of this experience that finds its way into her narratives is the presence of silence in the face of trauma, a characteristic that defines Martine's and, to a lesser extent, Sophie's experience when relating to this issue. Beside the traumatic reasons associated with this experience, another layer that composes this silence is the shame associated with sexual violence, in addition to the possibility of creating havoc in the family structure. Danticat comments on why it took her so long to address the issue in her writing, stating:

> While my parents were still alive, I was afraid to write about these nights. I was afraid that it might upset them. My parents had left Haiti in the middle of a thirty-year dictatorship during which most people were being terrorized. Women and girls being raped was not all that unusual. A girl could be walking down the street – she could be on her way to school alone – and if one of the dictator's henchmen decided he liked her, he could take her. My aunt and uncle managed to protect me from street threats – yet they were not aware that a different terror existed inside their home. (*The Art of Death* 43)

The act of testing becomes a secret in Sophie's life, as she would not share it with anyone, including Joseph, implying a sense of shame and guilt: "I did not tell Joseph what happened. He left for Providence and stayed away for five weeks. My mother still worked night shifts. She had no choice. However, she would test me every week to make sure that I was still whole" (*Breath, Eyes, Memory* 86). This invasion of privacy and abuse over the body creates a new set of worries for Sophie, who takes extreme actions to avoid the continuation of the practice. After relentless months of testing, Sophie who is clearly depressed, decides to mutilate herself in an effort to stop the testing. The narrator describes the act: "I was feeling alone and lost, like there was no longer any reason for me to live. I went down to the kitchen and searched my mother's cabinet for the mortar and pestle we used to crush spices. I took the pestle to bed with me and held it against my chest" (*Breath, Eyes, Memory* 87). This is the pronouncement of a desperate attempt to hold control over her body, as Sophie perpetrates on herself a kind of self-rape, with the intent of stopping the testing performed by her mother:

> My flesh ripped apart as I pressed the pestle into it. I could see the blood slowly dripping onto the bed sheet. I took the pestle and the bloody sheet and stuffed them into a bag. It was gone, the veil that always held my mother's finger back every time she tested me. My body was quivering when my mother walked into my room to test me. My legs were limp when she drew them aside. I ached so hard I could hardly move. Finally I failed the test. My mother grabbed me by the hand and pulled me off the bed. She was calm now, resigned to her anger. "Go," she said with tears running down her face. She seized my books and clothes and threw them at me. "You just go to him and see what he can do for you". (*Breath, Eyes, Memory* 88)

When the test fails to provide the answer Martine expects, Sophie is ordered to look for Joseph, as it seems that now he is responsible for caring for her. There is a clear transfer of responsibility in Martine's perspective, as now the male figure is the guardian of that body. The violence inflicted by Sophie upon her own body might be seen as a political act of self-determination and liberation, as pointed by Caldeira, in which the protagonist seeks to free herself from the patriarchal order, exemplifying unmistakably that the personal is also political (Writers as Citizens 214). Pin-Chia Feng, author of "'*Ou libéré!* Trauma and Memory in Edwidge Danticat's *Breath, Eyes, Memory*" comments on the issue through Julia Kristeva's abjection perspective, stating:

> This story of transformation teaches Sophie that she needs to change her own body in exchange for freedom. Here Sophie's self-abjection marks an attempt at demolishing her own corporeality to set up a boundary between the demanding mother and her withering selfhood. Thus Sophie frees herself from the nightly testing that dehumanizes her in her own eyes. (745)

Interestingly enough, before describing the self-inflicted act, like Martine and the *Marassas*, Sophie also resorts to a fable to explain her pain. The association of a legend to a moment that is clearly traumatic seems to be one of the strategies found by the author to deal with matters of trauma and violence representation, as the narration diverts from the act itself, and veers into a metaphoric instance. It may be argued that trauma is not contained into a realistic denotative language, and the metaphorical narrative offers more space for multiple interpretations for the representation of this experience, amplifying the possibilities of signification that are invariably restricted by language itself. In this fable a woman, who would be constantly bleeding out of her skin for twelve years, after not being able to find a cure with doctors, decides to consult with Erzulie, the healer virgin mother, to try and get a remedy for her illness. The goddess proposed that if she wanted to stop bleeding, she would need to give up her human form and turn into another creature. From all the animals she had known she decided to take the form of a butterfly, which Erzulie promptly tuned her into, and thus she no longer bled (87–88). Sophie might be seens as the bleeding woman of the story, who, after not finding any solution to her condition, resorts to a complete transformation in order to overcome it. The change in Sophie's life, after failing the test, completed her passage into adulthood. Erzulie might stand for Martine, as Sophie imagined her during her childhood, symbolizing the powerful mother figure who is able to modify a person's reality by her command. The transformation into a butterfly is indicative of the symbolic metamorphosis the protagonist must go through in order to thrive and leave her pain behind. After having experienced the negative consequences of sexual violence, Martine projects all the pressures of patriarchal purity onto Sophie's experience. Her trauma will extend out to Sophie, whose face brings back to Martine the haunting image of her rapist, will develop her own set of traumatic effects, disturbing her own sexuality in the long term, as she will come to feel uncomfortable with both her body and her sexuality. This distancing from her own sexuality may be understood as a means of creating some defense against unwelcomed touching and intimacy. Another negative consequences that are explored in the novel are the developing of bulimia in addition to issues related to body image.

Still feeling the physical pain of her self-injury, Sophie leaves the house following her mother's command, looking for Joseph as he becomes her lifeline and the possibility for a new beginning away from the reality of trauma that her mother imposed on her, escaping the dominance of Martine to marry Joseph:"[m]y body ached from the wound the pestle had made. I handed him my suitcase and the pinky ring he had given me. 'I am ready for a real ring,' I said" (88–89). Sophie asks Joseph to marry her at that very moment, with no priest or ceremony, desperately attempting to close the chapter of her life in which her mother had control over her body. This passage closes the second part of the novel, and the completion of Sophie's journey into adulthood. Left to her own devices, the protagonist hopes for a better future together with her husband. Sophie's religiosity is revised and evinced in this closure, as she both dismisses the dogmatic aspect of matrimony, and centers her hope in imagery that is metaphorically infused with the divine. They would get legally married not much later, and Sophie would come to recall these days by association to the physical as well as the psychological trauma she was going through, though it would never be fully understood by her husband that such an act of self-harm was also an act of self-empowerment:

> I looked at a small picture of Joseph's and my "wedding." The two of us were standing before a justice of the peace, a month after we had eloped. I had spent two days in the hospital in Providence and four weeks with stitches between my legs. Joseph could never understand why I had done something so horrible to myself. I could not explain to him that it was like breaking manacles, an act of freedom. (130)

The body becomes the arena in which trauma is performed in Sophie's story, a site in which generational trauma, as well as the self-inflicted liberation trauma take place. Andrea Queeley, in "Remembering the Wretched: Narratives of Return as a Practice of Freedom", explores the connections between memory and the body, stressing how much the black body is still a site of violence that needs to be addressed, as it is also the place that may enable practices of freedom:

Integral to these explorations of rupture and remembrance is the body and its designation as both the object of subjugation and the vehicle through which the subject engages in practices of freedom. In the highly racialized colonial/enslaved landscape, the body is not only an instrument of production, but is itself inscribed with an inferior status. Its characteristics distorted and despised as the incarnation of animalistic humanity and its reproductive capacity co-opted for the perpetuation of institutionalized domination, the Black body, and particularly the Black female body, has undoubtedly been the site of tremendous violation. (119)

Sophie's body is testament of this violation, and it bares the consequences of this historical oppressions. The disorders that would spark from this violation will also take place in her body, which will haunt her existence even far from the threat of testing.

3.2.4 Trauma and the Body

The body that Sophie inhabits as an adult is different from the one she envisions as normal, or even acceptable, making any reminder of the existence of this body a trigger for negative thoughts. These triggers take different shapes during the narrative, but the most powerful one is her husband's sexual desire, which forces Sophie to deal with her problematic sexuality. Sophie's rejection of her own body can be seen in many instances of the novel, an author's choice that clearly demonstrates that the protagonist's self-perceived dysmorphia became a chronic condition, one that has not allowed her to enjoy her life altogether given the inescapability of it. Sophie travels to Haiti as a way of figuring out how to deal with this problem, away from her husband and from Martine. Addressing Brigitte, her daughter, Sophie wonders about the consequences of this trip: "[a]re you going to remember all of this? Will you be mad at Mommy for severing you from your daddy? Are you going to inherit some of Mommy's problems?" (*Breath, Eyes, Memory* 110). Sophie demonstrates the awareness of the transmissibility of trauma, something that might be seen through the lenses of Marianne Hirsch's postmemory, a term that tries to deal with the relation of second generation subjects with the memory of traumatic events that took place even before their birth, but that are transmitted to them by their family members and artifacts in such a powerful way that these memories are finally perceived as also belonging to this second generation. Hirsh defines the term in "The Generation of Postmemory":

Postmemory is the term I came to on the basis of my autobiographical readings of works by second generation writers and visual artists. [...] it reflects an uneasy oscillation between

> continuity and rupture. And yet postmemory is not a movement, method, or idea; I see it, rather, as a *structure* of inter- and trans-generational transmission of traumatic knowledge and experience. It is a *consequence* o traumatic recall but (unlike post-traumatic stress disorder) at a generational remove. (106)

The trans-generational aspect of the transmission of trauma is made evident in Sophie's questions, and the awareness of this possibility already demonstrates that Sophie is better equipped to deal with her traumas than her mother ever was. While waiting for Atie, Sophie is met by Louise, one of her aunt's friends, a seller of drinks in the streets. Louise had become Atie's second love interest, a stance that is only obliquely explored by the author. Louise tries to sell Sophie a pig, in an effort to gather sufficient money to migrate to the United States by boat. Louise stands for the reality of those who rather risk what little they have at sea, in order to obtain some chance for a better life. Louise represents the desire to flee the dire reality of Haitian poverty and violence, stating that the voyage, though perilous, is not worse than their existing conditions. Poverty and persecution, as Danticat has stated in interviews, makes Haitian citizens refugees in their own country. Louise also comments on Sophie's body, stressing that she does not look like the post-pregnant women in Haiti, who are usually heavier, hinting at the bountiful reality of the American experience. Sophie's own body image differs from Louise's opinion, as she does believe she is heavier than she should be. This perspective about her own body is deeply connected to the trauma inflicted upon her by Martine, as pinpointed in the following passage:

> Even though so much time had passed since I'd given birth, I still felt extremely fat. I peeled off Joseph's shirt and scrubbed my flesh with the leaves in the water. The stems left tiny marks on my skin, which reminded me of the giant goose bumps my mother's testing used to leave on my flesh. (*Breath, Eyes, Memory* 112)

Louise also enquires about the precautions the protagonist might have taken during pregnancy to assure that the baby would be born with more desirable qualities: "[y]ou look very *meg*, bony. Not like women here who eat to fill a hole after their babies come out. When you were pregnant, you didn't eat corn so the baby could be yellow?" [...] "You should have eaten honey so her hair would be soft" (100). These pieces of advice stress the social hierarchy of colorism in the Haitian community, evidencing that the lighter, the better. The example given by the previously mentioned Martine's beauty routine also comes to mind. Brigitte, however, shares the dark complexion of her grandmother, as pointed by Atie: "'Who would have imagined it?' she said. 'The precious one has your manman's black face. She looks more like Martine's child than yours'." (101).

Martine, nonetheless, seems not to be a part of Sophie's life as an adult. She reports to Atie that though she tried to contact her mother, she was rejected multiple times. Atie was not aware of this distance, but brings news regarding Martine's mental state, hinting at a worsening of her PTSD (103). The estrangement between Martine and Sophie is an indicative consequence of the violence that existed in the act of testing, in addition to the morality that Martine imposed upon Sophie's body. The lack of desire to keep any contact with Sophie is telling of the depth of the betrayal Martine felt as her daughter decided to choose her own destiny by taking ownership of her sexual being, instead of conforming to tradition, and having chosen her own independent life instead of staying by her mother's side.

Grandmè Ifé is clearly the center of this matrilineal family, and she is the one who investigates Sophie's motives for her travel. She enquires about Sophie's life and family, meaning Joseph, and the granddaughter confides her difficulties as a married woman. This seems to be the kind of openness that she would not be able to share with her own mother, demonstrating the strength of the bond between Sophie and the family she left back in the island. In a very personal dialogue, Sophie and Grandmè Ifé talk about "marital duties". Sophie reveals that she has difficulties in her sexual life, and this amounts to the fear of how Joseph might interpret this, as Sophie makes clear that their intimacy would be one of the most important parts of their relationship. This assessment might be another vestige of the patriarchal education Sophie has received, as it would be the duty of a woman to perform sexually in a manner that would satisfy her partner, in addition to provide children. Grandè Ifé believes that Brigitte is the proof that to some extent Sophie is able to fulfill this role, and thus enquires if the problem is her husband's. Sophie responds that Joseph is not to blame as her grandmother encourages her to share her afflictions. Sophie finally confesses that her discomfort arises from her own body and a traumatic past, stressing that testing was central in negatively affecting her experience. Sophie acknowledges that she cannot perform because the act is too painful for her, and though Joseph is understanding of this, her own concerns regarding her body are too overwhelming. Ultimately, Sophie admits her struggles to Grandmè Ifé:

> "Secrets remain secret only if we keep our silence," she said. "Your husband? Is he a good man?"
>
> "He is a very good man, but I have no desire. I feel like it is an evil thing to do."
>
> "Your mother? Did she ever test you?"

"You can call it that."

"That is what we have always called it."

"I call it humiliation," I said. "I hate my body. I am ashamed to show it to anybody, including my husband. Sometimes I feel like I should be off somewhere by myself. That is why I am here". (123)

Sophie's confession to her grandmother seems to be one step in the direction of healing, as she discloses her feelings towards the practice of testing, revealing the traumatic consequences it had in her life. Her lack of connection to her own body is testament that the violence inflicted by her mother still bares consequences, even years later. Sarthou claims that "[...] the testing is part of a system that penalizes women for their sexuality and perpetuates the patriarchal valuation of women as 'virgins with ten fingers'" ("Unsilencing Défilés Daughters" 108), a reality that had been previously been made explicit in Atie's predicament.

Grandmè Ifé also interrupts the narrative during a moment in which the traumatic aspect of the conversation seems to peak by telling a story to some boys in her yard. Once again the strategy of using the fantastic or fable as metaphor is employed. Imaginative language is used once more as a rhetorical device during moments related to trauma and violence, serving to fill the gap in which descriptive language is insufficient, and as Grandmè Ifé puts it, the narrative aspect of this tale is capable of expanding the limitations of descriptive language, "bringing wings to their feet" (*Breath, Eyes, Memory* 123). This story, particularly, deals with a lark and a little girl, a fable in which the bird would give the child pomegranates in exchange for some favor, first just for looking at the girl, and then for a kiss, until they escalate to taking her to a faraway land. The frightened girl denies the bird this last request, stating she would miss her family, which prompts the bird to angrily respond and subdue the girl to his wishes. However, the intentions of the bird would soon be revealed:

As soon as the little girl got on the bird's back, the bird said to the girl, I didn't tell you this because it was a small thing, but in the land I am taking you to, there is a king there who will die if he does not have a little girl's heart. The girl, she said, I didn't tell you this because it was a small thing, but little girls, they leave their hearts at home when they walk outside. Hearts are so precious. They don't want to lose them. The bird, clever as it was, it said to the girl, You might want to return to your home and pick up your heart. It is a small matter, but you may need it. So the girl, she said, Okay, let us go back and get my heart. The bird took her home and put her down on the ground. He told her he would wait for her to come back with her heart. The girl ran and ran all the way to her family village and never did she come back to the bird. If you see a handsome lark in a tree, you had better know that he is waiting for a very very pretty little girl who will never come back to him. (125)

The fable metaphorically works towards representing the gender roles imposed by patriarchal society, as the girl is supposed to fulfill the bird's needs in exchange either for material goods or for appreciation. It also may be seen as a modified version of the Little Red Riding Hood, a tale that originally portrayed male sexuality as a threat, so as to curb female sexual development. Nonetheless, the lark's tale also reveals a strategy of female self-protection, in which the suppression of love/sexuality (leaving the heart at a safe place) is used as a form of empowerment, another take on the dissociation strategy. This choice, however, is questionable in the terms of developing a healthy relationship with other and with their own bodies, since by withholding their hearts, they make them unavailable to themselves and even to those who wish them no harm, a situation that could illustrate Sophie's relation to Joseph at this point. Sophie's metaphorical home as a safe place may be identified here as Atie and Grandmè Ifé, the two family members who have always nurtured her during her childhood and now as an adult woman.

When Sophie finally confronts her grandmother regarding the motive for the testing, Grandmè Ifé clearly states that honor is a key aspect in their culture:

> "The testing? Why do the mothers do that?" I asked my grandmother. "If a child dies, you do not die. But if your child is disgraced, you are disgraced. And people, they think daughters will be raised trash with no man in the house."
>
> "Did your mother do this to you?"
>
> "From the time a girl begins to menstruate to the time you turn her over to her husband, the mother is responsible for her purity. If I give a soiled daughter to her husband, he can shame my family, speak evil of me, even bring her back to me". (156)

The patriarchal aspect of this tradition is underlined by the fact that males would have the power to chastise the family that allows a girl that is no longer a virgin to be married, bringing shame both to the family and, more specifically, to the mother, who failed her duty of keeping the body of her offspring safe from the threat of other men. The comment on the perspective of the public sphere on the matter is also telling, as it denounces that a home without a man is subjected to the faith of producing women who do not meet the standards of patriarchal society. The centrality of men in this rationale decries the inequality that pervasively affects the lives of girls and women in this context in the name of honor and chastity. Sophie proceeds to question her grandmother about the reaction of her own daughters, to which she responds that she needed to keep them clean until they were married. Sophie points out that that do not have husbands, prompting an answer that illustrates the weight of tradition and honor to past generations: "[t]he burden was not mine alone" (156). When stressing that the burden was not solely hers, Grandmè Ifé is dealing with a twofold context, one that states that her daughters' lack of husbands is not her fault (accounting for Martine's mental state after the rape incident and for Atie's personal disillusionments), and that the burden of carrying the patriarchal traditions of upholding virginity and honor in detriment of protecting women is a practice that ran in the family for generations. Sophie proceeds then to confess her own feelings towards the tests that Martine performed, and the consequences that she feels are attributed to this practice:

> "I hated the tests," I said. "It is the most horrible thing that ever happened to me. When my husband is with me now, it gives me such nightmares that I have to bite my tongue to do it again."
>
> "With patience, it goes away."
>
> "No Grandmè Ifé, it does not."
>
> "Ti Alice, she has passed her examination."
>
> The sky reddened with a sudden flash of lightning. "Now you have a child of your own. You must know that everything a mother does, she does for her child's own good. You cannot always carry the pain. You must liberate yourself". (156–157)

Grandmè Ifé accepts that the testing bares traumatic consequences to the woman, and her comment on patience signals the naturalized reality of the practice in her experience. The retrieval of Ti Alice's testing, the girl she could hear in the distance, reinforces that testing continues to happen in many households of Haiti, stressing the permanence of the tradition. Grandmè Ifé stresses that whatever happened, it was all an effort in good intention. Sophie, however, as a young mother, a condition that elevates the protagonist in the family hierarchy, must now choose the best way to care for her own daughter, even if it means distancing from the practices that her family has carried out for generations. Finally, Grandmè Ifé pleads Sophie to liberate herself, letting the pain go away, so as to start a new cycle.

> She walked into her room, took her statue of Erzulie, and pressed it into my hand. "My heart, it weeps like a river," she said, "for the pain we have caused you." I held the statue against my chest as I cried in the night. I thought I heard my grandmother crying too, but it was the rain slowing down to a mere drizzle, tapping on the roof. (157)

The liberation that the grandmother referred to starts taking place at this moment, when Sophie is able to openly discuss with her family her feelings regarding her trauma. The apology that Sophie received from Grandmè Ifé functions as the recognition of her pain, and once acknowledged, she is able to move on. Erzulie, the mythical mother goddess, figures here as a token for a type of female-ness that allows itself to be ambiguous, both pure and sexual, in addition to symbolizing a return to Haiti as a place of belonging.

3.2.5 *Ou libéré?* Trauma Recognition, Working Through and Paths For Healing

Queeley examines narratives of return using the works of Franz Fanon and Michael Foucault as a basis for her analysis, claiming that there is a need to return to sites of trauma and memory, including the trauma of slavery, colonialism, or the anti-colonial struggle for liberation, as a means to resolve matters derived from these issues. Laying the theoretical foundation for her analysis, the author claims: "[t]hrough case studies of both victims and perpetrators of violence committed during the anti-colonial struggle in Algeria, Fanon reveals his belief that political violence cannot be bracketed, that it bleeds into the realm of the intimate" ("Remembering the Wretched" 110). Queeley links the violence of the anti-colonial struggle with a feeling of shame that permeates the life of the post-colonial subject, a view that is compared with the self-inflicted violence exercised by Sophie in order to escape her mother's testing, a violence that was seen as necessary for emancipation but caused a long lasting damage in her psyche. Commenting on Sophie's journey, the author states that the protagonist performs this investigation into her family's past in order to overcome her own history of trauma (115). In Haiti Sophie reencounters many characters from her past in the agitation of the market. While watching some women traders with their baskets, Sophie listens to a question that would become an important metaphor for her own burden:

> The female street vendors called to one another as they came down the road. When one merchant dropped her heavy basket, another called out of concern, *"Ou libéré?"* Are you free from your heavy load? The woman with the load would answer yes, if she had unloaded her freight without hurting herself. (*Breath, Eyes, Memory* 96)

During the third and forth part of the novel, Sophie tries to find ways of healing her body and her mind, and like the merchant, she tries to alleviate her heavy load so she can finally be free. This process starts with a visit of Martine to the island, in which she would try to solve two matters at once: organizing Grandmè Ifé's funeral, and making amends with Sophie, a request from the matriarch. The organization of the ceremony is planned up to the smallest detail, and as the oldest daughter, Martine is responsible for overseeing the arrangements. This is a trip that Martine avoided for a long time, as returning to Haiti would also be a return to the (site of) violence she desperately wants to leave behind. Upon meeting Sophie, Martine recognizes the different status that she occupies now as a mother, conceding that this new cycle in her life allows them to refigure their relation:

> "You and I, we started wrong," my mother said. "You are now a woman, with your own house. We are allowed to start again". The mid-morning sky looked like an old quilt, with long bands of red and indigo stretching their way past drifting clouds. Like everything else, eventually even the rainbows disappeared. (162)

The change in the weather also points to a modification in their lives, making explicit that this new phase is going to be a cycle of transformation for both characters. The acknowledgment of starting wrong is also a positive point towards the remaking of this relationship, as it validates Sophie's feelings of uneasiness around this mother figure. Martine, who had never returned to Haiti since her adolescence, has also noticeably changed, especially in matters of appearance. For one, her skin tone is now many hues above her usual tone, something that Grandmè Ifé was quick to notice (160). The lightening of the skin may be perceived as a clear adaptation to American beauty standards, as Martine bleached her face to appear more desirable in that context. By blaming the cold weather, a move to conceal her active efforts in bleaching her skin with creams and other products, Martine belies her true feelings towards the color hierarchy. Her reaction of shame to her mother's questions is also telling of a split consciousness, one that oscillates between adapting to the imposed model and accepting herself, and consequently her heritage and family through the appreciation of her natural skin tone. The allusion to Papa Shango, the orisha that manifests through the sun, also works towards the renovation of Martine, who will be able to return to her origins, regain her blackness, and stop identifying with the spectral reality of America.

Sophie and Martine's relation still carried the weight of the testing, and this transitional moment is one in which Sophie still grapples with the past, while trying to keep a positive attitude towards reuniting with her mother. During the first night Martine spends in Haiti, there is an episode in which Sophie's actions demonstrate how much she is still fearful of her mother's presence as she crossed her legs tightly everytime Martine would roam the corridors (162). Sophie's reaction may be characterized as an unconscious self-preservation response, as her body memory is triggered by Martine's presence, activating the drive to protect herself from any unwanted touching. This passage, however, also illustrates that Martine's mental state is clearly worse at this point, and returning to the island in which her own history of violence took place is certainly a stressor. Her tears might be interpreted as a demonstration of regret for having caused her daughter so much pain, and by crying even more as she sees her granddaughter, one might infer that she realizes that the trauma might be passed on to the next generation. During the days she spends there, Martine is not able to sleep properly, having constant nightmares any time her tiredness allows for some rest. Sophie feels compelled to accept Martine's apology, especially when seeing that the nightmares continue to devastate her mother's life. The remorse and negativity that plagued most of her adult life regarding her mother started to dissipate, and in a revelatory dialogue, mother and daughter are able to make amends and face a difficult conversation about their past, as Sophie confesses that she believed to be the source of her mother's night terrors. Interestingly enough, Sophie reveals that she sometimes was also afflicted by the nightmares, demonstrating a post-memory trait in her trauma, as she inherited her mother's own unresolved issues and they became her own. Later, Sophie would reveal some more aspects of these dynamics and the worsening of her own mental state as she lingered on suicidal ideation, exemplifying some other mechanisms of trauma in her life:

> After Joseph and I got married, all through the first year I had suicidal thoughts. Some nights I woke up in a cold sweat wondering if my mother's anxiety was somehow hereditary or if it was something that I had "caught" from living with her. Her nightmares had somehow become my own, so much so that I would wake up some mornings wondering if we hadn't both spent the night dreaming about the same thing: a man with no face, pounding a life into a helpless young girl. (193)

The conversation between Sophie and Martine then moves to the testing practice that runs in the family, and Sophie confronts her mother for the first time. This cathartic conversation serves the process of dealing with the issues that have affected their relationship as a means of facing them, so they can develop a healthier and happier connection from then on. Martine's ambiguous view on the testing are finally made explicit, as she reveals how much being tested also had a negative impact on her life:

> "Why did you put me through those tests?" I blurted out.

> "If I tell you today, you must never ask me again." I wanted to reserve my right to ask as many times as I needed to. I was not angry with her anymore. I had a greater need to understand, so that I would never repeat it myself.

> "I did it," she said, "because my mother had done it to me. I have no greater excuse. I realize standing here that the two greatest pains of my life are very much related. The one good thing about my being raped was that it made the testing stop. The testing and the rape. I live both every day". (170)

Sarthou comments on Martine's position, connecting her traumatic history with migration and its possibility of renewal, stating that: "[o]ne of the virtues of immigration is the ability to rewrite a person's life script", but Martine is too closed and too invested in the culture that brands her as unwholesome to be able to jettison those aspects of her birth culture that no longer make sense (109). Sophie's position is one of understanding, as she wants to learn how this violence was justifiable by her mother as a means to avoid doing the same to her own daughter, showing a clear break in the cycle of trauma that runs in the family. As Martine reveals her own feelings towards being tested, which are coupled with the rape incident, she demonstrates that the violence had not been normalized in her life either. In her perspective, the rape incident functioned as a blessing in disguise, as it freed Martine from her mother's testing. These negative experiences, however, continued to affect her life, since the traumatic night terrors kept reminding her of her two greatest pains, as she puts it. The adherence to the practice of testing, for Martine, was simply a question of keeping up with the family tradition inherited by her mother, a tradition of control over the female body. This explanation seems to serve the purpose of eliciting that unresolved trauma keeps on repeating itself, as the subject in unable to break from the cycle of violence and pain. Martine's clear avoidance of psychological treatment, as well as the absence of any other type of counseling, out of fear of being considered insane and therefore institutionalized, is telling of a bleak cycle of suffering that most often cannot be cracked alone. Sophie's choice to look for help, either through a therapist or with the holistic group she encounters, shows an opposite reality, demonstrating that healing is possible, if the matter that has caused the trauma is confronted and dealt with, as well as counting with the support of others.

The conversation between mother and daughter continues in the fourth part of the novel, in the plane ride back to the United States, as both characters head back to their lives. It is during this trip that Martine notices Sophie's eating habits, prompting the protagonist to disclose that she has developed bulimia as an adult, one of the consequences of the uneasiness Sophie feels in relation to her body. Martine seems to be unaware of what this condition means, and naively places the burden of the issue on Sophie, as if it were a conscious decision taken by the protagonist, who is fast in correcting the assumption, describing the routines of her condition.

The feeling of guilt related to the body and how much one's behavior is responsible for it is also explored by Gay in *Hunger*, in which she describes how much eating had become a coping mechanism after being raped at a young age by schoolmates. In contrast to Sophie's case, Gay chose to eat copious amounts of food so as to make her body undesirable, and therefore, outside the range of men who could possibly hurt her once again. Trauma and eating disorders are tied together, as the author ponders: "[i]s my body a crime scene when I already know I am the perpetrator, or at least one of the perpetrators? Or should I see myself as the victim of the crime that took place in my body?" (*Hunger* 18). This paradoxical relation between the traumatic event that took place over the body and the negative behaviors that have derived from it is the central aspect of Sophie's eating disorder. Gay had also developed bulimia when younger, shedding some light in its consequences, dynamics, and the overall rationale that has operated during this moment of her life:

> I have chronic heartburn because I used to make myself throw up after I ate. There's a word for this, "bulimia," but it always feels strange to use that word with regard to myself. For a time, I did try to become that girl I envy, the one with the discipline to disorder her eating. I didn't do it for that long, I tell myself. That's not really the truth. I did it for about two years, which isn't that long but is long enough. Or, maybe I don't want to use the word because it was so long ago, which is absolutely not the truth. I stopped making myself throw up about four years ago. And sometimes, I relapse. Sometimes, I just want to rid myself of all the food in my body. I want to feel empty. (177)

Gay exposes how much this eating disorder still touches her life, as purging episodes continue to take place, though not with the same frequency as when the author could be called bulimic. The search for a feeling of emptiness is somewhat relatable to Sophie's narrative, who wishes to shed not only weight, but also any possibility of being conceived as a sexual being, as she tries to make her body ever smaller. The practices of binging and purging are also clarified in Gay's memoir, as she describes:

> Like a thoroughly modern woman I consulted the Internet. I took that time to learn how to binge and purge and was both fascinated and appalled at the information I found. I learned that it helps to drink a lot of water right before you purge and that at the beginning of your binge you should eat carrots so you have a visual marker of when you've rid yourself of everything you've eaten. I learned that chocolate tastes the worst as it comes back up (and this would end up being absolutely true). I learned that my fingers might get cut from my teeth and that stomach acid would burn my knuckles (and these things were also true). (178)

Gay explores the different ways in which bulimia affects the body, aside from loosing weight, exemplifying the negative aspects if the disorder in its day to day occurrences. Though Sophie's narrative does not deal with this side of bulimia in detail, the readership is able to perceive that the disorder would have a pervasive effect in her life. Martine comments on the nature of Sophie's eating disorder, characterizing it as something outside the experience of Haitians, demonstrates the Americanness that is imbued in the identity formation of second-

generation immigrants, especially when in contrast with the previous generation. As Sophie points out that attributing the responsibility for the disorder to
herself is part of the problem, she is stressing the role of anxiety generation in
this specific case, as the affected person feels guilty for not being able to liberate
herself from this cycle, and therefore becomes even more entrenched in it. Gay
comments on guilt and shame as well, but from a different perspective, as the
practice of binging and purging seemed to momentarily give her a sense of control:

> [...] I rushed to my kitchen sink, gulped down three glasses of water, and stared into the
> aluminum basin as I shoved two fingers down my throat. It took a few jabs, but soon, I start-
> ed gagging. My eyes watered. And then I was heaving and vomiting all that food I had just
> eaten. When I was done, I turned on the water and the disposal and all evidence of what
> I had done slowly disappeared. For once, I did not feel shame after eating. I felt incredible.
> I felt in control. I wondered why it had taken me so long to try purging. (178)

This sense of control reported by Gay is something that Sophie desperately seeks
in her life, and bulimia seems to be one distorted way in which she could achieve
it. Sophie's defensiveness after Martine's comment may also be interpreted as a
sign of acculturation, as the mother sees it as utterly American:

> "You have become very American," she said. "I am not blaming you. It is advice. I want to
> give you some advice. Eat. Food is good for you. It is a luxury. When I just came to this
> country I gained sixty pounds my first year. I couldn't believe all the different kinds of ap-
> ples and ice cream. All the things that only the rich eat in Haiti, everyone could eat them
> here, dirt cheap. (*Breath, Eyes, Memory* 179)

The proactiveness in defending her position, a move towards a less passive
model of femininity, is seen as American, in opposition to a Haitian way of thinking and acting, that would demand obedience and complicity as Martine suggests. These different takes signal the already clear disparities between generations. Sarthou also comments that the migration aspect in Sophie's life, as
Haitian-American, has allowed her to have the freedom to speak about these issues, as well as a choice of questioning the traditions of her family, all because
of the distance from the motherland (108). The issue of unrequired advice also
surfaces in Gay's memoir, as her overweight body is the target of scrutiny by several members of her family, who frequently provided well-intentioned sugges-
tions:

> When you are overweight in a Haitian family, your body is a family concern. Everyone—sib-
> lings, parents, aunts, uncles, grandmothers, cousins—has an opinion, judgment, or piece of

counsel. They mean well. We love hard and that love is inescapable. My family has been inordinately preoccupied with my body since I was thirteen years old. (*Hunger* 49)

As with Gay, Sophie's body had been monitored and disciplined from a very early age. The concern over Sophie's body first concentrated on her purity, now turns to her body size and her eating disorder.

The material and economic differences between Haiti and the United States have had consequences in both mother and daughter, as Martine confesses that she also had dealt with body issues upon arriving. Their afflictions, however, are of opposite kinds, as Martine struggled to realize that she no longer needed to binge eat, given that food would be easily available. Haiti's poverty and scarcity had shaped the way the recently arrived immigrant perceived food and its consumption, and the sudden abundance of this resource created an imbalance in her life. Sophie, however, is reminded that upon meeting Martine in New York she was not sixty pounds heavier, but the opposite, looking thin and tired, prompting Martine to disclose her history of breast cancer and related loss of weight. She confesses that before the cancer she would eat uncontrollably. "When I first came, I used to eat the way we ate at home. I ate for tomorrow and the next day and the day after that, in case I had nothing to eat for the next couple of days. I ate reserves" (*Breath, Eyes, Memory* 179–180). Martine's description of her eating habits exemplify the extent to which poverty and hunger had affected her life in Haiti, and subsequently in the United States in its antithesis. Gay comments on the relation between Haitians and food, stating that: "Haitians love the food from our island, but they judge gluttony. I suspect this rises out of the poverty for which Haiti is too often and too narrowly known" (*Hunger* 49).

Sophie is able to draw a line of similarity in their lives, pointing that it is not unthinkable that she is going through her disorder, a lead that Martine takes to reiterate the difference between them, and a possible remark commenting on the Americanness of Sophie. Nonetheless, Martine wants to make clear that in this new phase she wants to be a positive influence in Sophie's life. Martine's effort to make her daughter comfortable and healthy may be perceived as simplistic at first; yet, it demonstrates the determination of the mother to make amends with her daughter. Sophie's strategy of only eating when she feels hungry is defeated by Martine's determination, showing once again that the daughter is also willing to accept her mother's effort. The next day, during breakfast, Martine shows once again that she is well intentioned, and Sophie needs to reaffirm that her disorder is more complex than having good food available, especially when these treats are associated to the feeling of guilt.

Sophie refers to eating in different moments of the novel, making explicit the connection between food and memory, as she would eat industrialized food, as well as dishes from different cultures, so as to avoid remembering Haiti and her ambiguous recollections: "I usually ate random concoctions: frozen dinners, samples from global cookbooks, food that was easy to put together and brought me no pain. No memories of a past that at times was cherished and at others despised" (*Breath, Eyes, Memory* 151). Another instance in which Sophie expresses uneasiness in relation to food happens soon after she is back with her husband, as he prepares a large amount of food to receive her and Brigitte. Though Joseph is not a skilled cook like Martine, presenting an amount of ready-made meals, Sophie is able to enjoy his effort. There is, however, an impulse to eliminate all she ate at the end of the meal, eliciting the complexities of food and its consumption in the life of a bulimic person (198).

Gay defines her body as "unruly" (*Hunger* 12), an adjective that describes both its dynamics, as well as being outside of the norm, and correspondingly to Sophie's story, the path towards healing involves accepting the very same body, which is the site of trauma, in a deconstructive effort of self-love. Gay comments on the issue:

> I have been living in this unruly body for more than twenty years. I have tried to make peace with this body. I have tried to love or at least tolerate this body in a world that displays nothing but contempt for it. I have tried to move on from the trauma that compelled me to create this body. I have tried to love and be loved. I have been silent about my story in a world where people assume they know the why of my body, or any fat body. And now, I am choosing to no longer be silent. I am tracing the story of my body from when I was a carefree young girl who could trust her body and who felt safe in her body, to the moment when that safety was destroyed, to the aftermath that continues even as I try to undo so much of what was done to me. (*Hunger* 21)

The body that Sophie created as a result of her trauma is one that she feels she cannot love and which cannot be loved fully. Yet, the steps in the direction of the aforementioned undoing of trauma are being taken, even if they are still unsure. Sarthou emphasizes that the openness to professional psychological treatment and for taking steps towards healing is a direct consequence of the hyphenated identity that Sophie was able to construct in the United States (109).

In a conversation with Joseph, after returning to Providence, Sophie discusses the impact of therapy in her life, and how she is not confident that she is getting the results she expected. Sophie worries that her incapacity to connect physically to her husband will be harmful for their relationship, and the therapy should help her "fix" this issue. Joseph, however, reassures the protagonist that she has his support, and that he will be by her side for as long as it takes for her to feel like herself again, stressing the positive influence of an understanding supportive network.

> Before returning to her home in Providence, Sophie is met with news that Martine is pregnant once again, a child resulting from her relationship with Marc. This piece of news, however, comes associated with the worsening of Martine's mental state, as nightmares are now more frequent and more severe. The fear of seeking treatment for her mental illness is present again, as Martine is terrified of having to face the memory of the rape and the rapist during therapy. There are then some other preoccupations too, as Martine believes that a baby would contribute to disrupt the tenuous stability she found in her relationship with Marc:

> "I know I should get help, but I am afraid. I am afraid it will become even more real if I see a psychiatrist and he starts telling me to face it. God help me, what if they want to hypnotize me and take me back to that day? I'll kill myself. Marc, he saves my life every night, but I am afraid he gave me this baby that's going to take that life away."

> "You can't say that."

> "The nightmares. I thought they would fade with age, but no, it's like getting raped every night. I can't keep this baby."

> "It must have been much harder then but you kept me". (*Breath, Eyes, Memory* 190)

Sophie's remark reiterates that Martine is in a better condition today than she was before, and that this child could be something positive in her life. Some other preoccupations of Martine are made evident too, as Sophie enquires if Marc intends to propose, to which Marine states: "[o]f course he wants to marry me, but look at me. I am a fat woman trying to pass for thin. A dark woman trying to pass for light. And I have no breasts. I don't know when this cancer will come back. I am not an ideal mother." (189). This statement makes evident all of Martine's anxieties in relation to her body, this pregnancy and her relationship, eliciting the different layers of inadequacy that in her perspective shape her experience. Sophie indirectly enquires about her mother's sexual life, and its relation to her nightmares, trying to piece together the dynamics of her relationship:

> "When you and Marc are together, do you have the nightmares then?"

> "I pretend; it is like eating grapefruit. I was tired of being alone. If that's what I had to do to have someone wake me up at night, I would do it. But never in my life did I think I could get pregnant."

> "You didn't use birth control?" She laughed through her tears". (191)

Once again Sophie is able to draw lines of similarity to Martine's life, as the mother confesses she is removed from the sexual life she has with Marc, seeing it as a duty in order to keep his company and his care. Like Sophie, Marine doubles during the sexual act, a situation she dissociatevely likens to eating a grape-fruit, a mundane act that has no direct impact in her life. Sophie, who feels that her own sexual life negatively impacts on her wellbeing, sees in her mother the pathway of untreated trauma. A child, who could be a positive influence in their lives, becomes another trigger for the trauma in Martine's life, which is threatened as the severity of her mental condition worsens.

Later, Martine would tell Sophie about the uncertainty of keeping this pregnancy, stressing that the baby is making her nightmares more vivid, and that now she sees the rapist not only in her dreams, but in the face of every man she meets. When considering an abortion, Martine heads to a clinic, in which she receives advice before having the procedure. She reports: "'I tried to get rid of it,' she said, 'Today. But they wanted me to think about it for twenty-four hours. When I thought of taking it out, it got more horrifying. That's when I began seeing him. Over and over. That man who raped me'." (*Breath, Eyes, Memory* 199). Martine sees the abortion as a way of avoiding to relive the mental instability she experienced during her first pregnancy, though the circumstances of conception are completely different.

Later, there is a description of an intimate moment in which Sophie, similarly to Martine, doubles, as she dissociates herself from the act itself, and lets her mind wander to a different setting. This episode takes place shortly after Sophie's visit to Haiti and, in this particular case, she concentrates on Martine and her problems, projecting a reality in which she could offer her mother support, protecting her from the nightmares, as well as reassuring her that the pregnancy she was now going through was a blessing and not a curse:

> He reached over and pulled my body towards his. I closed my eyes and thought of the Marassa, the doubling. I was lying there on that bed and my clothes were being peeled off my body, but really I was somewhere else. Finally, as an adult, I had a chance to console my mother again. I was lying in bed with my mother. I was holding her and fighting off that man, keeping those images out of her head. I was telling her that it was all right. That it was not a demon in her stomach, that it was a child, like I was once a child in her body. I was telling her that I would never let anyone put her away in a mental hospital, that I would take care of her. I would visit her every night in my doubling and, from my place as a shadow on the wall, I would look after her and wake her up as soon as the night-mares started, just like I did when I was home. I kept thinking of my mother, who now wanted to be my friend. Finally I had her approval. I was okay. I was safe. We were both safe. The past was gone. Even though she had forced it on me, of her sudden will, we were now even more than friends. We were twins, in spirit. *Marassas*. (*Breath, Eyes, Memory* 199)

It is significant to consider that Sophie chooses to identify with the Marassas, and ultimately with her mother, instead of focusing on Joseph, or in the body experience she is having. According to Queenley, "[t]here is a tension within the violated female body in which the temporary relief achieved through disassociation is matched by the necessity of being present to one's body in order for healing to occur" ("Remembering the Wretched" 119). In this passage, it is finally revealed how desperately Sophie needed the approval of Martine to resolve the issues in her own life, as the cycle of trauma and pain that existed between mother and daughter seems to comes to an end. Sophie symbolically retrieves the theme of the *Marassas* and both can finally be equals once again. The long and difficult confrontations Sophie went through in Haiti result in a better understanding of all the nefarious acts that were inflicted on her body and psyche, becoming the setting stone for better relationships with her family, as well as for her mental health.

While the issues related to her mother seem to have finally been settled, the ones regarding her own body and sexuality are still work in progress. The dialogue that ensues after this moment of intimacy reveals the extent to which Sophie's condition is still far from solved, as well as shading light into how distinct the experience was for both of them:

> "Can we visit my mother this weekend?" I asked Joseph.
>
> "Whatever you want." He was panting. "You were very good," he said. I kept my eyes closed so the tears wouldn't slip out. I waited for him to fall asleep, then went to the kitchen. I ate every scrap of the dinner leftovers, then went to the bathroom, locked the door, and purged all the food out of my body". (*Breath, Eyes, Memory* 199)

This passage also illustrates how much the rejection of her body and the bulimia are connected, as Sophie first holds in her true negative feelings related to the sexual act, while Joseph seems to consider that she had an enjoyable experience, and subsequently falls into the dynamics of binging and purging. This act might be read as an attempt to attain some satisfaction over the control of the body, as she was unable to feel pleasure during the previous moment in which she could not control her feelings of undesirability.

The sexual issue experienced by Sophie is further explored in a support group in the fourth part of the novel, as Sophie's therapy is approached and the readership gains access to what the protagonist calls the sexual phobia group. The participants in this support group have also experienced some form of violence that has affected their sexual life, gathering after their therapist had introduced them to one another. The forms of violence that have affected the women are different, ranging from female genital mutilation to continuous rape, the common ground being their incapacity to have a healthy and positive sexual life.

> There were three of us in my sexual phobia group. We gave it that name because that's what
> Rena—the therapist who introduced us—liked to call it. Buki, an Ethiopian college student,
> had her clitoris cut and her labia sewn up when she was a girl. Davina, a middle-aged Chi-
> cana, had been raped by her grandfather for ten years. We met at Davina's house. She was
> the only one of us with a place to herself. Buki lived in a college dorm and, of course, I lived
> with Joseph. (201)

The choice of choosing only immigrant women to figure in this group is relevant,
since it elicits the vulnerable reality of women in a transnational fashion, making explicit that violence against women is pervasive in all cultures and ages,
and that even in the United States, a country that sees itself as the definition
of development and equality, violence against women still takes place. In this
sense, Sarthou claims that "It [...] it is the transnational citizen that is able to recover memories and uncover traumatic events" (102). It is also interesting to
point out that all these women met in the United States, reiterating the immigrant character of its population. Historically founded by migrants, she continues to be populated by different people from different nations from all walks of
life who seek for better conditions of living. Sophie goes on to describe the rituals that this holistic group performs in order to try to overcome their sexual
phobia.

The wearing of white dresses and headscarves is symbolic of the purification
these characters seek in their lives, as the ritual assumes an almost religious
significance. The contributions of each one of the members is taken, displaying
a support network of women who, in their personal effort in healing, help one
another to make sense of their pain. The statue of Erzulie works as an offering
given by Sophie to this collective effort in healing, symbolizing the struggle
she faced with her mother, as well as the other women in her family. Queeley
also comments about the Erzulie statue and its significance, stressing the extent
to which returning to this symbol is crucial in her trauma healing process:

> The statue and spirit of Erzulie is the link between the past, present and future of the Caco
> women as it represents beliefs and practices ancient and alive. [...] Returning to the stories
> and spirits of Haiti, Sophie is in a place whose language and rhythm resonate with a part of
> her identity that had been sealed over in the process of Americanization. She is able to re-
> claim the power that is the source of the symbolic and is essential to being released from
> the prison of post-colonial, intergenerational trauma. (115–116)

The ceremony continues with a set of affirmations, which make explicit the
truths that these women are trying to live by (*Breath, Eyes, Memory*: 202). The affirmations touch matters of empowerment, in which the women clearly state that
despite what has happened in their lives, they are not diminished by it. The relational aspect of the affirmations is also relevant, since it makes explicit that

these women are supportive of each other and of other woman at large in the process of healing. In sum, they are able to become more resilient under the circumstances that they have been put through, helping one another in the process of trying to overcome this traumatic past. Another exercise that is displayed at this moment in the reading of a letter that Buki wrote to her grandmother, the perpetrator of the female genital mutilation:

> Buki read us a letter she was going to send to the dead grandmother who had cut off all her sexual organs and sewn her up, in a female rite of passage. There were tears rolling down her face as she read the letter. "Dear Taiwo. You sliced open my soul and then you told me I can't show it to anyone else. You took a great deal away from me. Because of you, I now carry with me an untouchable wound." Sobbing, she handed me the piece of paper. I continued reading the letter for her. "Because of you, I feel like a helpless cripple. I sometimes want to kill myself. All because of what you did to me, a child who could not say no, a child who could not defend herself. It would be easy to hate you, but I can't because you are part of me. You are me". (*Breath, Eyes, Memory* 202–203)

This exercise works towards the ability to express the grievances felt by these women, who must find words to make order out of their experience. By addressing the perpetrators they are able to finally communicate how much their actions have impacted their development, even if these perpetrators will not receive this message, such as in the case of Buki. When writing this letter, Buki expresses the complexity of the generation of violent cycles inside the family structure, an aspect that is comparable to Sophie's own struggle with her family. Though she realizes that her grandmother is responsible for her trauma, she is not capable of completely severing relations with this relative, since she understands that they are a continuation of each other. By admitting that she is also her grandmother, Buki infers the role of tradition and heritage, one that she reexamines as she tries to recover from the wounds, both physical and psychological, that were inflicted upon her.

Some other rituals takes place after the reading of the letter, in which all the women symbolically free themselves from their traumatic past by metaphorically destroying the name of the perpetrators of violence, in addition to Buki releasing a green balloon:

> We each wrote the name of our abusers in a piece of paper, raised it over a candle, and watched as the flames consumed it. Buki blew up a green balloon. We went to Davina's backyard and watched as she released it in the dark. It was hard to see where the balloon went, but at least it had floated out of our hands. (203)

These rituals are part of the holistic therapy that has helped Sophie in dealing with her personal history of trauma, as well as working towards the showcasing of a broader reality of trauma victims and the ways in which they share a path for healing. Sophie is able to put Martine's name in the fire not out of spite, but as a conscious effort to release herself from the succession of trauma and violence that took place in her family line. By doing so, she is breaking the cycle of trauma, assuring that Brigitte would not have to go though this process:

> I felt broken at the end of the meeting, but a little closer to being free. I didn't feel guilty about burning my mother's name anymore. I knew my hurt and hers were links in a long chain and if she hurt me, it was because she was hurt, too. It was up to me to avoid my turn in the fire. It was up to me to make sure that my daughter never slept with ghosts, never lived with nightmares, and never had her name burnt in the flames. (203)

Sophie's therapist appears in this section of the narrative, and their conversations illustrate the ways Sophie finds to deal with the matters that have haunted her for very long. She describes her therapist as someone who has a pluralistic approach, drawing from different traditions in her clinic: "My therapist was a gorgeous black woman who was an initiated Santeria priestess. She had done two years in the Peace Corps in the Dominican Republic, which showed in the brightly colored prints, noisy bangles, and open sandals she wore" (206). This description demonstrates the different ways that Sophie is able to connect with this healthcare professional, stressing that different discourses and practices aside from the medical establishment contribute in dealing with issues such as trauma. Similarly, Feng claims:

> The advices coming from Rena, to reclaim the mother line, to give a face to the rapist/father and to visit the spot where Martine was raped, combine professional psychoanalytical language and African folk wisdom. What is important is the creolized nature of this healing practice. Separately neither psychoanalysis nor folk wisdom can effectuate a cure. (Feng 746)

During the session they discuss how Sophie was able to finally confront her mother, as well as debating the discovery that all women in her family had been tested. Her feelings towards Martine are also investigated, giving Sophie the opportunity to showcase an alternative perception of temporality that would help her build a healthier relationship with her mother.

Sophie is willing to develop a new relationship with Martine, and to be able to do so she invokes the concept of the *Marassas* once again, stressing that the rationality that she would apply to make this new relationship possible is not limited by the understandings of western linear temporality. Another issue that is tackled during the session with the therapist regards the father figure that Sophie was never able to confront, as the therapist is able to draw a line connecting Martine's trauma with the memories she was never able to face:

"What about your father? Have you given him more thought?"

"I would rather not call him my father."

"We will have to address him soon. When we do address him, I'll have to ask you to confront your feelings about him in some way, give him a face."

"It's hard enough to deal with, without giving him a face."

"Your mother never gave him a face. That's why he's a shadow. That's why he can control her. I'm not surprised she's having nightmares. This pregnancy is bringing feelings to the surface that she had never completely dealt with". (*Breath, Eyes, Memory* 209)

Sophie's reluctance in attributing the status of father to the man who raped Martine also foments a culture of forgetting that would ultimately create negative consequences in her own life. By not dealing with the issue, Sophie would likely acquire the same symptoms of PTSD that Martine has so acutely developed. When the therapist urges Sophie to "give him a face" she is trying to stimulate a process in which the recognition of the perpetrator functions towards the development of a vocabulary to address the matter, thus trying to address the trauma it created. Ultimately, the therapist suggests that Sophie and her mother visit the grounds in which the incident happened, as a way of confronting this past:

> You and your mother should both go there again and see that you can walk away from it. Even if you can never face the man who is your father, there are things that you can say to the spot where it happened. I think you'll be free once you have your confrontation. There will be no more ghosts. (211)

When dealing with Sophie's sexual phobia, the therapist tries an exercise in which the protagonist must create a mental projection in which she can actively see Martine as a sexual being. This dynamic works towards an understanding of the limitations Sophie experiences in this aspect of her life, one the therapist believes she inherited from Martine (210). Her conversation with the therapist enlightens Sophie once again on the cyclical nature of trauma, as this traumatic past most likely would repeat itself if she did not take the necessary action to address it. It is in these conversations that Sophie realizes that she shares much of her life with Martine, as the trauma she lives was both inflicted on her, as well as inherited from her mother.

During the visit Sophie pays to Martine and Marc, in the company of Joseph, the family finally has the opportunity to get to know each other. It is at that moment that Martine is able to learn more about Joseph and his background, an occasion that serves to show the similarity they find in their diasporic origins. Joseph, who was born in Louisiana, uses his limited Creole to abridge the gap between their experiences, prompting Marc to claim that they were all Africans.

Martine confesses that upon arriving in the United States she would go to a
Southern church in Harlem, and that she felt she could have been Southern
too, and how she found solace in the Negro spirituals:

> "I feel like I could have been Southern African-American. When I just came to this country,
> I got it into my head that I needed some religion. I used to go to this old Southern church in
> Harlem where all they sang was Negro spirituals. Do you know what Negro spirituals are?"
> she said turning to Marc. Marc shrugged.
>
> "I try to get him to church," my mother said, "just to listen to them, but he won't go. You tell
> him, Joseph. Tell this old Haitian, with his old ways, about a Negro spiritual."
>
> "They're like prayers," Joseph said, "hymns that the slaves used to sing. Some were happy,
> some sad, but most had to do with freedom, going to another world. Sometimes that other
> world meant home, Africa. Other times, it meant Heaven, like it says in the Bible. More
> often it meant freedom". (214–215)

Marc then listens to Joseph perform a spiritual, one called "Oh Mary, don't you
weep", an experience he immediately recognizes as similar to a Vaudoo song,
"Erzulie, don't you weep". Martine then sings her favorite Negro spiritual,
"Sometimes I feel like a motherless child", an emblematic choice that encapsu-
lates the history of the Caco women. The comparison between the Negro spirituals e and the Haitian Vaudoo songs works as a kind of cultural translation, in
which different sides are able to better understand each other due to the similarities they can draw from this exchange. This moment certainly helps Martine better understand and accept Joseph, who had so far been a stranger in the family. It
also makes the distance between these two cultures to be diminished, as Marc
posed, they were all Africans, in different diasporic contexts, but still, sharing
a motherland.

This afternoon of bonding disguises Martine's growing apprehension, as
the PTSD symptoms are becoming more severe with the development of the pregnancy. After returning home, mother and daughter have a conversation on the
phone, in which Martine reports that she is now able to listen to the baby
speak to her in a male voice. She claims: "[e]verywhere I go, I hear it. I hear
him saying things to me. You *tintin, malpròp.* He calls me a filthy whore.
I never want to see this child's face. Your child looks like Manman. This child,
I will never look into its face" (217). This experience compels Martine to opt
for the termination of the pregnancy, a matter that is not settled by the end of
their call.

Joseph suggests that Sophie help her mother by practicing some of the tech-
niques of the sexual phobia group, such as a release ritual, as she is unwilling to
reach for psychological help on her own. Joseph questions if Marc is aware of the
pregnancy, referring to him as his mother's lover, a term Sophie clearly dislikes

as she finds it too sexual to be applied to Martine. Joseph states: "[t]oo sexual to be linked with your mother? I think you have a Madonna image of your mother. Part of you feels that this child is a testimonial of her true sexuality. It's a child she conceived willingly. Maybe even she is not able to face that" (220). Although Sophie was able to work on her relationship with Martine, the mother's mental health was clearly worse. After returning home from another private ritual in the sexual phobia room, in which she could see the Erzulie statue crying in the flickering light of the candles, Sophie is met by a message left by Marc on the answering machine, in which he asked her to return the call as soon as possible. After some hours trying to get ahold of her mother and Marc, Sophie is able to talk to him and discovers that Martine had committed suicide that night by stabbing her stomach seventeen times. Sophie questions Marc about the circumstances of the act, to which he responds:

> "She was still breathing when I found her," he said. "She even said something in the ambulance. She died there in the ambulance."
>
> "What did she say in the ambulance?"
>
> "*Mwin pa kapab enkò*. She could not carry the baby. She said that to the ambulance people". (224)

Both characters seem to be stoic during this conversation, a choice that Danticat would comment later on, and that tries to mimic a reaction of shock and paralisys in the face of death:

> During the phone call, Sophie tries to concentrate on anything but what her mother's boyfriend is telling her, which is how my younger self imagined I might react if I were getting similar news. Sophie remains calm. The mother's boyfriend is blunt and passes on information coldly, in part because he is in shock. At this point both Sophie and the boyfriend are in shock. To convey this, I stripped down the language as much as I could, even reducing dialogue tags. [...] I was trying not to be sentimental and melodramatic, not to overdramatize the reaction to a death that already seemed over-the-top. Still I wish now that I had included more confrontation, so that It wouldn't seem as though Sophie had immediately accepted her mother's death. (85–86)

Language, exemplified here, is never adequate in the face of trauma and violence, and by choosing to strip language of sentimentality, Danticat was able to achieve one level of representation that focused on the inability people have of dealing with these traumatic events. Martine's death, although unexpected, given the context of finally being in peace with Sophie, was somehow "in character" when considering her past of mental instability. The apparent acceptance of Martine's death by Sophie is finally a byproduct of the limitation of language in the face of trauma and violence, something the author could realize years later, as pointed in the previous passage.

The worsening of the nightmares, and the feeling of possession by the rapist memory work towards the depiction of the consequences of trauma, in which the victim is unable to distinguish between memory and reality, as well as between present and past, in an ever consuming (re)experience of the event. Martine's suicide may be seen as the ultimate result of untreated trauma, in which the subject is incapable of dealing with the reality of her existence, as many triggers transform her life into an impossible experience. Martine's last words reveal the reason she opted to commit this act against herself, as she was unable to relive the experience of motherhood, even if this one was a desired child. On the other hand, Sophie demonstrates that there is still a possibility of existence in trauma if it is assessed and dealt with in a timely manner, with the help of a supportive network. Rosello comments on the contribution of the novel in regards to rape representation, stating:

> [...] the novel does not simply suggest that the rape victim loses the ability to tell her story, but suggests that as readers we become incapable of listening to her because the parameters that we look for in a story are precisely those that have been destroyed, not so much by the rape but by the consequences of the rape on both the subject and the story. By portraying a character who is incapable of organizing a story and who dies of that syndrome, *Breath, Eyes, Memory* makes a specifically literary contribution to the representation of rape because the novel invites us to reconsider the distinction not between facts and fiction, but between the narrator who can choose to make rape the subject of a story and the body defined by an event that usurps the place that the storyteller should normally occupy. (Rosello 122)

Although trauma and its consequences will never allow for a complete healing, Sophie's experience, especially when contrasted with Martine's, shows that one may learn to live with this kind of wounds.

It is now Sophie's duty to prepare her mother's funeral, and after going to Brooklyn and having a conversation with Marc, she decides do fly her mother's body to Haiti, so she could rest in the same place as her ancestors. Emblematically, she chooses the outfit in which Martine will be buried, making evident her opinion of this mother figure:

> I picked out the most crimson of all my mother's clothes, a bright red, two-piece suit that she was too afraid to wear to the Pentecostal services. It was too loud a color for a burial. I knew it. She would look like a Jezebel, hot-blooded Erzulie who feared no men, but rather made them her slaves, raped them, and killed them. She was the only woman with that power. It was too bright a red for burial. If we had an open coffin at the funeral home, people would talk. It was too loud a color for burial, but I chose it. There would be no osten-

tation, no viewing, neither pomp nor circumstance. It would be simple like she had wanted, a simple prayer at the grave site and some words of remembrance. (*Breath, Eyes, Memory* 227)

Although Sophie sees her mother as a strong female figure, it is invariably a
characterization that is deeply linked with violence. The choice of the red
dress also echoes the family surname, Caco, a bird that becomes intensely scar-
let at the moment of its death. Marc conservatively ponders on the choice, which
prompts Sophie to respond vehemently: "'Saint Peter won't allow your mother
into Heaven in that,' he said. 'She is going to Guinea,' I said, 'or she is going
to be a star. She's going to be a butterfly or a lark in a tree. She's going to be
free.' He looked at me as though he thought me as insane as my mother"
(228). Sophie's comments on where her mother will find her final rest is in
clear opposition to Marc's catholic view, and the exchange elicits the creation
of an abysmal line between them, as Sophie turns to a kind of religiosity that
differs from the imposed Christianity that erased her traditional beliefs. Another
layer of difference in this exchange regards the control of the female body, even
in death, by the patriarchal order, as the color chosen by Sophie to dress her
mother would not allow for redemption under the male gaze of the gatekeeper
of heaven, resulting in the relegation of both female characters to the realm of
madness, thus dehumanizing them.

After arriving in Haiti, Sophie spends the waking night with the other Caco
women, who gather in the yard. While singing traditional songs and playing
games, Sophie realizes that the mother-daughter theme was not something related to her family specifically, but something altogether Haitian:

> Listening to the song, I realized that it was neither my mother nor my Tante Atie who had
> given all the mother-and-daughter motifs to all the stories they told and all the songs they
> sang. It was something that was essentially Haitian. Somehow, early on, our song makers
> and tale weavers had decided that we were all daughters of this land. (230)

As daughters of this land, these women and their intertwined histories and stories are emblematic of the transmission of trauma in its different versions, an
also of its healing either directly or indirectly. The primordial trauma of slavery
and colonialism becomes the paradigm from which all these daughters derive,
even if they are in their paths towards healing. The stories of Martine and Sophie,
though fictional, work towards a more complex understanding of trauma and
violence, shedding some light on the processes in which these forces take over
lives. The land and its characteristics are described once again towards the
end of the narrative, as Danticat summarizes many of the motifs that were
used during the novel:

> There is always a place where women live near trees that, blowing in the wind, sound like music. These women tell stories to their children both to frighten and delight them. These women, they are fluttering lanterns on the hills, the fireflies in the night, the faces that loom over you and recreate the same unspeakable acts that they themselves lived through. There is always a place where nightmares are passed on through generations like heirlooms. Where women like cardinal birds return to look at their own faces in stagnant bodies of water. (234)

Danticat reiterates the importance of memory in Haitian culture, figuring here as trauma in this passage, as well as restating the importance of fictional narratives in the construction of the identity of these subjects who, as previously mentioned, are a people obsessed with memory. Sophie, describes her mother as one of the daughters of this land, a characteristic she finally attributes to herself, making evident the cyclical nature of their existence: "[m]y mother was as brave as stars at dawn. She too was from this place. My mother was like that woman who could never bleed and then could never stop bleeding, the one who gave in to her pain, to live as a butterfly. Yes, my mother was like me" (234).

Martine's service takes place on the hills where the members of the Caco family are buried, a simple ceremony that gathered a small number of mourners who followed along the funeral procession. Sophie is overcome with emotions and narrates that it was impossible to see her mother being buried, an image that would symbolize the end of a cycle in this mother-daughter relation. The cane field situated between their house and the hills is the place in which Martine was first raped, the site that symbolizes the inception of Sophie's life and one of the origins of her trauma. Following her therapist's advice, Sophie decides to confront the site, as she attacks the canes, which literally and metaphorically fight back during this cathartic moment:

> There were only a few men working in the cane fields. I ran through the field, attacking the cane. I took off my shoes and began to beat a cane stalk. I pounded it until it began to lean over. I pushed over the cane stalk. It snapped back, striking my shoulder. I pulled at it, yanking it from the ground. My palm was bleeding.
>
> The cane cutters stared at me as though I was possessed. The funeral crowd was now standing between the stalks, watching me beat and pound the cane. My grandmother held back the priest as he tried to come for me.
>
> From where she was standing, my grandmother shouted like the women from the market place, "*Ou libéré?*" Are you free?
>
> Tante Atie echoed her cry, her voice quivering with her sobs.
>
> "*Ou libéré!*". (233)

Sophie finally is able to liberate herself from the trauma that has haunted her life in this releasing and spontaneous ritual. Grandmè Ifé's question is testimony that she understood what was happening in Sophie's mind during this outburst, demonstrating that the healing of this psychological wound was even more clear to those who do not share the hegemonic discourse of western medicine. Moreover, Atie's sobs are evidence that Sophie's liberation was also a form of liberation for all the Caco women, who could finally see the new generations as free from this specific history of trauma, which had impacted them directly and indirectly. It is also interesting to notice that all male characters in this passage seem to relegate Sophie's actions to madness, in similar fashion to Marc's reaction regarding the dress, while the female ones seem to be understanding and sharing. Conclusively, Sophie is able to walk away from the site of trauma, eventually free.

Healing, overcoming trauma, or simply being able to feel free, is a process that seems never to be completed. Gay describes the different stages in her path towards healing, stating that the process towards liberation is an everyday occurrence, in which managing PTSD symptoms is a constant endeavor, as much as self-care and self-love.

> Years ago, I told myself that one day I would stop feeling this quiet but abiding rage about the things I have been through at the hands of others. I would wake up and there would be no more flashbacks. I wouldn't wake up and think about my histories of violence. [...] That day never came, or it hasn't come, and I am no longer waiting for it. A different day has come, though. I flinch less and less when I am touched. I don't always see gentleness as the calm before the storm because, more often than not, I can trust that no storm is coming. I harbor less hatred toward myself. I try to forgive myself for my trespasses. (*Hunger* 258)

Sophie's story, though fictional, serves the same purpose than Gay's, as it demonstrates the extent to which a life is altered in the presence of sexual violence, as well as demonstrating that there are paths towards healing, and that even though trauma can never be obliterated, living is still possible. Summing up, healing is not a clear process, as stated by Gay (258). The process of healing is not clear from the onset of trauma, however, it becomes gradually clearer as experiences are shared, and as support networks are formed, and stories are told. Similarly, Sarthou concludes that "[a]s the stories of the Caco family illustrate, if trauma is to be mediated for these Haitians, it will be through free and open speech." ("Unsilencing Défilés Daughters" 110). Similarly, Shaw states:

> Sophie embodies the hopes and dreams of the next generation of Haitian women, those who can revolt, speak out, and fill in the collective lacunae of their individual and national identity. Sophie also represents a "connection between history and storytelling, between

Haitian women's lives and the ways in which narrative enables Haitian women writers to preserve those lives." Through the testimony of Sophie's journey and her process of healing, *Breath, Eyes, Memory* becomes a story of hope and healing for Haitians as well. ("Textual Healing" 8)

Moreover, the sharing of stories helps subjects to better understand each other in their complexity, as well as help diminish the relational aspect of their invariably unique experience, traumatic or otherwise, abridging the abysmal line between them, to use Santos' terminology ("Além do Pensamento Abissal" 2). Danticat states in *The Art of Death:* "[p]oems, essays, stories, and novels can help fill depth gaps in a way that numbers and statistics can't. One person's well described life and death can sometimes move us more than the mere mention of thousands of deaths" (50). Grandmè Ifé closes the narrative in *Breath, Eyes, Memory* by reinstating the importance of stories and storytelling, as words, in their limited power, are able to unleash new possibilities, especially when there is a possibility for liberation.

As a woman in her own right in Haitian terms, Sophie now occupies the place of keeper of the traditions of her family, having the right to not perpetuate any practice she does not find suitable to maintain in her life and in her family. Healing becomes now a possibility, not in the sense of restoring life to a previous state in which the trauma had never taken place, but to one in which living is not made impossible because of the presence of trauma. Sophie is able to finally break the cycle of trauma in the life of the Caco women, as she seems to have finally been able to heal from the primordial act of violence that has created her.

4 Morrison and hooks

"What you do to children matters. And they might never forget" (*God Help the Child*)

"Only grown-ups think that the things children say come out of nowhere. We know that they come from the deepest part of ourselves" (*Bone Black*)

Toni Morrison and bell hooks are the authors that are going to be analyzed in the current chapter, investigating different kinds of violence in their coming-of-age stories from the ones treated earlier, as the authors explore matters related to colorism, abuse, poverty, and domestic violence. These authors also investigate matters related to the mother-daughter relationship, raising questions related to the choices mothers must make to raise their daughters in a hostile environment. Black girlhood is seen in different contexts in the United States, as the characters learn to undo the hatred that was instilled in their psyches. Audre Lorde provides a clear account of this condition of hatred in the lives of black girls in "Eye to Eye: Black Women, Hatred, and Anger", stating:

> Little Black girls, tutored by hate into wanting to become anything else. We cut our eyes at sister because she can only reflect what everybody else except momma seemed to know – that we were hateful, or ugly, or worthless, but certainly unblessed. We were not boys and we were not white, so we counted for less than nothing, except to our mommas.
>
> If we can learn to give ourselves the recognition and acceptance that we have come to expect only from our mommas, Black women will be able to see each other much more clearly and deal with each other much more directly. (159)

The complications arise, though, when the mothers are not capable of supporting and accepting their daughters as they are, a matter that is present in both narratives selected for this chapter. hooks comments in *Salvation – Black People and Love*, that ever more often the presence of these abusing mothers can be noticed in literature:

> Any black woman who reads contemporary fiction by African-American women finds there narrative after narrative of mothers emotionally shaming and wounding their daughters. As a girl I was always disturbed by the old saying "Black women raise their daughters and love their sons." It suggested not only that girls did not matter but that the only role our mothers played in relationship to us was to keep us in check, to discipline and punish us how to conform to a woman's lot, showing us how to be subordinate and servile. (108 – 109)

To some extent, this trait in mother-daughter relationships was seen in the previously analyzed texts, yet, in this chapter, these issues will be made more evi-

https://doi.org/10.1515/9783110752755-006

dent. Similarly to the previous chapter, the narratives that were selected for this part of the investigation are diverse in their form, since Morrison's *God Help the Child* is a novel and hooks' *Bone Black – Memories of Girlhood* is a memoir. It may be argued, however, that the limits that separate these genres are permeable, and that the authors are able to explore the interstices between the representation of a fictional reality, the lived experience, and the ways in which literature provides a locus for the flexilibization of conventions. This is possible as the authors appropriate the *Bildungsroman* genre, transforming it in their effort to account for the presence of violence and trauma in their stories, experimenting with different forms of organization of the plot, as well as dealing with different forms of (avoiding) closure. While Morrison offers a narrative that is infused with a form of magic realism, as her protagonist mysteriously feels she is tracing her journey back to childhood in the aftermath of the release of suppressed traumatic memories, hooks deals with a post-modern kind of organization, recollecting different episodes of her childhood in short vignettes, that may or may not be chronologically structured, rupturing the sense of teleology, while the age of the narrator/protagonist is more often than not unclear to the reader.

Ultimately, both authors explore the ways in which childhood is stolen from black girls. They must learn how to deal with an unsympathetic and violent reality from a very early age, as they trade some part of the innocence of their tender years in the name of survival, as pointed by Lorde:

> A piece of the price we paid for learning survival was our childhood. We were never allowed to be children. It is the right of children to be able to play at living for a little while, but for a Black child, every act can have deadly serious consequences, and for a Black girl child, even more so. [...] Sometimes it feels as if I were to experience all the collective hatred that I have had directed at me as a Black woman, admit its implications into my consciousness, I might die of the bleak and horrible weight. Is that why a sister once said to me, "white people feel, Black people do?". (*Sister Outsider* 171)

4.1 Toni Morrison

> "We die. That may be the meaning of life. But we do language. That may be the measure of our lives" ("Nobel Lecture").

Toni Morrison was born Chloe Wofford in Lorain, Ohio, in 1931. She was an editor at Random House for years before attempting her first novel, *The Bluest Eye* (1970), at the age of thirty-nine. Morrison is arguably one of the most relevant voices of American contemporary times, being the receiver of a number of prizes and awards, cementing her place among the greatest writers of our times in the world. She is a Nobel Laureate, receiving the prize in 1993, and the author of *Beloved* (1987), which was the Pulitzer Prize winner in 1998. She is also the author of several novels such as, *Sula* (1973), *Song of Solomon* (1977), *Tar Baby* (1981), *Jazz* (1992), *Paradise* (1997), *Love* (2003), *A Mercy* (2008), *Home* (2012), and finally, *God Help the Child* (2015), which will be the focus of the following analysis. Morrison's fiction deals with a large span of American history, from pre-colonial Virginia in *A Mercy*, to slavery in *Beloved*, reaching the 20th century with *The Bluest Eye, Sula*, and *Home*, and finally reaching contemporary times in *God Help The Child*, her latest fictional work, as the author passed on August 5, 2019.

Morrison writes about the black experience, in its complexity and ordinariness, creating fictional universes in which racism is central to their construction. Rachel Kaadzi Ghansah writes for the *The York Times Magazine*, commenting on Morrison's production in an interview with the author titled "The Radical Vision of Toni Morrison", stating:

> On one level, Morrison's project is obvious: It is a history that stretches across 11 novels and just as many geographies and eras to tell a story that is hardly chronological but is thematically chained and continuous. This is the project most readily understood and accepted by even her least generous critics. But then there is the other mission, the less obvious one, the one in which Morrison often does the unthinkable as a minority, as a woman, as a former member of the working class: She democratically opens the door to all only to say, "You can come in and you can sit, and you can tell me what you think, and I'm glad you are here, but you should know that this house isn't built for you or by you." Here, blackness isn't a commodity; it isn't inherently political; it is the race of a people who are varied and complicated. This is where her works become less of a history and more of a liturgy, still stretching across geographies and time, but now more pointedly, to capture and historicize: This is how we pray, this is how we escape, this is how we hurt, this is how we repent, this is how we move on. It is a project that, although ignored by many critics, evidences itself on the page.

Morrison makes black life – regular, quotidian black life, the kind that does not sell out concert halls or sports stadiums – complex, fantastic and heroic, despite its devaluation. As Ghansah aptly points out, Morrison approaches blackness in a different manner than it is usually seen in American literature, in which the author usually must assume the position of both explorer and explainer of a culture; Morrison however deals with race in a fashion that values aesthetics without ever leaving aside the political, without falling pray to the need of demanding validation of others for her experience, but instead, wishes to complicate it. Morrison does not intend to alienate a wider audience, but to valorize her experience and referents making them central to her work. She remarks in the same interview:

> What I'm interested in is writing without the gaze, without the white gaze [...]. In so many earlier books by African-American writers, particularly the men, I felt that they were not writing to me. But what interested me was the African-American experience throughout whichever time I spoke of. It was always about African-American culture and people – good, bad, indifferent, whatever – but that was, for me, the universe.

This instance is very expressive of her counter-discourse, claiming black experience to be universal. Morrison's audience is primarily African-American, which did not impede her from achieving mainstream notoriety as one of the most important American writers in history, despite the criticism that her positioning in relation to Black people as her target audience has generated. Though African Americans continue to be perceived as nonreaders, they comprise a sizeable share of her audience, demonstrating that stories written by black authors for a black audience can exist in a market and a culture that continually intends to ignore their presence. Her ideas have been challenged many times by the conservative establishment which has attempted to ban her books from the curriculum of high schools all across the United States on the basis of their explicit sexual content and violence. Most notably in this regard are the titles *The Bluest Eye* and *Beloved*. Her editorial work also contributes to her relevance, as she propped the voices of many relevant African-American figures such as Angela Davis, Toni Cade Bambara, Gayl Jones, Henry Dumas, Huey R. Newton, and even Muhammad Ali. Ghansah comments:

> Morrison wanted to not only broaden the tastes of the industry, she also wanted to change the fate of a literary culture that had to either diversify or die. She told me that the books she edited and wrote were her contribution to the civil rights movement. By publishing black geniuses, she was also forcing the ranks of the big publishing houses and the industry to become more hospitable to her point of view, to the idea that a black writer could write for a black audience first and still write literature. She was more humanist than nationalistic, more visionary than didactic, but to some extent her editorial work was political.

Her stances on the political are not surmised in the editorial work, as her fiction shows clearly the different realties inhabited by African-American subjects. Her eleventh fictional work surely contributes to her overall project, centering her tale on the black experience withoug having to explain it, exploring how childhood trauma shapes the existence of subjects for the time of their life, especially when it is not properly addressed.

4.1.1 *God Help the Child*

God Help The Child, published in 2015, presents the story of Bride, former Lula Ann Bridewell, a successful self-made business woman in the cosmetic industry, who must reckon with her troubled past which involves witnessing child abuse, the accusation and following imprisonment of a teacher, in addition to deep rejection from her mother on account of her extremely dark skin. Bride might be compared to Jadine in *Tar Baby,* a professional woman that uses her beauty and professional life as coping strategies of counteraction against racism and emotional damage, though Bride's relation to race is more complex in a different way. She does not feel alienated from her race like Jadine did, but consciously uses it as a commodity to her own benefit. Bride also shares something with Hagar, from *Song of Solomon,* as her fascination with cosmetic products makes evident the need to believe in their promise of a better life based on appearance.

After being rejected by her boyfriend, Booker, Bride attempts to contact Sofia, the ex-felon, the one who had been accused of child abuse, and is severely attacked by her. Subsequently Bride is rescued by her assistant and must go through plastic surgery, leading her to a period of solitary recovery. During this time the protagonist realizes she knows very little about her former partner and decides to track him in Northern California, ending up in a car accident. This quest seems to be motivated more for her own satisfaction in discovering why she had been suddenly abandoned than out of love for Booker. She is saved by a couple of hippies who live outside the grid of modern technology, and during this time she meets Rain, a child victim of sexual abuse, and they develop a close relationship during her short stay. After the healing of her injuries from the car accident she is able to continue her journey, subsequently finding her boyfriend and meeting his closest family, finally learning more about his own traumatic past. During this moment Bride is able to make sense of her own past and also his, finding a way to work through her trauma. The plot seems to be contrived in some aspects, as different subplots pile up, and some characters seem to be underdeveloped.

The novel has received mixed reviews, in which the critics have generally argued that due to the brevity of the text the themes approached in the narrative are not duly explored. Razia Iqbal writes for *The Independent:* "Surprisingly for Morrison, some of the characters, though their stories have stayed in my mind, are too didactic on the page: prototypes for an idea rather than real people". Ibarrola-Armendariz writes in "Too Huge a Theme for Too Slight a Treatment: Toni Morrison's *God Help the Child*":

[...] despite Morrison's unquestionable narrative skill and her audacity in terms of form—
with constant shifts in language and point of view—, one must conclude that the themes
she wants to delve into in this slim novel prove far too complicated to be properly explored
in this short span. (76)

Roxane Gay also writes about the novel in "*God Help the Child* by Toni Morrison":
"incredibly powerful'", she says, but also comments on its flaws, stressing that
the short length of the novel is to blame for the lack of development of many
characters, as well as for the shortcomings of some of its subplots:

> *God Help the Child* is the kind of novel where you can feel the magnificence just beyond
> your reach. The writing and storytelling are utterly compelling, but so much is frustratingly
> flawed. The story carries the shape of a far grander book, where the characters are more
> fully explored and there is far more at stake. [...] Yet still, there is that magnificence, burn-
> ing beneath the surface of every word. [...] In *God Help the Child* we have a coming-of-age
> story for an adult woman in arrested development. (Gay)

God Help The Child revisits some of the topics first approached in *The Bluest Eye*,
although the setting of the story is very different. While in *The Bluest Eye* the
main character lives in Lorain, Ohio, during the 1940s, facing problems concerning beauty standards, racism and sexual violence, in *God Help The Child* Morrison explores these issues and their complexities focusing on the trauma inflicted
upon a child in a story set in the 1990s, in Los Angeles, California. Her first and
latest fictional books to date are intertwined in their interest in the role of ideas
of (self)worth and prejudice inside the black community, the importance of the
mother figure in the nurturing of a healthy sense of self, and the consequences of
the violence in the lives of children and the people who surround them. Violence
and trauma in their different forms play a very important part in both novels,
as violent events in the lives of the characters become determinant in their development. Paula Martín Salván writes in "Secrets, Lies and Non-Events: The
Production of Causality and Self-Deconstruction in Toni Morrison's *God Help
the Child*":

> Trauma narratives like *God Help the Child* emphasize the determining force of original
> events, by underscoring the causal relations between events in a character's life trajectory
> from a psychological perspective. Through analepsis, the events from what Freud called the
> "prehistoric period of childhood" [...], anterior to the diegetic present, should be reconsti-
> tuted in order to provide an explanation as to the characters' adult selves. In Morrison's
> novel, the two main characters, Bride and Booker, may be said to have become the adults
> they are because of their respective original traumas. (71)

Morrison explores child abuse and trauma in *God Help the Child*, demonstrating the ways in which pain that is inflicted during infancy unrevels in the lives of the characters. Child abuse is a fenomenon that is difficult to be precisely quantified, since reports often are not made by the victims. However, there are some figures that express to some extent the pervasiveness of child abuse and violence in our times. Data reported by the National Center for Victims of Crime claims that in the United States over sixty percent of all children deal with some form of violence, from ages zero to seventeen, while thirty-eight percent are witnesses of violence in their childhood. These numbers rise even higher when the children in question are African-American: black children and adolescents are three times more likely to be victims of reported child abuse or neglect, as well as to be victims of robbery, in addition to being five times more likely to be homicide victims (homicide is the leading cause of death among African-American youth between the ages of fifteen and twenty-four[1]. Moreover, this survey also claims that one in five girls are victims of sexual abuse, while for boys the rate is one in twenty[2]. According to Childhelp in 2017, more than 3.6 million child abuse referrals are received every year in the United States, involving more than 6.6 million child victims[3]. This reality, however, is not confined to the United States, being present all across the globe. In Portugal, for instance, in the first semester of 2018, at least five cases related to sexual violence involving children have been reported daily[4].

Morrison has explored the theme of child abuse in many other novels, focusing in different specificities according to the plot. Most notably, *The Bluest Eye* approaches questions of beauty standards, racism, and sexual violence. The story of Pecola Breedlove, the child who most wished to have Shirley Temple's blue eyes so she could finally be loved, and who was despised by her own mother on the account of her ugliness, is exemplary of the treatment Morrison has given to matters that relate childhood, trauma and violence. *Beloved* is also relevant here, since the unresolved trauma of slavery is embodied in the returning ghost of Sethe's daughter. In *Tar Baby* Michael is abused by his mother, Margareth, while in *Home* Frank Money suffers his share of injustice during his traumatic childhood. In *A Mercy* Sorrow embodies the character of a battered child, with-

1 http://victimsofcrime.org/our-programs/other-projects/youth-initiative/interventions-for-black-children%27s-exposure-to-violence/black-children-exposed-to-violence#fn1
2 https://victimsofcrime.org/media/reporting-on-child-sexual-abuse/child-sexual-abuse-statistics
3 https://www.childhelp.org/child-abuse-statistics/
4 https://observador.pt/2018/09/12/todos-os-dias-cinco-criancas-sao-vitimas-de-crimes-sexuais-em-portugal/

out a clear origin, and abused by those who surround her, as well as Florens, the protagonist, who is sold as an agreement between white colonizers as a request from her mother, so the child could escape the threat of sexual abuse by her master and have a better chance in life. Kusumita Mukherjee points the similarities between novels in "Politics of Selfhood and Magic Realism in Morrison's *The Bluest Eye* and *God Help the Child*": stating that Pecola and Bride are loathed by their mothers because of their appearance. While Pecola internalizes her mother's rejection and wants to feel accepted through the means of obtaining blue eyes, Bride overcomes her mother's attitude and is capable of fashioning a meaningful sense of self (Mukherjee 497). Fatoumata Keita summarizes the main themes approached in the novel in "Conjuring Aesthetic Blackness/ Abjection and Trauma in Toni Morrison's *God Help the Child*", stating that:

> Lula Ann Bridewell's experience illustrates the idea that childhood trauma or sins return like lingering ghosts to visit and haunt their subjects in adult life. This return of the repressed shores up the idea of circularity and circling back that has been identified as an aesthetic hallmark of *Beloved* [...] (44)

Sweetness Bridewell is the narrator of the first chapter in *God Help the Child*, an African American passing for white trying to hold on to her dignity in the aftermath of her blue-black daughter's birth. Sweetness feels betrayed by Lula Ann's darkness, as the promise of social uplifting and belonging granted by her own fairness is dismantled by her daughter's skin tone, resulting in a fraught relationship between mother and daughter. It is possible to claim that even Bride's professional field, cosmetics, is also a reflex of Sweetness preoccupation with her image, as well as a form of resistance to her mother's rejection, as pointed by Manuela López Ramírez in "'Racialized Beauty': The Ugly Duckling in Toni Morrison's *God Help the Child*" (181).

Lula Ann is anxious for her mother's love, as Sweetness would even avoid touching her dark-skinned daughter and forbidding that the child called her by anything other than her name, the protagonist resorts to desperate measures to achieve any sign of appreciation and love. She describes how far she would go to feel her mother's touch, devising mischievous actions just to be physically punished, which rarely happened, as the mother resorted to chastisements that did not involve any contact. Sweetness aversion to her daughter's skin was present in the most mundane circumstances, as Bride narrates: "Distaste was all over her face when I was little and she had to bathe me. Rinse me, actually, after a halfhearted rub with a soapy washcloth. I used to pray she would slap my face or spank me just to feel her touch" (*God Help the Child* 31). Sweetness' harsh education sparks not only from the colorist prejudice she feels toward her child, but also from the fearful consequences that might be brought by having such a dark skin. Sweetness confesses her regrets concerning Lula Ann's education believing that the ways she found to instruct Lula Ann would finally serve the purpose of protecting her, highlighting the aspects of a racist structure that would impact her child's life. Sweetness acknowledges the reasons for her attitudes:

> Oh, yeah, I feel bad sometimes about how I treated Lula Ann when she was little. But you have to understand: I had to protect her. She didn't know the world. There was no point in being tough or sassy even when you were right. Not in a world where you could be sent to a juvenile lockup for talking back or fighting in school, a world where you'd be the last one hired and the first one fired. She couldn't know any of that or how her black skin would scare white people or make them laugh and trick her. (41)

Although Sweetness intentions are to protect her daughter, the colorist component of this education creates scars that would finally drive the daughter away from her mother, as well as shape the reality of her life. Ramírez writes about the skills taught by African-American parents to their children that help foster a positive identity in "'What You Do to Children Matters' – Toxic Motherhood in Toni Morrison's *God Help he Child*", commenting on Sweetness' contrasting style of child rearing. Sweetness' choices for protecting her daughter were more concerned with creating a docile and subservient subject than with empowering her daughter through self-acceptance and a sense of belonging, which would ultimately help her challenge any racist struggle she would surely face:

> To help black children cope with racism, their parents teach them special skills (self-reliance, self-defense, dealing with pain and disappointment), however, Sweetness' motherhood only seeks absolute and uncontested obedience. She does not foster a positive racial identity in her daughter so she can resist racist practices, conversely, she imposes on her the societal cultural norms, values and expectations of the dominant culture. (114)

Meanwhile, during Lula Ann's childhood, the need to please her mother produces the circumstances that would create the central aspect of the narrative: the false confession that would spur even more trauma in the protagonist's life. Lula is summoned as a witness during a trial regarding child abuse, accusing one teacher of her school, together with some other children. Her performance on the stand grants her the mother's touch she had craved for so long:

> Outside the courtroom all the mothers smiled at me, and two actually touched and hugged me. Fathers gave me thumbs-up. Best of all was Sweetness. As we walked down the courthouse steps she held my hand, my hand. She never did that before and it surprised me as

much as it pleased me because I always knew she didn't like touching me. (*God Help the Child* 31)

The accused is sentenced to fifteen years in prison, and Lula Ann finally receives some recognition from her mother for performing her duty as witness and bringing a white child molester to justice. Later on in the novel, Bride confesses that the accusation she made was false, and that the only reason she did it was to receive some measure of appreciation from her mother.

Sofia Huxley, the accused, did her time in prison and on the day of her release she is contacted by Bride, who wishes to gift her a large amount of money and a free plane ticket, so she could start over, possibly a move towards the atonement of her guilty conscience. The indeterminacy of Sofia's guilt is an interesting choice made by Morrison, as the readership can only confirm that Bride's testimony was false, but not the other children's. Salván writes: "[t]he question of whether Sofia Huxley was ultimately innocent of child molestation remains unanswered, for even if she acts as narrator in two chapters in the novel, she never claims her innocence or discusses the sentence she was given" (77). The protagonist, however, undoublty witnesses an episode of child sexual abuse in the building she lives in, an episode that is going to be further approached later on, adding another contributing factor to her childhood trauma.

Notwithstanding, Bride is affected by the release of Sofia, as the diegetic present of the narrative commences with the attempt of contact made by the protagonist to make amends for her troubled past. The ambivalence of Sofia's trial is telling of a reconfiguration of the *Bildungsroman* genre, since this pivotal event in the narrative seems to be rendered ambiguously open, but still carrying consequences in the protagonist's development, as this instance, in conjunction with her boyfriend's rejection, sparks Bride's reverting to her childhood traumatic past, in a deconstructive unmaking of her hard-earned womanhood. Thus, this novel might be understood as a form of the *Bildungsroman*, but one that differs from the canonical structure and premises, due to the account of trauma. The apparent regression of Bride's body into a state of pre-pubescent development subverts the premise of progress and growth that is preconized by the canonical *Bildungsroman*, delivering instead a subject in process of deconstruction, one that is finally reversed only when the working through of trauma takes place. Another difference from the traditional *Bildungsroman* is that the narrative comprises multiple narrators, as the voices of different characters recount their plights, creating a complex story in their polyphony, in addition to a third-person narrator, who is able to connect these life stories. To some extent, most of the characters in the novel have been victims of some kind of violence during their childhood (physical, sexual, psychological), a strategy used by the author to make evident the pervasiveness of violence in children's lives throughout the United States, across races, classes, and sexes.

As is typical of Morrison, the reader arrives in the narrative *in medias res*, and at a moment of confusion, as pointed by John Updike in his review of *A Mercy*, "Dreamy Wilderness – Unmastered Women in Colonial Virginia", published in *The New Yorker:* "Toni Morrison has a habit, perhaps traceable to the pernicious influence of William Faulkner, of plunging into the narrative before the reader has a clue to what is going on". Commenting on the writing of her openings, Morrison explains that comfort is not a priority in this first contact between the reader and the text. The writer explains in "Unspeakable Things Unspoken" (160) that when writing the opening sentences of *Beloved,* the experience of the reader should be figuratively likened to that of the slave while being captured, with no preparation, just abrupt confusion that yearns for meaning.

> No native informant here. The reader is snatched, yanked, thrown into an environment completely foreign, and I want it as the first stroke of the shared experience that might be possible between the reader and the novel's population. Snatched just as the slaves were from one place to another, from any place to another, without preparation and without defense. No lobby, no door, no entrance – a gangplank, perhaps (but a very short one). (161)

Regarding *God Help the Child,* the experience is no different, as Sweetness confesses her dread and her innocence regarding a crime the readership is not yet privy to. Her rejection of her baby's skin color shocks the readership from the first page, creating an experience that is complex and inevitably engaging. Salván comments on the structure of the novel, mapping the ways in which a revelatory pattern shapes the narrative around a central event from which the narrative events of the plot unfold, organizing the different diegetic times that are comprised in the text in the process of unveiling it, also helping to foster a sense of foreignness for the reader (67).

Sweetness inaugurates the tale with a strong affirmation that promptly inserts the reader in a story that was already in motion: "It's not my fault. So you can't blame me. I didn't do it and have no idea how it happened. It didn't take more than an hour after they pulled her out from between my legs to realize something was wrong. Really wrong" (*God Help the Child* 3). The revelatory moment of the narrative is centered on the fact that this mother-daughter relation is plagued by abjection and prejudice, which will shape the realities of both characters, as their choices will be always conditioned by this revelatory circumstance throughout the narrative. This revelatory moment will guide both Sweetness' education of Lula Ann, as well as the protagonist's relation to her body and racial identity, as well as her relationships as an adult.

Lula Anne Bridewell actively becomes Bride during her adolescence, demonstrating she took the necessary steps to distance herself from the ideation imposed by her mother. In an exercise of self-invention, she sheds her name piece by piece, carefully fashioning her new identity. Lula Ann is clearly a past identity she has outgrown, one that could not contain all the attributes she intended to display as Bride.

> Lula Ann was a sixteen-year-old-me who dropped that dumb countrified name as soon as I left high school. I was Ann Bride for two years until I interviewed for a sales job at Sylvia, Inc., and, on a hunch, shortened my name to Bride, with nothing anybody needs to say before or after that one memorable syllable. (11)

Bride seems to be both an identity and a brand the protagonist wishes to propel. This construction is also based on the acceptance and celebration of her skin tone, in a reversal of the years of inferiorization she lived before. This inferiorization is demonstrated as Bride recalls the insults she suffered at school, as her color became a mark of inferiority. The name-calling ranged from the usual epithets of a Jim Crow era embedded in stereotypical representations of blackness to sheer animalization, an image that recalls the historic photographs of the first black students in segregated schools, such as Dorothy Counts-Scoggins and Ruby Nell Bridges Hall. Bride narrates her experiences of discrimination:

> Just like later in school when other curses—with mysterious definitions but clear meanings—were hissed or shouted at me. Coon. Topsy. Clinkertop. Sambo. Ooga booga. Ape sounds and scratching of the sides, imitating zoo monkeys. One day a girl and three boys heaped a bunch of bananas on my desk and did their monkey imitations. They treated me like a freak, strange, soiling like a spill of ink on white paper. [...] So I let the name-calling, the bullying travel like poison, like lethal viruses through my veins, with no antibiotic available. Which, actually, was a good thing now I think of it, because I built up immunity so tough that not being a "nigger girl" was all I needed to win. I became a deep dark beauty who doesn't need Botox for kissable lips or tanning spas to hide a deathlike pallor. And I don't need silicon in my butt. I sold my elegant blackness to all those childhood ghosts and now they pay me for it. I have to say, forcing those tormentors—the real ones and others like them—to drool with envy when they see me is more than payback. It's glory. (56–57)

Lula Ann's endurance of all the prejudice she experienced from birth is transformed in the resilience that the protagonist capitalizes as Bride, by treating the same features that were used to diminish her humanity into attributes that are defined as desirable, something that is only possible due to a shift in the public perspective of white mainstream America regarding race and color. Though figures such as Grace Jones and Iman were actively involved in the fashion industry since 1973, and *Vogue* had displayed the first African-American supermodel, Beverly Johnson, in its cover in 1974, they were largely seen as exotic features in an industry that continues to be predominantly white. The 1980s and 1990s would cement the place of black subjects in the industry, with prominent figures such as Naomi Campbell and Tyra Banks. Other artists, singers, actors, and athletes also helped to change the public perception of beauty during this era, exemplary are Tina Turner, Whitney Huston, Janet Jackson, Serena and Venus Williams, among others. The trend continues to present times, when figures such as Beyoncé, Rihanna, and Nicky Minaj dictate the standards of beauty and desirability in the United States and abroad.

Lula Ann must deal with the implications of the prejudice that is embedded in her life from the moment she is born, as her dark skin is read as a sign of evil, backwardness and adultery, and later on is used as a powerful tool of self-invention. As Bride, the extreme darkness of her skin is used as an advantage. The protagonist creates a new image for herself, wearing only white to accentuate her darkness, and finally takes control of the narrative of her color. She learns how to frame her extreme blackness in her favor, directing the other's gazes to her exotic look as a positive difference. Heeding the advice of an image consultant, Bride creates the impression she desires to achieve by carefully crafting her looks and attitudes.

> "You should always wear white, Bride. Only white and all white all the time." Jeri, calling himself a "total person" designer, insisted. Looking for a makeover for my second interview at Sylvia, Inc., I consulted him. "Not only because of your name," he told me, "but because of what it does to your licorice skin," he said. "And black is the new black. Know what I mean? Wait. You're more Hershey's syrup than licorice. Makes people think of whipped cream and chocolate soufflé every time they see you." That made me laugh. "Or Oreos?"
>
> "Never. Something classy. Bonbons. Hand-dipped". (33)

Bride transforms the idea of inferiority inculcated by her mother into an asset, and the metaphor of delicious and luxurious food is telling of the kinds of desire that Bride intends to emulate by her look. Her blackness becomes then a commodity she sells to her advantage. After her transformation, in her second interview for the cosmetic company she intended to work for, Bride is able to reap the benefits of this elaborate transformation:

> [...] walking down the hall toward the interviewer's office, I could see the effect I was having: wide admiring eyes, grins and whispers: "Whoa!" "Oh, baby." In no time I rocketed to regional manager. "See?" said Jeri. "Black sells. It's the hottest commodity in the civilized world. White girls, even brown girls have to strip naked to get that kind of attention". (36)

4.1.2 How else can we hold on to a little dignity? – Colorism as Violence

In *God Help the Child* Morrison explores how colorism is a nuanced form of oppression that takes place both inside the African-American community and outside it, as the interaction between different groups takes place. Although colorism is the term that is used most often when referring to this kind of discrimination, there are other options such as shadism, color bias, and chromatism[5]. This practice of discrimination is present since the transatlantic slave trade, as pointed by Nina G. Jablonski in *Living Color – The Biological and Social Meaning of Skin Color:*

> Colorism is a type of skin-color bias that involves systematic discrimination against the darker-skinned members of a particular group. It is primarily a product of the skin-color hierarchy that became entrenched and institutionalized with the transatlantic slave trade, and it has been collectively reinforced ever since. During the era of slavery in the United States [...], people of African descent who were lighter-skinned enjoyed advantages: light skin, in the words of one historian, became their "most precious possession." [...] (172)

In "From Color Line to Color Chart: Racism and Colorism in the New Century", Angela Harris points that the elimination of race and an emphasis on color would not undo the structural oppression performed by racism, but rather, would weaken the language created to address this issue. The author claims that "[...] in the new millennium traditional racism is indeed disappearing, but only to be slowly supplanted by colorism, in which the color of a person's skin will take on more importance in determining how she is treated by others than her ancestry" (Harris 53–54). Looking at colonial history, Harris claims that the creation of a mixed-race slave, as the result of sexual violence, shaped the attitudes toward this more light-skinned subject in two different manners, determining our understanding of colorism to this day:

> In the United States, African chattel slavery, which relied heavily on white sexual violence against blacks, led to at least two American practices said to be important for the evolution of colorism: the house slave/field slave dichotomy, and the emergence in some regions of a

---

5 Ashcroft *et al* describe the phenomenon of color discrimination using chromatism, stating its definition as such:"From 'chromatic' (1603) meaning 'of or belonging to colour or colours' (*OED*), this term is used to refer to the essentialist distinction between people on the basis of colour. It is sometimes used in conjunction with the term 'genitalism', a distinction between men and women based on the obvious biological difference between male and female. Both terms are employed to indicate the fallacy of making simplistic and stereotypical distinctions of race and gender and to suggest that the range of difference within these categories is a matter of representation and discursive construction" (37).

"mulatto" buffer class between whites and blacks. The house slave/field slave theory is that because house slaves were subject to rape and forced procreation, their children were often of mixed heritage, and thus preferred by slaveholders. [...] [Trina] Jones also notes that in the antebellum Lower South, "mulattoes" were recognized as a distinct social class, and that these persons of mixed white and African ancestry often had lighter skin than people of purely African descent. (55–56)

The two practices reveal firstly the colonial violence that created these subjects, in addition to revealing that the privileges associated with lighter skin are historically ingrained in the African-American community. The dynamics of the relationships among blacks, whites, and mixed-raced subjects was recorded by in 1944 by Gunnar Myrdal in his historically relevant contribution *An American Dilemma: The Negro Problem and Modern Democracy*:

The mulattoes followed the white people's valuation and associated their privileges with their lighter color. They considered themselves superior to the black slave people and attributed their superiority to the fact of their mixed blood. The black slaves, too, came to hold this same valuation. The white people, however, excluded even the fairest of the mulatto group from their own caste – in so far as they did not succeed in passing – and the mulattoes, in their turn, held themselves more and more aloof from the black slaves and the humbler blacks among the free Negroes; thus the mulattoes tended early to form a separate intermediary caste of their own. Although they were constantly augmented by mulatto ex-slaves, they seldom married down into the slave group. In such cities as New Orleans, Charleston, Mobile, Natchez, and later Washington, highly exclusive mulatto societies were formed which still exist, to a certain extent, today. Color thus became a badge of status and social distinction among the Negro people. (969)

Harris also points out that lighter skin tones is not a phenomenon that is restricted to the Americas and the New World, but a practice that takes places all around the globe. Lighter-skinned subjects are also preferred throughout Asia for instance, there subject from Japanese or Chinese descent enjoy privileges that South Asians with distinctively darker skins from countries like India, Indonesia, or Sri Lanka, do not. In addition to that, one can also consider the colonial whitening campaigns that took place in South America as a reflex of a colorist ideology, such as in the Cases of Brazil and Argentina. Skin color, as well as hair texture, physiognomy, and behavior, among others, also factor as markers of difference in colorist discrimination.

Colorism may be seen through the lenses of abjection in this novel, a term that is commonly associated with the work of Julia Kristeva, but that will be understood here through the definition provided by Judith Butler in *Bodies That Matter*:

> The abject [...] designates here precisely those "unlivable" and "uninhabitable" zones of so-
> cial life which are nevertheless densely populated by those who do not enjoy the status of
> the subject, but whose living under the sign of the "unlivable" is required to circumscribe
> the domain of the subject. This zone of uninhabitability will constitute the defining limit of
> the subject's domain; it will constitute that site of dreaded identification against which—
> and by virtue of which—the domain of the subject will circumscribe its own claim to au-
> tonomy and to life. (3)

Lula Ann's blackness functions in Butler's terms of "unlivable", defining what
the mother, as a passing for white high-yellow African American no longer considers herself to be. Sweetness reveals from the inception of the tale that she is
not to blame for whatever caused her daughter to look the way she does, and for
this mother, the child is proof that something inherently negative is correlated
with her complexion. "She was so black she scared me. Midnight black, Suda-
nese black. I'm light-skinned, with good hair, what we call high yellow, and so
is Lula Ann's father " (*God Help the Child* 3). Lula Ann's crime was to look a different caste, one her mother thought she had overcome, either by her own complexion, or by marrying into a more desirable one. What ensues is then a process
in which the mother describes this baby completely as the Other, not able to
identify any similarity between her and the person who had shortly been born
out of her own body. "Ain't nobody in my family anywhere near that color. Tar
is the closest I can think of yet her hair don't go with the skin. It's different—
straight but curly like those naked tribes in Australia" (*God Help the Child* 3).
One of Sweetness' first reaction to her baby is shame, as Lula Ann's difference
marked both the child and the mother, who would have to explain to onlookers
the reason for the disparity between them. The mother even goes to the extreme
of contemplating the killing of her child in the face of such stark abjection:

> I hate to say it, but from the very beginning in the maternity ward the baby, Lula Ann, em-
> barrassed me. Her birth skin was pale like all babies', even African ones, but it changed
> fast. I thought I was going crazy when she turned blue-black right before my eyes.
> I know I went crazy for a minute because once—just for a few seconds—I held a blanket
> over her face and pressed. But I couldn't do that, no matter how much I wished she
> hadn't been born with that terrible color. (*God Help the Child* 4–5)

Lula Ann is seen as the proof of the lack of privilege enjoyed by passable African-Americans, and instead of protecting this vulnerable child, Sweetness' instinct is to reject her, not being able to abdicate of her own privilege. Michelle
Dreiding in "Inaugurating Ambivalence – Toni Morrison's *God Help the Child*"
comments on the chasm that is created by the birth of this character, stressing
the ways in which Lula Ann represents a reality of struggles that have not yet
been resolved, both inside the African-American community and outside it:

Lula Ann stands [...] at the beginning of a text which is situated in the discursive reality that wants to find victorious historical closure. But Lula Ann is a "wounding encounter" [...], embodying the co-presence of the wish of a historically progressing trajectory (for the African American this means that blackness should be gradually decolorized within the logic of the family narrative) and a painful contemporary reminder of the traumatizing biography of the African-American subject (she is a "throwback"). Textually, then, Lula Ann does not figure as the absolute beginning of the text. Instead, she is inscribed in a discursive anteriority which specifically struggles with irresolution. (132)

For Sweetness, her blue-black daughter is a reminder of a past of subjugation and discrimination she wishes she had forgotten, contrasting with the empowering aspect of her passability as white. Light-skin privilege was a reality that guaranteed a life that was less harsh to many individuals and families, as better opportunities were offered to those with traits that were mistook for whites. Lula Ann does not symbolically belong to Sweetness lineage due to her color, a break in the narrative of "progress" enjoyed by her light-skinned family, and therefore is bound to the categories of abjection and foreignness. The making of hierarchies based on colorist notions as a phenomenon that took place inside the private realms of sociability might be seen as a reflex of the colorist institutional reality of the United States, a reality that can be perceived as Jablonksi points out through the records of the census. The evolution of the classification by color that started during times of slavery developed to accommodate the changes in society that comprised more nuance due to the mixing of the races:

Classification of people by skin color became an ineradicable part of American government as well as business and social life. In the first decades of the United States census, from decades of the United States census, from 1790 to 1850, the only categories recorded were "white" and "black (Negro)," with black divided into categories of "free" and "slave." [...] As the population of individuals of mixed ancestry swelled in the years around the Civil War, a column titled "color" was added, and attempts were made to classify Africans and their descendants as "blacks" and "mulattoes". (Jablonski 150)

Jablonsky also claims that after the civil war lighter-skinned men resulting from a mixed background continued to enjoy a higher status, as better educational prospects were available to them. Even in historical black colleges and universities, created mostly during the 19th century, these individuals had better chances of being accepted, and therefore were set on a path towards the possibility of a better life[6], as "[...] some school administrators considered it a waste of time to

---

6 Harris indicates some more studies that back this perspective: "'Vema Keith, using data about African-American women collected in 1979–1980, identified a positive relationship between

educate dark-skinned men and women for career paths that would be closed to them" (Jablonski 172).

God Help the Child is set during the 1990s, inferring that Sweetness ancestors lived during the 1940s and 1950s. During this era, the proliferation of mixed race subjects, in their complex heritage of different hues due to miscegenation, would force the institutional classification to be refined. By 1880, different categories were created in the the census so they could accommodate the different type of ancestry of black subjects, and these categories were based on the quantity of African blood a subject might carry, as pointed by Jablonski:

> Careful instructions were issued to "be particularly careful to distinguish between blacks, mulattoes, quadroons, and octoroons. The word 'black' should be used to describe those persons who have three-fourths or more black blood; 'mulatto,' those persons who have from three-eighths to five-eighths black blood; 'quadroon,' those persons who have one-fourth black blood; and 'octoroon,' those persons who have one-eighth or any trace of black blood [...] (150–151)

Passing for white, however, was possible if the person in question had physiognomic traits that could be considered ambiguously associated to whiteness: fair skin, smooth hair, and generally perceived European features. At this time, the amount of blackness that was attributed to each person based on their ancestry did not put in check white passing subjects who might have had some distant black ancestor, or even white people who might have some long distant black relative, something Sweetness aptly points out: "Can you imagine how many white folks have Negro blood running and hiding in their veins? Guess. Twenty percent, I heard." (*God Help the Child* 4). Jablonski comments on the phenomenon of passability, stating: "The United States after the Civil War became fixated on skin color, especially concerns that black could 'pass' as whites. People with darkly pigmented skin bore a visible badge of inferiority, but those with lighter skin tones passing for white were considered evils in disguise" (Jablonski 154). The categorization of different skin tones created a clear hierarchy of access to

---

lighter skin color and educational attainment, occupational standing, and family income'. Taunya Banks notes that 'a more recent study of 2000 men in Los Angeles found that race, skin tone, and the existence of a criminal record are major factors in determining whether men with similar educational backgrounds are employed. According to the study, being black and dark-skinned reduced a man's odds of working by 52 percent.' [...] As Trina Jones has explained, these phenomena do not fit the standard paradigm of racial discrimination, in which all persons recognized as African-American face similar discrimination and all persons recognized as white enjoy privilege regardless of their individual features [...]. Rather, colorism reveals hierarchies of privilege and disadvantage within racialized groups" (Harris 55).

power, and therefore to better conditions of living, which allowed for the rise of a mixed-race elite during the late 19th century. The initial scale based on the proportion of African blood multiplied, as a more complex range of possibilities became more evident.

> Skin color defined personal choice and prospects, and the names associated with skin colors created what was, effectively, a caste system: high yellow, high yella, crème-colored, ginger, saffron, octoroon, quadroon, bronze, mulatto, red-bone, light brown, black as tar, coal, blue-veined, café au lait, pinkie, blue-black. Girls and young women, especially, were encouraged not to play in the sun because darkened skin would reduce their chances of attracting a light-skinned husband and having light-skinned children. Some light-skinned individuals who could pass usually did so. (Jablonski 172–173)

However, the categorization changed radically in the 20th century, as the infamous "one-drop rule" sentenced that any child prevenient from a mixed background to be categorized as black automatically, despite their passability. The passing of the 1924 "Racial Integrity Act" in Virginia declared that there should be a racial description registered at the time of birth. The possible categories for this registration were simply "white" or "colored", the latter serving to all those who did not fit the first, including Native Americans. This change in law was propelled by the rise of eugenics and the racist pseudo-scientific debate that took place during the period, assuring a more thorough discriminatory apparatus that would protect the supremacy of whites, by preventing interracial marriages, and the possible enfranchisement of passable subjects. Jablonski claims that The Bureau of the Census decided to no longer adopt the subdivisions of race in 1900 basing the decision on the principle of hypodescent: any "black blood" in a person's automatically made them irrevocably black, making the previous categorizations obsolete (Jablonski 150–151).

As *God Help the Child* is set during the 1990s, the racial dynamics of this specific time are portrayed in the narrative through mother-daughter relationship, revealing a period of racial unrest that may be epitomized in the Rodney King episode, and that is reenacted in the novel through the difficult relationship between Sweetness and Lula Ann. Dreiding contends:

> [...] it could be argued that the difficulty of identification that is dramatized at the beginning of *God Help the Child* reproduces the particular medialization of the historical context of the 1990s, that is the irresolution of two narratives that cannot mutually explain each other, that exist alongside, simultaneously, that pull in different directions, but that cannot be resolved into one. The historical logic of the 1990s that the novel rewrites in 2015 painfully resonates with the beating of Rodney King in 1992. It thus becomes an actualized resurgence of that unfinished history, threatening to remain unfinishable, returning compulsively, symptomatically to the American psyche – the American psyche, which is confined

> to the neurotic repetition of its historical structure, not yet having reached the stage of working through that which it has been repressing. (135)

The two narratives that are antagonistic here, one of progress and social mobility symbolically imbued in Sweetness' light skin, and another of a past rooted in discrimination, violence and inhumanity, realities that are reminded in Lula Ann's deep blackness, compete in their irresolution in turbulent racial and social unrest. The impossibility of the working through that Dreiding refers might be seen in the continuous subjugation of African Americans and other racialized minorities in the United States until the current times, in which multiple narratives on race participate in their unresolved tension, and that seem bound to be repeated until they are fully addressed. The contrasting realities may also be seen in the following disparate realities: though the carceral population of African-Americans and Latinos has never been larger and the violence against racialized subjects is rampant, mixed-race people are perceived positively by the media, as hip and fashionable tokens of racial democracy. Harris claims that artists such as Tiger Woods, Mariah Carey, and The Rock Dwayne Johnson are able to discuss their mixed background proudly, seeing it as an asset rather then a setback. President Barack Obama is an example of the approachability granted to mixed-race subjects, appealing to both sides of his heritage. The author indicates that the future of race in the United States is destined to assume a "multiracial matrix", as supposedly are the cases of Brazil, Cuba, and Puerto Rico (Harris 53). This kind of discourse that heeds toward a future of racial democracy, eliminating the discourse on race, as a different type of "color-blindness" or "post-racial era", fails to combat the practice of racism, as implicit bias continues to determine the lives of people. Harris also notices that although, sometimes, it may be useful to conceptually differentiate racism from colorism, race and color are not different, as both forms of oppression represent related ways of attributing stigma and status. "Traditional racism places a higher value on ancestry than colorism; traditional racism assigns people to discrete racial categories, while colorism assigns people to places along a spectrum from dark to light, indigenous or African to European" (61). Thus, although colorism creates a more complex racial setting, in which different tensions are negotiated, it does not equate to a less racist environment.

Sweetness, abjection of the darker hues derives from the sense of superiority resulting from the light skin privilege she enjoys. The narrator addresses the readership later in the narrative, as a means to try to expose the reasons why there is this division inside the racialized groups, enumerating the benefits that lighter skin are able to provide:

Some of you probably think it's a bad thing to group ourselves according to skin color—the lighter, the better—in social clubs, neighborhoods, churches, sororities, even colored schools. But how else can we hold on to a little dignity? How else can you avoid being spit on in a drugstore, shoving elbows at the bus stop, walking in the gutter to let whites have the whole sidewalk, charged a nickel at the grocer's for a paper bag that's free to white shoppers? Let alone all the name-calling. (*God Help the Child* 4)

The brutal and dehumanizing racist treatment given to African Americans in the United States summarizes the reasons for the creation of this caste system inside racial lines, creating a fracturing reality of discrimination and trauma inflicting as a means of cyclical repetition of oppression, in the name of holding on to "a little dignity", and finally partaking in full citizenship. Sweetness describes the ways in which passability could affect their lives in the simplest ways, granting light-skin African-Americans the possibility of enjoying citizenship in a manner that almost fulfilled their fantasies of belonging, especially during the segregation period:

I heard about all of that and much, much more. But because of my mother's skin color, she wasn't stopped from trying on hats in the department stores or using their ladies' room. And my father could try on shoes in the front part of the shoestore, not in a back room. Neither one would let themselves drink from a "colored only" fountain even if they were dying of thirst. (5)

This internalization of the racist logic of white ideals and white standards, ramified in the colorist practice, seems to be one effect of white supremacy over racialized subjects, who see this discrimination as a way to side with the oppressors, appeasing their own concerns with the racial tension they are able to avoid. Sweetness' family had a history of passing for white, starting with her grandmother, who had severed any ties she had with her children, so she could protect this privilege from any suspecting eyes: "You should've seen my grandmother; she passed for white and never said another word to any one of her children. Any letter she got from my mother or my aunts she sent right back, unopened. Finally they got the message of no message and let her be" (*God Help the Child* 3). Passing granted these subjects the possibility of emancipation when compared to subjects who were undeniably black. This practice was enacted frequently, as Sweetness points out: "[a]lmost all mulatto types and quadroons did that back in the day—if they had the right kind of hair, that is" (3). Her mother, however, chose not to pass for white, which brought about consequences. On the occasion of her marriage, Lula Mae was prompted to lay her hand on the Bible that was designated for black people in the courthouse, a memory Sweetness carried with ambiguous shame:

> The other one was for white people's hands. The Bible! Can you beat it? My mother was housekeeper for a rich white couple. They ate every meal she cooked and insisted she scrub their backs while they sat in the tub and God knows what other intimate things they made her do, but no touching of the same Bible. (4)

Sweetness considers her mother to be equal to whites in many regards, but her choice to embrace her racial identity as black becomes a form of inferiority in the eyes of her daughter, as the mother would dispose of any privilege she might have obtained otherwise. Lula Ann's birth signals another blow to the made-up equality granted by passability, as pointed by Dreiding:

> God Help the Child [...] is the story about a daughter who must transgenerationally negotiate the traumatic historicity that she is discursively born into. Her birth is an infraction to a fantasy of a rehabilitated African-American subjectivity which has appropriated the cipher of skin color as a token of belonging. (132)

The fantasy that Dreiding refers to is related to the possibility of passability a fair-skinned African American could achieve, and thus profit from, a reality that escaped the trials and tribulations associated with racism in the United States, finally being able to fully participate as a citizen in American society.

The formation of colorist hierarchies inside this family is also shared by Lula Ann's father, Louis, who barely figures in the novel, since his sense of abjection for her surpasses that of her mother. For him her color is proof of Sweetness, infidelity. As a porter, Louis spends much of his time out of town, which fueled the suspicion that such a dark baby could not possibly be his, leading to their divorce. For Louis, Lula Ann was not simply a stranger in the nest, she was, as Sweetness puts it, "an enemy" (God Help the Child 5). The utmost expression of his feeling of betrayal and disgust was that he never touched the child. The breaking point for the couple happened when Sweetness claims that their daughter's blackness must have come from his side of the family: "[w]e argued and argued till I told him her blackness must be from his own family—not mine. That's when it got worse, so bad he just up and left and I had to look for another, cheaper place to live" (6). As a single mother, Sweetness experiences the intersectional hardships that her class, sex, and her daughter's color create. When looking for a new apartment, Lula Ann's color would raise questions, as their relation would be denounced: either Sweetness was the baby's mother, and therefore black (dissolving any pretense of ambiguousness), or she was white and mothered a child from a black man. Her ambiguity could be maintained had their skin tone been the opposite, as pointed by Sweetness: "I could have been the babysitter if our skin colors were reversed" (6), as black subject tending to a white baby would be a normalized role in a white supremacist society. Other strategy the mother found to maintain her ambiguous status when in public was instructing the child to call her only by her name.

> I told her to call me "Sweetness" instead of "Mother" or "Mama." It was safer. Being that black and having what I think are too-thick lips calling me "Mama" would confuse people. Besides, she has funny-colored eyes, crow-black with a blue tint, something witchy about them too. (*God Help the Child* 6)

The description of the too-thick lips infers once again the characteristics of blackness that are hierarchically inferior in the eyes of the mother. Lula Ann's eyes, however, belong to a realm of strangeness, almost granting the child a supernatural status, given her alien aspect. Sweetness denies Lula Ann the proximity of a healthy mother-daughter relationship in detriment of her privilege, creating a distance between them that is especially harmful given the age of the child. Ramírez states in "'Childhood Cuts Festered and Never Scabbed Over': Child Abuse in Toni Morrison's *God Help the Child*": "Sweetness's habitual emotional and psychological maltreatment severs the mother-daughter bond. Lula Ann misses her unavailable and unresponsive mother". (Ramírez 152)

In addition to that, the mother expressed that she avoided taking the baby outside in general, fearing the reactions of people when facing such a light-skin mother and her blue-black baby, and especially during the apartment hunt. "I knew enough not to take her with me when I applied to landlords so I left her with a teenage cousin to babysit" (*God Help the Child* 6). Housing discrimination has been outlawed on the occasion of the passing of the 1968 Civil Rights Act, Title VIII, which prohibited any discrimination on the basis of race, color, sex, disability, familial status or natural origin in the terms and conditions of sale or rental of a residence. These laws, however, are difficult to be enforced, which creates the possibility of extortion of vulnerable parts that belong to minorities. Although this part of the plot takes place during the 1990s, Sweetness claims that the landlord for the apartment she was able to find would charge her more than what had been advertised, demonstrating the ways in which racial bias surpasses the legislation that was created to protect these subjects. In confessional tone, Sweetness narrates the hardships of being a black woman, even if a passable one, especially being a single mother in a position of economic vulnerability:

> It was hard enough just being a colored woman—even a high-yellow one—trying to rent in a decent part of the city. Back in the nineties when Lula Ann was born, the law was against discriminating in who you could rent to, but not many landlords paid attention to it. They made up reasons to keep you out. But I got lucky with Mr. Leigh. I know he upped the rent

seven dollars from what he advertised, and he has a fit if you a minute late with the money. (6)

Later, Lula Ann witnesses Mr. Leigh sexually abusing a boy, something she promptly reports to Sweetness. Her reaction was different from the trial held against Sofia Huxley, as the mother feared being evicted if she made a denunciation. "I know now what I didn't know then—that standing up to Mr. Leigh meant having to look for another apartment. And that it would be hard finding a location in another safe, meaning mixed, neighborhood" (54). Bride tells Booker about the abuse:

> Me hearing a cat's meow through the open window, how pained it sounded, frightened, even. I looked. Down below in the walled area that led to the building's basement I saw not a cat but a man. He was leaning over the short, fat legs of a child between his hairy white thighs. The boy's little hands were fists, opening and closing. His crying was soft, squeaky and loaded with pain. The man's trousers were down around his ankles. I leaned over the windowsill and stared. The man had the same red hair as Mr. Leigh, the landlord, but I knew it couldn't be him because he was stern but not dirty. (54)

Lula Ann's witnessing of this crime figures as another traumatic moment in her childhood. The protagonist's reaction of disbelief at the identity of the perpetrator adds another layer of relevance to this even in her development, as it shows that the perpetrator of sexual violence usually is not a stranger, but someone who is known by the victim. Although Lula Ann is not the direct victim of this episode, it illustrates that she could have been, which would weaken her trust in adult figures who were supposed to care for her, as a vulnerable subject. The economic power dynamics are evidenced in Sweetness, reaction, and her silencing of her daughter amounts to the lack of support her mother continually demonstrated during her childhood.

> So when I told Sweetness what I'd seen, she was furious. Not about a little crying boy, but about spreading the story. She wasn't interested in tiny fists or big hairy thighs; she was interested in keeping our apartment. She said, "Don't you say a word about it. Not to anybody, you hear me, Lula? Forget it. Not a single word". (54)

Lula Ann's experience with this molester was not surmized by the witnessing of a crime, as her mother's silencing prevented her from sharing what ensued. Once the perpetrator realizes he is being watched by Lula Ann, she is met by violent reproach and racist name calling:

So I was afraid to tell her the rest—that although I didn't make a sound, I just hung over the windowsill and stared, something made the man look up. And it was Mr. Leigh. He was zipping his pants while the boy lay whimpering between his boots. The look on his face scared me but I couldn't move. That's when I heard him shout, "Hey, little nigger cunt! Close that window and get the fuck outta there!". (55)

As an adult, Bride revisits this moment of her life in the light of Sofia's trial and her false accusation (56) as both events seem to be connected in some way, as if justice could have been made in a different setting, since the crime that was performed was similar, though the perpetrators were different. The racism that was inflicted upon her can be seen as a coupled feature of child abuse in Lula Ann's perspective, one that is somewhat corroborated by Sweetness, when she states after Sofia's trial: "It's not often you see a little black girl take down some evil whites" (42). It seems that for them racial justice was someway achieved in this conundrum, despite being served to the apparently wrong evildoer.

The economic hardships that are overdetermined in the female African-American experience are also demonstrated by Sweetness and were not limited to housing discrimination, as she finds herself in a precarious financial situation being the sole provider and a fulltime mother. Louis, although absent, financially contributed a little every month after discovering Sweetness' new address. Before that, her work was not sufficient to support her family, and the mother had to resort to welfare assistance so she could make ends meet. The welfare reform that took place during the 1990s greatly affected the lives of those who depended on it, as it became more restrictive than it had previously been. The racial stereotype of the welfare queen employed frequently by Ronald Reagan instigated discussions for a reform of the system, culminating in Bill Clinton's extinction of the Aid to Families with Dependent Children program, and the implementation of the Temporary Assistance to Needy Families program, under the Personal Responsibility and Work Opportunity Act, passed in 1996[7].

Sweetness provides a critique of the welfare system, and the ways in which it vilifies the people who are in need of assistance, and how racism is an intrinsic force in this dynamic:

---

[7] The usual perception is that racialized minorities are most often the recipients of this social aid, however, as pointed by Robert A. Moffitt and Peter T. Gottschalk, in *America Becoming: Racial Trends and Their Consequences:*"Many analysts have noted that the general popular perception that minority racial and ethnic groups dominate the welfare rolls has been historically incorrect, for minorities have historically accounted for no more of the welfare caseload than White families. Ethnic minorities do, however, have higher rates of participation in the welfare system than does the majority White population, given their lesser total numbers. Thus, the popular perception has some basis in fact, if interpreted to mean that minorities have higher propensities to make use of the welfare system (152).

> His [Louis'] fifty-dollar money orders and my night job at the hospital got me and Lula Ann off welfare. Which was a good thing. I wish they would stop calling it welfare and go back to the word they used when my mother was a girl. Then it was called "Relief." Sounds much better, like it's just a short-term breather while you get yourself together. Besides, those welfare clerks are mean as spit. When finally I got work and didn't need them anymore, I was making more money than they ever did. I guess meanness filled out their skimpy paychecks, which is why they treated us like beggars. More so when they looked at Lula Ann and back at me—like I was cheating or something. (*God Help the Child* 6–7)

The change in nomenclature is also telling, as relief provided a more humanistic dimension to this king of aid, as referred by Sweetness. The discrepancy between their skin tones creates another layer of distrust, as it seemed that Sweetness was partaking on some kind of scheme so she could benefit from the financial aid in the eyes of the clerks who were presumably white.

Although the financial situation for this family was stabilized, Sweetness foresaw the trials and tribulations that Lula Ann would go through in her life due to her dark skin, leading the mother to adopt a kind of education that was extremely severe, as a means to prepare the child for the violence she would invariably encounter. This kind of understanding, when combined with the abjection that the mother felt in relation to her child, created a childhood that was extremely traumatic for Lula Ann. Sweetness closes the first chapter with the reiteration of her first sentence:

> Things got better but I still had to be careful. Very careful in how I raised her. I had to be strict, very strict. Lula Ann needed to learn how to behave, how to keep her head down and not to make trouble. I don't care how many times she changes her name. Her color is a cross she will always carry. But it's not my fault. It's not my fault. It's not my fault. It's not. (*God Help the Child* 7)

Sweetness reiterates in her closing remarks that the burden of Lula Ann's existence, her color, is not something that she can be blamed for, still perceiving it to be a sin and a fault. However, her attitudes as a parent are reprimandable. As a mother she was able to guarantee the basic material conditions for the development of her child, yet she could not provide the nurturing necessary for the establishment of a positive self-image. Her lack of affection carried consequences in Bride's life that would resurface.

4.1.3 What you do to children matters – Children's trauma Representation

"I'm scared. Something bad is happening to me. I feel like I'm melting away. I can't explain it to you but I do know when it started. It began after he said,

'You not the woman I want'" (*God Help the Child* 8). Although the protagonist is aware that some change is taking place, she does not know what it is. What she is certain is the moment when it all started. Booker's parting declaration triggers the repressed memories of loneliness and lovelessness that were created during Bride's childhood. The reader meets Bride in the aftermath of her breakup of a relationship that compensated for much of what Bride could not experience in her childhood. Although Bride realizes she knew very little about Booker's past at the moment he left, he was the one partner she had as an adult who would listen to her without objectifying her. Booker cherished both her personality and her appearance, and Bride was even able to confide him secrets she had never shared with anyone, namely the witnessing of the abuse performed by Mr. Leigh. The change that was taking place concerned her body, as it started a process of deconstruction as a consequence of a number of factors. Salván comments:

> The novel begins just as Sofia Huxley, the woman accused and convicted for child moles-tation, is about to be released from prison after a fifteen-year sentence. This is signaled in the text as the trigger for Bride's identity crisis, which from the psychoanalytical perspective can be read in terms of the return of the repressed, as she associates her visit to Sofia to feelings of guilt, vulnerability and what appears to be a psychosomatic acting out of an ac-tually unresolved trauma, as she seems to be physically returning to prepubescence. (68)

The fantastic is employed in this narrative as a means to represent how trauma is present in the body of this character. Bride slowly perceives changes in her body, which seems to be going back to its former childhood form. This change has been interpreted through the perspective of magic realism[8], used in the sense

---

8 The history of the terminology is clarified by Ashcroft *et al:* "This term, which has a long and quite distinctive history in Latin American criticism, was first used in a wider post-colonial con-text in the foundational essay by Jacques Stephen Alexis, 'Of the magical realism of the Haitians' (Alexis 1956). Alexis sought to reconcile the arguments of post-war, radical intellectuals in fa-vour of social realism as a tool for revolutionary social representation, with a recognition that in many post-colonial societies a peasant, pre-industrial population had its imaginative life root-ed in a living tradition of the mythic, the legendary and the magical. The term became popular-ized when it was employed to characterize the work of South American writers widely translated into English and other languages, such as Gabriel Garcia Marquez. It tended to be used indis-criminately during the 'Boom' period of the 1960s and 1970s by some critics who saw it as a de-fining feature of all Latin American writing, in stark contrast to its older, more specific usage in Latin American criticism, a usage that differed in marked ways from the recent rather loose and generalized use of the term (Zamora and Faris 1995). However, its origins in the 1950s lay in the specific need to wed Caribbean social revolution to local cultural tradition" (Ashcroft *et al* 132). Ashcroft *et al* also caution that the usage of this terminology has become so popular it may now

of the insertion of fantastic, mythic, or imaginary aspects in an otherwise realis-tic narrative. Bride's body regression is perceived solely by the protagonist, and not by any other character that interacts with her, emphasizing the psychological aspect of trauma in this representation, as her self-image is deconstructed, and the features that separate the woman from the girl are gradually taken away. Firstly, Bride notices that she has lost her pubic hair, symbolically demonstrating that her womanhood is in check. "It was when I got dressed for the drive I noticed the first peculiar thing. Every bit of my pubic hair was gone. Not gone as in shaved or waxed, but gone as in erased, as in never having been there in the first place" (*God Help the Child* 12–13). The change is telling since it is character-ized not as a cosmetic effort that was taken in a perfomative act of traditional female beauty, but as something that was absent because it had not yet existed, hinting once again that her prepubescent self is claiming her body. Shortly after, the hair in her armpits also vanishes (52). Another change in her body is her ear-lobes, which are no longer pierced. Bride recalls the special circumstances in which she got her first earrings, small golden hoops Sweetness gave her after So-fia's trial, marking her childhood as one of the few positive gestures that her mother displayed towards her. They may also be seen as another sign of her ma-turity, demarcating different stages in her development. Her regressing body, however, displays no sign of their existence (51).

Later in the narrative, after her car accident in California, Bride realizes her breasts also vanished. When she was healed enough totake a bath, she faced her nakedness for the first time in weeks, realizing their absence: "It was when she stood to dry herself that she discovered that her chest was flat. Completely flat, with only nipples to prove it was not her back. [...] I must be sick, dying, she thought" (92). Bride's suspicion is only physiological, not considering the psy-chological side of these changes. Evelyn, one of her hosts in California, does not comment on Bride's appearance when seeing her naked, leaving room for doubt and ambiguity. In addition to the disappearance of her breasts, Bride no-tices she is several sizes smaller, as it she feels her body is reducing, and the adult clothes she is handed do not fit her:

"Please, do you have something I can wear?"

"Sure," said Evelyn, and after a few minutes brought Bride a T-shirt and a pair of her own jeans. She said nothing about Bride's chest or the wet towel. She simply left her to get

---

serve as a misnomer for everything that does not comply with realistic fiction: "Although the term has been useful, its increasingly ubiquitous use for any text that has a fabulous or mythic dimension has tended to bring it into disrepute with some critics who suggest that it has become a catch-all for any narrative device that does not adhere to Western realist conventions" (133).

dressed in private. When Bride called her back saying the jeans were too large to stay on her hips, Evelyn exchanged them for a pair of Rain's, which fit Bride perfectly. When did I get so small? She wondered. She meant to lie down just for a minute, to quiet the terror, collect her thoughts and figure out what was happening to her shrinking body, but without any drowsiness or warning she fell asleep. (92–93)

A third-person narrator interrogates the nature of her apparent sickness, questioning: "Nothing hurt; her organs worked as usual except for a strangely delayed menstrual period. So what kind of illness was she suffering? One that was both visible and invisible. Him, she thought. His curse" (95). The reference to her delayed period is another indication of her body's regression to its childhood form. Bride must resort to wearing Rain's clothes, as the child's size is a more appropriate fit. Rain is characterized by being small, either by her age, which cannot be precisely assessed in the narrative, or by her malnutrition after being mistreated by her biological mother and living a homeless life. Rain seems to be another character that is afflicted by childhood trauma, which is expressed in magic realistic fashion. After being rescued by Steve and Evelyn, it seems that the child has not developed since, seemingly remaining a child. Mukherjee states that "[t]his magical stasis of Rain's body is perhaps because she is too scared to grow-up having been witness to the ugliest side of adulthood" (503). While talking with Evelyn about Rain's origins, Bride has a revelation about her own predicament:

> "You said she was about six when you found her?" asked Bride. "I guess. I don't really know. She never said and I doubt she knows. Her baby teeth were gone when we took her. And so far she has never had a period and her chest is flat as a skateboard." Bride shot up. Just the mention of a flat chest yanked her back to her problem. Had her ankle not prohibited it, she would have run, rocketed away from the scary suspicion that she was changing back into a little black girl. (*God Help the Child* 97)

Finally, Bride attributes the reason for the changes to Booker, and the explicit reference to a curse corroborates the interpretation of the phenomena through the perspective of magic realism. These fantastic changes amount to a reflex of her feeling of lovelessness and loneliness after Booker's departure, and his curse is related to her intentions of helping Sofia. As he does not give her a sensible motive for the breakup, something she is only able to get after confronting him, her body is the site in which her childhood trauma resurfaces. In addition to this, her plan of redemption for the false statement she made falters, as helping Sofia get a new beginning only caused her physical trauma and a feeling of inadequateness.

In her quest to find him she ends up in Whiskey, the small town in which Booker has the only relative he still keeps in touch with, his aunt Queen.

First, she is able to locate the aunt, who straightaway judges Bride's appearance, shattering her already compromised confidence: "Come on in. You look like something a raccoon found and refused to eat" (144). Queen, however, is not mean spirited in her comments, as she invites Bride inside for a meal, and immediately sooths the protagonist's worries about Booker, reveling his troubled past. Queen simply remarks that it seems that Bride looks too thin, almost sickly. This is a stark contrast to the opinions Bride was previously able to get, when in control of her body and life, as the successful manager of a cosmetic line. Queen's comment contributes to the ambiguity of Bride's change, adding an external perspective to her body and appearance. Her opinion, though, is not conclusive, as the readership cannot precisely grasp if her comments reflect the changes that Bride reports, or if they simply assess her weight loss, especially after living in a hippie household for some weeks while recovering from her car accident, without the comforts she used to enjoy in Los Angeles. Queen's comment confines Bride once again to the skin of Lula Ann:

> Bride swallowed. For the past three years she'd only been told how exotic, how gorgeous she was—everywhere, from almost everybody—stunning, dreamy, hot, wow! Now this old woman with woolly red hair and judging eyes had deleted an entire vocabulary of compliments in one stroke. Once again she was the ugly, too-black little girl in her mother's house. Queen curled her finger. "Get in here, girl. You need feeding".(144)

Booker reveals he left Bride due to her sympathy for a child molester, something that triggered his own childhood trauma. Her plans to help Sofia convinced Booker that Bride was not the person he considered her to be, prompting him to journey back to his family. The reversal of the curse takes place when Bride is able to reach Booker and confront him for having left her. At the same time she finally admits her testimony as a child was false. She regains her body when she is able to address her trauma, acknowledging her wrongdoing and her guilt, providing Booker with an explanation for her behavior (helping Sofia), and finally creating an understanding between the couple. The first physical reversal takes place after the conversation with Booker, being noticed as they try to save Queen from her house fire. When trying to put out the fire from her hair, Bride removes her shirt, trying to extinguish the flames with it, revealing the return of her breasts to all around her:

> Suddenly, a spark hiding in Queen's hair burst into flame, devouring the mass of red hair in a blink—just enough time for Bride to pull off her T-shirt and use it to smother the hair fire. [...] As the ambulance parked, the crowd became bigger and some of the onlookers seemed transfixed—but not at the moaning patient being trundled into the ambulance. They were focused, wide-eyed, on Bride's lovely, plump breasts. However pleased the onlookers were, it was zero compared to Bride's delight. So much so she delayed accepting the blanket the

medical technician held toward her—until she saw the look on Booker's face. But it was hard to suppress her glee, even though she was slightly ashamed at dividing her attention between the sad sight of Queen's slide into the back of the ambulance and the magical return of her flawless breasts. (165–166)

Bride's delight might be read as the satisfaction to finally regain her self, her womanhood, her image and her sense of assertion in the world. The pleasure derived from the gaze of others is indicative of Bride's sense of empowerment that comes from her beauty, finally regaining control over the narrative of color that had haunted her for all of her childhood. While caring for Queen in the hospital, Bride and Booker have the time to focus on somebody else other than themselves, which allows room for their own internal healing, cementing their relationship: "[n]either one spoke during those ablutions and, except for Bride's occasional humming, the quiet served as the balm they both needed. They worked together like a true couple, thinking not of themselves, but of helping somebody else (167). The next sign of her recovery are her earlobes, which regain the small piercings given by her mother's first earrings:

> Suddenly, as though he'd forgotten something, Booker snapped his fingers. Then he reached into his shirt pocket and took out Queen's gold earrings. They had been removed to bandage Queen's head. All this time they had been in a little plastic bag tucked in the drawer of her bedside table. "Take these," he said. "She prized them and would want you to wear them while she recovers." Bride touched her earlobes, felt the return of tiny holes and teared up while grinning. (168–169)

The reclaiming of her body is completed in the closing episode of the novel, as Bride announces she is pregnant. This is relevant because, in addition to demonstrating that her adult body is certainly restored to its normality, it also serves the purpose to finally offer the opportunity for this character to develop a healthy mother-daughter relationship. Booker figures here as the bridge to this reality, offering his support symbolized as the hand she had never been offered:

> Bride took a deep breath before breaking into the deathly silence. Now or never, she thought. "I'm pregnant," she said in a clear, calm voice. She looked straight ahead at the well-traveled road of dirt and gravel." [...] "Then he offered her the hand she had craved all her life, the hand that did not need a lie to deserve it, the hand of trust and caring for—a combination that some call natural love. (174–175)

The setting of an open road, with its dirt and gravel may be seen as a metaphor for motherhood, a path that clearly displays hardships, but one that Bride is willing to travel with Booker by her side. The pregnancy also offers the opportunity of showcasing that the cycle of childhood trauma inflicting might have a chance to stop being perpetuated, as a new child offers a clean slate to start over. "A child. New life. Immune to evil or illness, protected from kidnap, beatings, rape, racism, insult, hurt, self-loathing, abandonment. Error-free. All goodness. Minus wrath. So they believe" (*God Help the Child* 175).

Wrath might be a word to describe Rain's behavior, as she stands for the child who is deeply affected by childhood trauma, seemingly refusing to grow up. Her story of violence and abandonment illustrates the despicable things that can be done by those who are supposed to protect children. Rain's behavior is understood as a consequence of the life she was forced to live before being rescued by Evelyn and Steve. When asked, Evelyn is not capable of giving Bride a full account of Rain's past, only providing the story of how she was found in the rain while returning from a protest. The girl was standing in the rain, something the couple attributed to a minor mishap, such as being locked out of her home. She did not provide a name when they asked her, and at the moment Steve tried to reach to touch her shoulder, she escaped. The couple decided to give up and leave, as the rain got heavier, parking at a nearby diner to find some safety from the weather. They found the girl once again after the rain subsided, standing by the trash outside the building. Evelyn describes her rescue as a tumultuous affair, one in which the child is obviously terrified of being in the presence of adults (96–97).

Later, Bride develops a relationship with the child. Rain's identification with Bride while convalescing is indicative of the power of acceptance, as the protagonist is able to listen to her story without judging it, something her well-intentioned saviors could never do. In the only chapter narrated by the child, she reveals her special relationship with Bride, and the ways in which it functions as a safe place for the telling of her side of her story:

> I don't know who I can talk to. Evelyn is real good to me and so is Steve but they frown or look away if I say stuff about how it was in my mother's house or if I start to tell them how smart I was when I was thrown out. Anyway I don't want to kill them like I used to when I first got here. But then I wanted to kill everybody—until they brought me a kitten. She's a cat now and I tell her everything. My black lady listens to me tell how it was. Steve won't let me talk about it. Neither will Evelyn. (104)

Rain's violence is seen here in this confession, as it extents to murder desires which are appeased when the girl is gifted with something to care for. Her cat functions as a device that channels her anger and transforms it into something positive, as she is able to communicate with it. Interestingly, Rain, who calls Bride "my black lady", likens the protagonist to her cat, once again reiterating her affection for these figures: "[t]hat time I saw her stuck in the car her eyes scared me at first. Silky, my cat, has eyes like that. But it wasn't long before I began to like her a lot" (104).

The conversation that takes place between Rain and Bride takes place as the protagonist invites the child for a walk, one that serves the purpose simply of distracting Bride from the tedious process of recovering. The invitation is met with suspicion, which turns into acceptance: "'What for?' By the tone of her voice it was clear the ants were far more interesting than Bride's company. 'I don't know,' said Bride. That answer seemed to please. She jumped up smiling and brushing her shorts. 'Okay, if you wanna'" (100). The open-endness of the invitation allures the child, and this interaction might be seen as a sign that Bride has no wish to control her, and therefore, though Bride is an adult she does not present a risk. Finally, Rain discloses her side of the story of her rescue, in a dialogue that is not hostile, but that demonstrates how Rain perceives her own story, as she tells Bride she has been stolen by Evelyn and Steve since they never asked if she wanted to go. She is not resentful, since they provided her with food, care, and shelter. When asked if she wanted to leave, Rain promptly responds: "Oh, no. Never. This is the best place. Besides there's no place else to go" (101). This last statement encourages Bride to ask more specific questions about Rain's past and her family, which would reveal the traumatic events that have shaped her childhood. When asked about her home, Rain claims that she used to have one, but her mother lives there, which lead Bride to ask if Rain had run away:

"No I didn't. She threw me out. Said 'Get the fuck out.' So I did."

"Why? Why would she do that?" Why would anybody do that to a child? Bride wondered. Even Sweetness, who for years couldn't bear to look at or touch her, never threw her out.

"Because I bit him."

"Bit who?"

"Some guy. A regular. One of the ones she let do it to me. Oh, look. Blueberries!" Rain was searching through roadside bushes. "Wait a minute," Bride said. "Do what to you?"

"He stuck his pee thing in my mouth and I bit it. So she apologized to him, gave back his twenty-dollar bill and made me stand outside." The berries were bitter, not the wild sweet stuff she expected. "She wouldn't let me back in. I kept pounding on the door. She opened it once to throw me my sweater." Rain spit the last bit of blueberry into the dirt. As Bride imagined the scene her stomach fluttered. How could anybody do that to a child, any child, and one's own? (101)

Bride is shocked to discover that Rain was prostituted in her own home by her mother, and that this mother was the same who had expelled this child from home, and forced her to roam the streets. When Bride compares her childhood in its traumatic stance, she feels that Sweetness was not as bad, when in comparison to Rain's mother. Although their experiences are different, it is precisely because her trauma happened during her tender years that Bride is interested in understanding Rain's account, possibly trying to help the young child to work-through this negative experience, so she can finally be released from it. When asked what she would like to say to her mother, Rain's response is direct: "Nothing. I'd chop her head off" (102). Rain's anger at her mother continues as she fantasizes about the ways in which she could hurt her, demonstrating that the child continues to be deeply affected by the violence that was implicated on her. After this revelation, Bride encourages Rain to share her story, and the child elaborates on the time she spent on the streets (102–103).

Rain reveals the strategies she had to develop so she could survive during this period, as she needed to struggle for the simplest of actions. Her comments on the relations she developed are also telling, since they inform the readership of the different dangers that afflict such a vulnerable subject in these circumstances. Her connection with the prostitutes is also interesting in the perspective of a sense of sisterhood, as they try to warn her of the different dangers that men might bring, being either civilians or agents of the law. The explicit reference to her own experience with pain and hurt culminates her report, demonstrating a more complex representation of her mother figure, who, at some point, would show some willingness to protect her, but who was the one responsible for putting her own child in this unforgivable position of vulnerability. Rain's reaction regarding Steve's touch during her rescue comes to light in their conversation, as Rain admits that men continue to be a source of fear to her, confirming that the previously reported abuses she had suffered in her mother's care continue to affect her life deeply, impeding her of developing healthy relations with all masculine figures.

> Men scared her, Rain confessed, and made her feel sick. She had been waiting on some steps at the Salvation Army truck stop when it began to rain. [...] That's when Evelyn and Steve came along, and when he touched her she thought of the men who came to her mother's house, so she had to run off, miss the food lady and hide. Rain giggled on occasion as she described her homeless life, relishing her smarts, her escapes, while Bride fought against the danger of tears for anyone other than herself. Listening to this tough little girl who wasted no time on self-pity, she felt a companionship that was surprisingly free of envy. Like the closeness of schoolgirls. (*God Help the Child* 103)

Their closeness is confirmed in the last episode narrated by Rain, in which Bride is the victim of a racist attack that culminates into violence. As Bride and Rain are returning home, after the conversation about Rain's past, the pair is approached by some young men in a pick up truck. Rain narrates: "One of them hollered 'Hey, Rain. Who's your mammy?' My black lady didn't turn around but I stuck out my tongue and thumbed my nose at him" (105). Bride, used to the racial slurs, is not moved by the attack; however, Rain, seeing her newfound friend, retorts in the most childlike manner. It is not clear if Rain is aware of the meaning of the offense that Bride was the target to, yet, she had perceived that it was not something positive. The young men return to persecute them with guns, and Bride is quick to protect Rain from any harm:

> The driver, an older boy, turned the truck around so they could come after us. Regis pointed a shotgun just like Steve's at us. My black lady saw him and threw her arm in front of my face. The birdshot messed up her hand and arm. We fell, both of us, her on top of me. I saw Regis duck down as the truck gunned its engine and shot off. What could I do but help her up and hold on to her bloody arm as we hurried back to our house as fast as her ankle would let her. [...] My heart was beating fast because nobody had done that before. I mean Steve and Evelyn took me in and all but nobody put their own self in danger to save me. Save my life. But that's what my black lady did without even thinking about it. She's gone now but who knows maybe I'll see her again sometime. I miss my black lady. (105–106)

Bride's gesture, something expected from a responsible adult, is seen as the redemption of all the wrongs performed by those who were supposed to care for the life of this child. Bride's interaction with Rain is short, only the weeks she spends recuperating during her quest for Booker; however, the impression she left in Rain is lasting and positive, and works in the path of the possible addressing of trauma. Rain's realization that Bride had risked her life to save her demonstrates a step in the direction of healing for the child, as she begins to understand the possibility of trusting adults once again.

Another character that shares her story of trauma is Brooklyn, Bride's assistant. Brooklyn figures as a frustratingly minor character in the novel, who offers little of her own story to the plot, but that nonetheless helps the narrative by providing Bride with a frame to display the protagonist's development. Brooklyn's story of trauma, which involves sexual abuse from a family member, is briefly described by the character as she compares her own story of overcoming to Bride's. Just as Bride left her world and name behind when she was sixteen, reinventing herself and taking the reigns of her own narrative, Brooklyn also abandoned her family and set out for a change in her life (140).

Brooklyn's and Bride's sense of sisterhood seems tenuous, at least from Brooklyn's perspective, as she characterizes the protagonist as a "[b]eautiful dumb bitch" (139). Her commentary on Bride's life is often condescending, and though she recognizes Bride's beauty, she does not seem to see her as her equal, especially when it comes to Bride's personal life. She is described as the polar opposite to Bride physically, extremely pale and with blond dreadlocks, and is summoned anytime the protagonist must deal with complications, either being the rescue and treatment after the altercation with Sofia, or when Bride suddenly decides to chase Booker in Whiskey, leaving the professional responsibilities of Sylvia Inc. in her care. She also figures as a character that does not condone Bride's relationship with Booker, as she considers his unknown past to be suspicious, in addition to censuring what little she knows of him, calling him a conman, and a panhandler. Though it seems that her rejection of Booker is born completely out of spite, as he somewhat rejected her sexual advances, something that Bride is completely unaware of. Brooklyn seems to fit the archetype of the self-made character, working for Bride as her personal assistant in the beauty product company, but always striving for something better in her career. Little is known about Brooklyn's background, but she recounts about her dysfunctional family, stressing the ability she developed to predict people's behavior, which would protect her from the predatory advances of an abusive uncle. Brooklyn describes her past, stressing how the sexual abuse had an enormous impact on her life, as well as the strategies she managed to develop so she could resist:

> [...] my uncle started thinking of putting his fingers between my legs again, even before he knew himself what he was planning to do. I hid or ran or screamed with a fake stomach-ache so my mother would wake from her drunken nap to tend to me. Believe it. I've always sensed what people want and how to please them. (139)

The first revelation highlights that the abuse was not something that had happened only once, which would lead the readership to think that this kind of violence had taken place from an early age, of that at least it had happened multiple times. The dysfuctionality in the family continues to be described by the adjectives used to characterize the mother, revealing an environment in which a child was clearly vulnerable, and lacking the supportive network that would protect her from this violence. Finally, the ability to predict people's behavior is what makes Brooklyn such a useful and proactive assistant to Bride.

Sofia's storyline and the indeterminacy of her crimes play a more important part than her past at the hands of strict religious parents. Sofia discloses little about her upbringing, after telling about not having received letters, nor visits, from her parents while serving her sentence, though they would send care packages on Christmas and her birthday. She describes her parents as following:

> They were always hard to please. The family Bible was placed on a stand right next to the piano, where my mother played hymns after supper. They never said so, but I suspect they

were glad to be rid of me. In their world of God and Devil no innocent person is sentenced to prison. (68)

It seems that Sofia's parents are religious fundamentalists who discipline their children, upholding the word of the Bible as the model for morality. Sofia also comments that her life was devoid of literature, something that she could only find inside the prison as a form of distraction. Though she pursued an education major, finally becoming a schoolteacher, she was not required to deal with any literature in her formal education. "Anything other than religious tracts and the Bible were banned in my family's home" (*God Help the Child* 68). While describing the home, she compares it to the church she used to frequent with her parents, an austere environment that though comforting, was depleted from warmth. While comparing the two spaces, the aspect of her home that is stressed also highlights the disciplinary aspect of her upbringing, as she remembers the wallpaper colors and patternsin the dining room corner better than her own facial features, as she was often grounded in this specific space:

> Roses, lilacs, clematis all shades of blue against snowy white. I stood there, sometimes for two hours; a quiet scolding, a punishment for something I don't remember now or even then. I wet my underwear? I played "wrestle" with a neighbor's son? (78)

The strictness of Sofia's upbringing is revealed more due to her "mistakes" than by the way she is punished, as the character displays the innocence and irrelevance of her crimes as a child. Sofia also confesses that she wanted to escape this strict environment, resorting to the only way out for a woman in a deeply religious context: marriage. Her relationship with this man seems to be a continuation of her parents' rules, as she reveals the corner to which she was confined, a metaphor for her own house, became larger:

> I couldn't wait to get out of Mommy's house and marry the first man who asked. Two years with him was the same—obedience, silence, a bigger blue-and-white corner. Teaching was the only pleasure I had. I have to admit, though, that Mommy's rules, her strict discipline helped me survive in Decagon. (78)

The realization that her mother's education was useful to her time in prison elicits the formative character of this mother, who is never given a voice in the novel. The internalization of this discipline, a bitter aspect of her childhood, is what grants Sofia the ability to resist the fifteen years inside the prison, in its physical and psychological violent dynamics. The cathartic moment for Sofia takes places after leaving the institution, as Bride tries to meet her. Sofia describes the physical violence inflicted upon Bride as the moment she was finally free from all this terrible past: "I beat up that black girl who testified against me. Beating her, kicking and punching her freed me up more than being paroled. I felt I was ripping blue-and-white wallpaper, returning slaps and running the devil Mommy knew so well out of my life" (77). When remembering the confrontation with Bride, Sofia frames it in a sense of redemption, not as a form of revenge in the violence itself, but as a form of traumatic past release, which allowed her to finally express the feelings she had managed to conceal since her childhood. The aftermath of the physical altercation is described as such:

> As soon as I threw her out and got rid of her Satan's disguise, I curled up into a ball on the bed and waited for the police. Waited and waited. None came. If they had bashed in the door they would have seen a woman finally broken down after fifteen years of staying strong. For the first time after all those years, I cried. Cried and cried and cried until I fell asleep. When I woke up I reminded myself that freedom is never free. You have to fight for it. Work for it and make sure you are able to handle it. Now I think of it, that black girl did do me a favor. Not the foolish one she had in mind, not the money she offered, but the gift that neither of us planned: the release of tears unshed for fifteen years. No more bottling up. No more filth. Now I am clean and able. (70)

After this moment, the realization of this release comes full circle as Sofia works in a hospice during her parole. The act of caring for ageing bodies, with their delicate features, serves as the locus in which Sofia enacts some form of compensation for having destroyed Bride's beautiful body, acknowledging the release gift received at this moment:

> When I tend to my patients—put their teeth back in their mouths, rub their behinds, their thighs to limit bed sores, or when I sponge their lacy skin before lotioning it, in my mind I am putting the black girl back together, healing her, thanking her. For the release. (77)

Sofia's story of trauma is explored in a manner that takes into account the ambiguous status of either being innocent or guilty, but focuses on the consequences of the early years of her life, though they are only briefly described in two short chapters. Both the discipline of her mother, and the sense of morality informed by her religious upbringing, help her understand and cope with the experience of imprisonment, and the subsequent working through of her traumatic past.

Booker's own story of trauma is also relevant, since he portrays the struggle against trauma that is both direct and indirect: he experiences the loss of his brother, figuring as his personal history of tragedy, and lives through the tribulations of his brother's experience through the investigation and subsequent prosecution of the perpetrator of violence, resulting in the complications of a life that has been lived in the shadow of pain. To clarify these different experiences, Booker's storyline is best understood in the perspective of a *secondary trauma*, as he suffers the traumatization upon learning about the experiences that his brother suffered, rather than under the scope of *vicarious trauma*, which would imply a cumulative experience of traumatization by frequent retelling and reliving of the traumatic event, such as in the case of the Holocaust and their survivors, or even the experience of slavery, for instance.

Charles R. Figley and Rolf J. Kleber discuss the terminology of secondary trauma in "Beyond the 'Victim' Secondary Traumatic Stress" stating: "We prefer the term secondary traumatic stress [...] because it combines and integrates the many aspects mentioned in the other concepts. It is the exposure to knowledge of a traumatizing event experienced by a significant other that is associated with posttraumatic stress symptoms" (79). The authors further claim that:

> We would like to define a secondary traumatic stressor as the knowledge of a traumatizing event experienced by a significant other. For people who are in some way close to a victim, the exposure to this knowledge may also be a confrontation with powerlessness and disruption. Secondary traumatic stress refers to the behaviors and emotions resulting from this knowledge. It is the stress resulting from hearing about the event and/or from helping or attempting to help a traumatized or suffering person. This conceptualization of primary and secondary traumatic stress describes the distinction between those "in harm's way" and those who care for them and become impaired in the process. (Figley and Kleber 78)

It might be considered that the telling of Booker's trauma is different even in the narratological aspect of the novel, since Booker's trials with the past are told not by the character, as in the case of Bride, Sweetness, Sofia, Rain, or Brooklyn; instead, they are told by a third person narrator who shares the past that has burdened Booker's life since his childhood, posing the question later in the narrative: "[h]ow long had childhood trauma hurtled him away from the rip and wave of life? His eyes burned but were incapable of weeping." (*God Help the Child* 173–174). Booker incapability to shed tears is representative of the difficult relation that the character develops with the human bonds he created since his childhood, as building trust and believing in an optimistic take on life is nearly impossible, as long as he carries his brother's death as his personal burden.

Booker's story of trauma starts when his brother Adam goes missing, and the subsequent discovery of his body months later, inside a culvert, only half dressed and already decomposing. The last time Adam was seen by his brother was while he was skateboarding, a description that eternalizes Adam almost as a mythical figure, who just like the leaves that had not yet faded into goldens and browns, promised to live forever.

> The last time Booker saw Adam he was skateboarding down the sidewalk in twilight, his yellow T-shirt fluorescent under the Northern Ash trees. It was early September and nothing anywhere had begun to die. Maple leaves behaved as though their green was immortal. Ash trees were still climbing toward a cloudless sky. The sun began turning aggressively alive in the process of setting. Down the sidewalk between hedges and towering trees Adam floated, a spot of gold moving down a shadowy tunnel toward the mouth of a living sun. (*God Help the Child* 115)

This stance as an eternal figure in Booker's memory is reinforced by his aunt Queen, who seems to be the only adult that is capable of perceiving Booker's particular grief, as he used to have a special connection with his older brother. Booker had a twin brother who had died during the delivery, and Adam figured as a substitute twin, though he was older. Queen advises Booker: "'[d]on't let him go,' she said. 'Not until he's ready. Meantime, hang on to him tooth and claw. Adam will let you know when it's time.'" (117). This idea of fixation, of hanging on to the memory of his brother, continued for years, causing a set of difficulties for this character's development.

His family life, before the incident of his brother, demonstrates a kind of household in which the development and safety of children was of utmost importance for the parents, who cared for the psychological wellbeing of their children by instating a routine that allowed them to voice their concerns and troubles, as well as to share things they had learned. This politics of care helped foster an environment in which parents had access to their children's inner lives without ever being intrusive, creating bonds of trust that certainly would help them develop as psychologically healthy people. The description of this rountine forum is given as follows:

> Every Saturday morning, first thing before breakfast, his parents held conferences with their children requiring them to answer two questions put to each of them: 1. What have you learned that is true (and how do you know)? 2. What problem do you have? Over the years answers to the first question ranged from "Worms can't fly," "Ice burns," "There are only three counties in this state," to "The pawn is mightier than the queen." Topics relevant to the second question might be "A girl slapped me," "My acne is back," "Algebra," "The conjugation of Latin verbs." Questions about personal problems prompted solutions from anyone at the table, and after they were solved or left pending, the children were sent to bathe and dress—the older ones helping the younger. Booker loved those Saturday morning conferences rewarded by the highlight of the weekend—his mother's huge breakfast feasts. (112–113)

This environment of care and harmony would be disrupted by Adam's abduction, as the family routine fell to the background in the presence of the uncertainty that was brought about by his disappearance. The narrator points clearly at the police's neglect and racism, as the family is under scrutiny after they reach the authorities, a clear contrast to the caring environment that was previously described (114).

The idea of this wholesome family, however, is made more complex as the readership learns about their family history, particularly as their grandfather's past is brought forward. Mr. Drew made his fortune as a slumlord, charging exorbitantly high rent from his tenants. The possibility for the healthy household in which Booker and Adam lived existed because their mother had severed her connections with this figure. This is an example of how Morrison is capable of delivering a complex account of the African-American experience, depicting it without ever being simplistic and unambiguous. By placing Booker's story in such a complex web of circumstances, she restores once again the undeniable humanity of her characters, diverting from any Manichean representation. Mr. Drew returns to the lives of the Starbern during Adam's funeral, an episode that demarcates the rift between the patriarch and the rest of his family (116). This character is relevant, since he enables Booker to live independently later on in the narrative, as he inherits Mr. Drew's fortune, who leaves his grandchildren all he had, cutting out his own children in the process. This inheritance does not come free from a moral impasse that serves the point of once again reaffirming the complexity of Morrison's characters. Booker's righteousness comes into question as he accepts the material compensation that has been created though violence and extortion, and this circumstance is aggravated as the money had come most possibly from poor racialized families.

> Mr. Drew had died and to everyone's surprise he had included his grandchildren—but not his own children—in his will. Booker was to share the old man's constantly-bragged-about fortune with his siblings. He refused to think about the greed and criminality that produced his grandfather's fortune. He told himself the slumlord money had been cleansed by death. (*God Help the Child* 130)

The centrality of money for Booker's understanding of society is a trait that defined much of his interests during his upbringing, something that seems to be at odds with the acceptance of the tainted inheritance left by Mr. Drew. Booker's years in college after his brother's death are characterized as a period in which the character develops a detached attitude in relation to all that surrounds him, resorting to irony as coping mechanism, peaking in hopeless depression. "All he did from freshman year through sophomore was react—sneer, laugh, dismiss, find fault, demean—a young man's version of critical thinking. [...] It was as a junior that his mild cynicism morphed into depression" (*God Help the Child* 121). During this period, his academic interests are described as broad, finally focusing on economics, the discipline that seemed able to respond to most of the central questions of his existence as a black subject.

> Four years ago, as an undergraduate, he'd nibbled courses in several curricula, psychology, political science, humanities, and he'd taken multiple courses in African-American Studies, where the best professors were brilliant at description but could not answer any question to his satisfaction beginning with "Why." He suspected most of the real answers concerning slavery, lynching, forced labor, sharecropping, racism, Reconstruction, Jim Crow, prison labor, migration, civil rights and black revolution movements were all about money. Money withheld, money stolen, money as power, as war. (*God Help the Child* 110 – 111)

This understanding is complicated by the moral implication imbued in the mistreatment of black subjects for profit, as Booker conjectures that it is white hate and violence that would propel the material gains derived from a slave system of exploitation, concluding that money would explain all the different forms of oppression that have torn humanity asunder. His own personal history of trauma is added to the realization of his position in the world, and more specifically as a black subject in a capitalist United States:

> Where was the lecture on how slavery alone catapulted the whole country from agriculture into the industrial age in two decades? White folks' hatred, their violence, was the gasoline that kept the profit motors running. So as a graduate student he turned to economics—its history, its theories—to learn how money shaped every single oppression in the world and created all the empires, nations, colonies with God and His enemies employed to reap, then veil, the riches. (111)

Booker's disillusionment with the world is best explained as he returns to the questions he used to answer during his childhood, before Adam's disappearance, and the contrast demonstrates the extent to which this character has his life unraveled since the moment of Adam's disappearance: "[…] Booker replayed those questions posed by his parents during those Saturday conferences on Decatur Street: 1. What have you learned that is true (and how do you know)? 2. What problem do you have? 1. So far nothing. 2. Despair" (122). The evaluation of Booker's progress in his own terms demonstrate the different ramifications of his initial childhood trauma, as his disinterest in life in general might be traced back to the moment his sadness turned into bitterness and anger. His unresolved trauma continues to symbolically be repeated, as he is not able to create with anyone else deep and meaningful connections, such as he had with Adam. His identification with the history of economics might also be read as a sign of this reality, since money in its abstraction serves as the perfect symbol for the impersonal. Upon receiving his master's degree, Booker is invited back home, which allows the narrator to reveal a trait of his that was still unclear, as he gives up the idea of taking an old ex-girlfriend, feeling she might judge the Starberns. The narrator is privy to his reasoning, claiming: "[h]e thought about asking Felicity, his on-again, off-again girlfriend, to accompany him, but decided against it. He didn't want an outsider judging his family. That was his job" (123). Thus, it is possible to claim that Booker's detachment extends to his family. The Starberns seem to have worked through the trauma of losing a child, something that feels like betrayal as Booker continues to cling to the memory of Adam. Upon visiting the Starberns, Booker is able to see the changes made in the house, which are testament of the family's movement toward a position of overcoming of the trauma. Though it seems that the house once again lives a moment of harmony and positivity, Booker's true feelings are revealed as he sees the changes that have been made during his absence. The material modifications in his previous bedroom, which he used to share with his older brother, are symbolic of the family's working through, and Booker feels the family is too easily setting aside the memory of Adam, disrupting the harmonious environment that had been restored (123–125).

Booker sees in his brother's effacement from the physical reality of the house as an effacement of himself from his family, his detachment reaching its most acute form, as he feels he does not belong even in the house in which he grew up. His abrupt leaving is described by the narrator as something that should have happened years before, stressing once again the distance created between this character and his family as the result of unresolved trauma.

> When he visited his and Adam's old bedroom, the thread of disapproval he'd felt during his proposal of a memorial became a rope, as he saw the savage absence not only of Adam but of himself. So when he shut the door on his family and stepped out into the rain it was an already belated act. (*God Help the Child* 125)

Booker's sense of detachment might also explain his own distance from Bride, who knows very little about his life before meeting her. Finally, Queen remains the only family member with whom Booker keeps in touch, as revealed in the last chapters of the novel, and because of her intervention Booker is able to start working through his trauma, as she confronts him regarding his attachment to the memory of Adam (156–157). Queen's questions point directly at Booker's predicament, providing him with the clarity he needed to understand his own condition. Once again the argument of the moral superiority is invoked by one of Booker's relatives, and after its refutation, Queen indicates the centrality of Adam's figure in Booker's life, as it has taken proportions that are not healthy for him, since the memory of his brother overtakes the reality that Booker inhabits and prevents him from enjoying the present. Booker's recollection of seeing Bride for the first time is revisited in his memory as the moment he could finally be free from the memory of Adam. The cyclical nature of trauma is implied in Queen's response, as she indicates the ways in which Booker continues to summon the presence of Adam in his life, even if unintentionally, demonstrating through metaphor the ways in which this presence has a negative effect in him. The use of medical terms, such as "formaldehyde", and even "cadaver" (157) point once again to the experience of dealing with the finding of Adam's body, and the subsequent need for the identification of the remains. Queen, however, is trying to salvage the remains of Booker's existence, as he still has a chance to live the life that Adam would never have the chance to experience. Queen's opinion of Booker's predicament is made clear as she ends the conversation: "Booker and Queen stared at each other for a long time until she stood up and, not taking the trouble to hide her disappointment, said, 'Fool,' and left him slouched in his chair" (157). Her thoughts are exposed later as she returns home, evaluating the situation and depicting the clear idea of repetition of trauma:

> They will blow it, she thought. Each will cling to a sad little story of hurt and sorrow—some long-ago trouble and pain life dumped on their pure and innocent selves. And each one will rewrite that story forever, knowing the plot, guessing the theme, inventing its meaning and dismissing its origin. What waste. (158)

Booker seems to be able to break the cycle of trauma after understanding Bride's reason to have helped Sofia, and thus being able to accept her love as genuine. Queen's intervention and later death are also motivators for change in his life, as she was the only family member who seemed to have a grasp on his feelings, and their final interactions were able to make him address his unresolved trauma. Bride's dedication to the care of Queen in hospital also helps toward the creation of an environment that is less egotistical for Booker, who is finally able to see beyond his personal history of tragedy and finally be integrated in the small community of his affects.

4.1.4 The nicest man in the world

Besides exploring matters of secondary trauma, Booker's story also serves the purpose of dealing with the representation of the abuser, as he interrogates the figure of the "nicest man in the world", a common characterization of perpetrators of violence before they are exposed as such. Some other characters in Morrison's fiction have occupied the place of the sexual abuser, namely Cholly and Mr. Henry in *The Bluest Eye,* and Frank Money in *Home.* These characters, in their specific contexts, also performed the difficult task of complexifying the narrative about pedophilia, providing humanizing views that have contributed towards a broader understanding of this specific kind of violence, and their victims. Sabine Sielke writes in *Reading Rape – The Rhetoric of Sexual Violence in American Literature and Culture 1770–1990:*

> African American women's writing expresses a certain sympathy with the disempowered perpetrator. *The Bluest Eye*, for instance, depicts the father's sexual aggression against his daughter as resulting to a considerable degree from his own troubled first sexual experience. "Overseen" and enforced by white spectators, thus embedded in a scene that visualizes the denial of privacy and the racialized construction of black sexuality, this experience breeds not love but illicit family relations. In Morrison's depiction of Pecola's rape, rendered through the perspective of a powerless aggressor, the reader even senses a certain degree of empathy for this victim-turned-violator. (152)

In *God Help the Child,* however, Morrison explores the issue more obliquely: in the character of Sofia and the indeterminacy of her crime, in Sweetness' landlord Mr. Leigh, through Lula Ann/Bride's perspective, in Rain's account of domestic violence and prostitution, and later in Booker's brother abuser, Mr. Humboldt, who indirectly figures in the story both through the perspective of a third-person narrator and through Booker's account. This move towards a more oblique narrativization of sexual violence might be understood in the sense of empowering the victims of sexual violence, focusing on their impressions instead of trying to comprehend the reasons why such violence takes place, a trait that can be seen in *The Bluest Eye* and *Home.*

In *God Help the Child,* though we are not granted an inside perspective of the abuser through Sofia in relation to the act itself and its motivations, the readership is granted a glimpse at the reality of the accused (and in this case the convicted). Sofia's treatment by the other inmates in prison is telling of the social perception of child abusers, who are relegated to the bottom of the social hierarchy and are despised even by all the other criminals. Sofia also tells the story of Julie, her cellmate who was also sentenced on child abuse charges, revealing the ways in which this kind of crime is seen as the utmost form of inhumanity, even by those who have committed actions that would harm and impair the lives of others, even the lives of children. Sofia narrates:

> For the first two years we two, sentenced for child abuse, were avoided in the cafeteria. We were cursed and spit on, and the guards tossed our cell every now and then. After a while they mostly forgot about us. We were at the bottom of the heap of murderers, arsonists, drug dealers, bomb-throwing revolutionaries and the mentally ill. Hurting little children was their idea of the lowest of the low—which is a hoot since the drug dealers could care less about who they poison or how old they were and the arsonists didn't separate the chil-

dren from the families they burned. And bomb throwers are not selective or known for precision. (*God Help the Child* 66)

The opinions about the other child abusers in the novel are not different, especially Booker's perspective, whose rage and anger toward his brother's abuser
is described in detail. Although the readership is only certain that Bride's accusation was false, and does not know if Sofia committed the crimes she accused
of, the allegation of this crime is sufficient to unmake Sofia's humanity in the
eyes of the public, as narrated by Sweetness during the weeks of the trial:
"[f]or weeks, crowds of people with and without children in the school yelled
outside the courthouse. Some had home-made signs saying, KILL THE FREAKS
and NO MERCY FOR DEVILS" (42). The public reaction demonstrates the extent
to which the humanity of the perpetrators of this kind of violence is suspended,
as they are characterized as misfits and as evil personified. Moreover, there is
also a gender bias that is not fully explored in this novel, as female abusers
are perceived as even more disturbing in a society that places the work of nurture
and care in the hands of women.

To achieve the unmaking of personalization in the other child abuser characters, in addition to presenting them in a more indirect manner, Morrison in-
vests in the figure of "the nicest man in the world", a normalized trope that describes the presumption of innocence of abusers, and that also expresses the
pervasiveness of sexual abuse, since the majority of abusers are known, familiar, seemingly harmless people in the victims' lives. Mr. Henry in *The Bluest Eye* is
a clear example of this kind of abuser, unsuspected, showering the children with
compliments, only to later lure Frieda to his room and touch the child inap-
propriately.

Adam's murder was performed by a seemingly innocent man who used to
work in the neighborhood as a repairman, announcing his trade in a van with
his name. The description of the perpetrator by the narrator elicits an idea of affable homeliness, someone who would be granted access to the privacy of homes
without causing disturbances:

Another feature some remembered was his smile, how welcoming, attractive, even. Otherwise he was fastidious, capable and, well, nice. The single other thing people remembered
most about him was that he always traveled with a cute little dog in his van, a terrier he
called "Boy". (118–119)

The dog was used as a lure with which the nicest man would attract his victims
to his van, as discovered later. The identity of Adam's killer was only learned several years after his disappearance, and the subsequent discovery of his body. The

resolution of the investigation happened because Adam was seen inside a van, playing with a dog. The car was marked with the name of the abuser, as he advertised it with the American flag colors. The narrator describes the process of the investigation, making evident the familiarity of the community with the criminal:

> A central witness, an elderly widow, remembered that she had seen a child in the passenger side of his van laughing and holding a little dog up to his face. Later, after seeing the missing-child posters displayed in store windows, on telephone poles and trees, she thought she recognized a face as that of the laughing boy. She called the police. Of course they knew the van. It advertised in red and blue letters its promise: PROBLEM? SOLVED! W.M.V. HUMBOLDT. HOME REPAIR. (119)

The investigation that lead to his arrest uncovered a series of crimes that were performed by Mr. Humboldt, all involving children of different ethnicities, as commented by the narrator, pointing that the crime did not seem to be driven by race, but by the age of his victims. The comments on the tattoos and the tattoo artist are relevant since they display another way in which the community might have been attentive, and could have prevented more crimes from taking place, evidencing once again that the perpetrators of violence, especially sexual violence, are present and among us, wearing the social disguise of "commonplace folk".

> Six years later [...] the nicest man in the world was caught, tried and convicted of SSS, the sexually stimulated slaughter of six boys, each of whose names, including Adam's, was tattooed across the shoulders of the nicest man in the world. Boise. Lenny. Adam. Matthew. Kevin. Roland. Clearly an equal-opportunity killer, his victims seemed to be representative of the We Are the World video. The tattoo artist said he thought they were the names of his client's children, not those of other people. (*God Help the Child* 118)

The investigation of the latest crime performed by Mr. Humboldt displays the cruel aspects of sexual violence, and though the reports are not directly connected to Adam, the readership contiguously perceives that the atrocities inflicted upon the other victim(s) might have been similarly applied to him. The families are described to be justly and painfully taken aback by the findings of the investigation, as the reality of the violence merges with the imagination of what could have happened, in addition to the long six years of unanswered questions:

> There was not much left of Adam when he was found, but the details of the more recent abductions were Gothic. Apparently the children were kept bound while molested, tortured and there were amputations. [...] When Mr. Humboldt's house was searched a dirty mattress sporting dried blood was found in the basement along with an elaborately decorated candy

tin that held carefully wrapped pieces of dry flesh, which, on not very close inspection, turned out to be small penises. (119)

Morrison explains the cruelty that might be inflicted upon children's bodies as a means to surmise the main idea in the text: "What you do to children matters" (43). The care for these vulnerable beings is a task that primarily belongs to their caretakers, family, parents, grandparents, and is extended to the community, who must protect them regardless of affiliation, a task that must be seen as a moral duty. The reaction from the public to the findings of the investigation are also telling, as they inherently adopt a rhetoric of violence, which stands for some kind of cathartic righteousness. Booker joins these manifestations of violence, as some form of his pain is addressed in the imagination of worse-than-death consequences for the perpetrator of violence (120). His demand for violence grows during the trials for his brother's case, in which the justice offered by the system did not seem enough to mitigate his need for revenge. His anger, turned into torture wish thinking, would culminate in a fantasy in which the complete and slow destruction of the perpetrator's physical body, in addition to the obliteration of his moral entity, would suffice as justice.

> What he wanted was not the man's death; he wanted his life, and spent time inventing scenarios involving pain and despair without end. Wasn't there a tribe in Africa that lashed the dead body to the back of the one who had murdered it? That would certainly be justice—to carry the rotting corpse around as a physical burden as well as public shame and damnation. (120)

While attending college, Booker faces another molester, but this time he is able to intervene in the situation. The chapter regarding Booker's background begins with a description of this encounter, in which Booker uses violence to counteract the upcoming threat of a child abuser near his prey.

> Blood stained his knuckles and his fingers began to swell. The stranger he'd been beating wasn't moving anymore or groaning, but he knew he'd better walk away quickly before a student or campus guard thought he was the lawless one instead of the man lying on the grass. He'd left the beaten man's jeans open and his penis exposed just the way it was when he first saw him at the edge of the campus playground. Only a few faculty children were near the slide and one was on the swing. (109)

Booker notices the man observing the children while masturbating on the hedges, a description that is completed with the subtle details that the narrator adds little by little through Booker's perspective, never granting the reader access to the mind of the pedophile. The relation between the innocent unaware children and the man pleasuring himself at their sight is described in terms that are sensual, to a point of discomfort to the readership.

> None apparently had noticed the man licking his lips and waving his little white gristle toward them. It was the lip licking that got to him—the tongue grazing the upper lip, the swallowing before its return to grazing. Obviously the sight of the children was as pleasurable to the man as touching them because just as obviously, in his warped mind, they were calling to him and he was answering their plump thighs and their tight little behinds, beckoning in panties or shorts as they climbed up to the slide or pumped air on the swing. (109)

The discomfort arises from the sexualization of the children's bodies, which are seen in terms that are infantile, but that nonetheless insert them in a universe of desired objects for this man. This description can be seen as a conscious choice by Morrison, who tries to portray the desires of the pedophile, irredeemably wrong as they are, as something that apparently is difficult to be controlled. Booker's reaction is instantly to attack the man, a reaction that is clearly associated with his traumatic background, trying to protect the lives of these children as well as he can by eliminating the threat. It is not clear if the man is dead when Booker leaves him in the ground, but the narrator makes it seem like the threat is no longer present.

> Booker's fist was in the man's mouth before thinking about it. A light spray of blood dappled his sweatshirt, and when the man lost consciousness, Booker grabbed his book bag off the ground and walked away—not too fast, but fast enough to cross the road, turn his shirt inside out and make it to class on time. (109–110)

Booker's reaction might even be seen as understandable given the context of his loss, as his first instinct is resorting to violence. The description of Booker's parting with this man underlines the ways in which violence is multifaceted in this episode, since he disguises any sign of the encounter as a way of protecting himself from being perceived as a violent threat. Booker's black body is already inscribed in a narrative of violence in the United States, in which his presence may be set in two different directions: either he is the aggressive black man who assaulted a presumably innocent (white) person; or he is the victim of a racist policing institution that more often than not presumes his guilt due to his skin color.

Morrison addresses in this episode a complex issue: should the perpetrators of sexual violence (pedophiles especially) be treated with empathy? Is violence the (only) way in this potential threat be dealt with? Morrison investigates through this unnamed pedophile character the commonness and ubiquitousness of this kind of threat in the lives of children, as Booker describes the man in similar terms to those used by the community to describe Mr. Humboldt, making evident his average aspect:

> Bald. Normal-looking. Probably an otherwise nice man—they always were. The "nicest man in the world," the neighbors always said. "He wouldn't hurt a fly." Where did that cliché come from? Why not hurt a fly? Did it mean he was too tender to take the life of a disease-carrying insect but could happily ax the life of a child? (111)

His averageness contributes to the idea that the perpetrators of (sexual) violence do not belong to a specific race, or class, national origin, or any other markers of difference besides sex (96% of reported abusers are male, and 76.8% of abusers are adults[9]). By using the common saying "not hurt a fly" as an adjectifying element Morrison once again inscribes this average man in popular speech. By making Booker question such characterization she successfully denormalizes both the presence of the perpetrator and the culture that the perpetrator inhabits. The fly, which might be seen as an innocuous and innocent insect at best, is described in its most repugnant aspect, as a disease carrier, spreading filth and possibly death. The children, who are compared by Booker to the insect due to their fragility, or even simply due to their size, are the victims of this presumably sensitive man. It might be said that by questioning the cliché Booker ends up reversing the analogy, placing this man as the disease-carrying agent, spreading a different kind of sickness, one no less virulent and deadly.

Finally, Sweetness closes the narrative by giving advice on child rearing, since Bride writes her a letter to let her know she is pregnant. The mother, now estranged in a nursing home all paid for by Bride, demonstrates how much Sweetness is not an active part in Bride's life. After noting the signature of the brief letter, signed Bride and not Lula Ann, Sweetness comments on the apparent happiness of the mother and the absence of any mention of the child's father, conjecturing about his possible skin tone and the consequences of this in the life of the child:

> I reckon the thrill is about the baby, not its father, because she doesn't mention him at all. I wonder if he is as black as she is. If so, she needn't worry like I did. Things have changed a mite from when I was young. Blue blacks are all over TV, in fashion magazines, commercials, even starring in movies. (176)

---

9 Data provided by the U.S. Department of Justice in a statistical study by Snyder, H. N. (2000) "Sexual Assault of Young Children as Reported to Law Enforcement: Victim, Incident and Offender Characteristics" available at https://www.bjs.gov/content/pub/pdf/saycrle.pdf.

Sweetness acknowledges the change that has taken place since she became a mother, recognizing that the racism that plagued her generation continues to exist, though taking shape in different ways, as black people are also celebrated. It is interesting to notice that the examples used by Sweetness to demonstrate this change are all encompassed in the media, revealing in her final words that blackness, especially the extreme kind denoted by the categorization "blue blacks", has become a commodity of a different kind, something that already had been presented as a positive idea by Bride's image consultant.

Sweetness examines her relation with Bride, conceding that the distance created between them is her fault, as she tried her best to protect her child. The regret of the mother continues to be described in this last chapter, as she revisits her attitudes concerning the upbringing she was able to offer, giving the readership a different perspective on the traumatic episodes that have marked Bride's childhood. Sweetness enumerates the instances that she regrets, claiming once again that she did the best she could to bring up a child in the context in which they lived, Though she claims she is regretful, this character does not seem able to confront the negative impact that her actions had in her child's life, representing a person who is aware of her attitudes, but has not been able to break the cycle of trauma that had been imposed upon her, excusing herself without ever apologizing for it. Sweetness narrates:

> If I sound irritable, ungrateful, part of it is because underneath is regret. All the little things I didn't do or did wrong. I remember when she had her first period and how I reacted. Or the times I shouted when she stumbled or dropped something. How I screamed at her to keep her from tattling on the landlord—the dog. True. I was really upset, even repelled by her black skin when she was born and at first I thought of... No. I have to push those memories away—fast. No point. I know I did the best for her under the circumstances. When my husband ran out on us, Lula Ann was a burden. A heavy one but I bore it well. (177)

Sweetness moves to her final considerations on raising children, as now Bride must fulfill the role of the mother in her own life. Sweetness' worldview is superimposed once again in her advice, as she warns Bride that motherhood is a much heavier load than what she might expect, hinting that the criticism that she faced might as well be something Bride will have to confront. Through these comments the readership is able to perceive that Sweetness presupposes that when Bride assumes the role of the mother, she will be able to better perceive the reality that surrounds her, the inherent racism that permeates the African-American experience, and implicitly, Sweetness might as well have a chance to be forgiven and redeemed in the eyes of her child. Her final affirmation summarizes the main idea of the novel, which finally becomes its title.

> Now she's pregnant. Good move, Lula Ann. If you think mothering is all cooing, booties and diapers you're in for a big shock. Big. You and your nameless boyfriend, husband, pick-up—whoever—imagine OOOH! A baby! Kitchee kitchee koo!
>
> Listen to me. You are about to find out what it takes, how the world is, how it works and how it changes when you are a parent.
>
> Good luck and God help the child. (178)

Finally, Bride has a chance of breaking the cycle of violence, discrimination and trauma, as she is set to raise her own daughter in her own terms. Though the worlds that these two mothers inhabit are different, though no less racist and violent, Bride's self transformation and working through of her traumatic past grant her a better chance of creating life opportunities that will empower her child, rather than repeating her mother's and foremother's mistakes.

4.2 bell hooks

bell hooks was born in Hopkinsville, Kentucky, in 1952. She is a renowned author and academic pertaining to the fields of feminist studies, African-American studies, critical pedagogy, social activism, among others. The author's informal style of writing, in addition to her prolific production, positions her as one of the most influential feminist thinkers of current days. Her most celebrated titles are: *Ain't I a Woman? Black Women and Feminism* (1981), *Feminist Theory: From Margin to Center* (1984), *Talking Back: Thinking Feminist Thinking Black* (1989), *Yearning – Race, Genderm and Cultural Politics* (1990), *Teaching to Transgress: Education as the Practice of Freedom* (1994), *Outlaw Culture: Resisting Representation* (1994), *Killing Rage: Ending Racism* (1995), *Bone Black: Memories of Girlhood* (1996), *All About Love: New Visions* (2000), *Where we stand: Class Matters* (2000), *Feminism is for everybody: Passionate Politics* (2000), *Salvation: Black People and Love* (2001), *Communion: The Female Search for Love* (2001), *We Real Cool: Black Men and Masculinity* (2004), *Writing beyond race: Living Theory and Practice (2013)*, among others.

Daughter of a Southern working-class couple, hooks would experience in the early years of her life the poverty and the sexism that were engrained in the social environment, as well as the racism that was rampant, before becoming a notorious feminist author and theorist. As a child of the 1950s, hooks would also experience the outcomes of the Civil Rights Movement during her teenage years, such as the integration of previously segregated schools and the passing of the 1965 Civil Rights act. hooks' original name is Gloria Jean Watkins, a name she omitted from her public life as a writer and lecturer. The history related to her penname, adopted when publishing her first book *And there we wept – poems* (1978), is deeply connected to her family history. The author reveals the origin of the adopted name in *Talking Back: Thinking Feminist, Thinking Black*, first published in 1989:

> One of the many reasons I chose to write using the pseudonym bell hooks, a family name (mother to Sarah Oldham, grandmother to Rosa Bell Oldham, great-grandmother to me), was to construct a writer-identity that would challenge and subdue all impulses leading me away from speech into silence. I was a young girl buying bubble gum at the corner store when I first really heard the full name bell hooks. I had just "talked back" to a grown person. Even now I can recall the surprised look, the mocking tones that informed me I must be kin to bell hooks—a sharp tongued woman, a woman who spoke her mind, a woman who was not afraid to talk back. I claimed this legacy of defiance, of will, of courage, affirming my link to female ancestors who were bold and daring in their speech. Unlike my bold and daring mother and grandmother, who were not supportive of talking back, even though they were assertive and powerful in their speech, bell hooks as I discovered, claimed, and invented her was my ally, my support. (28)

By reclaiming the name of this figure, hooks also reclaims the matrilineal resistance that she knew was present in her family, even though the family she knew seemed to be subservient to patriarchal hierarchies of subordination, which tried to stifle the outspoken streak of her personality, deeming that it was not a child's place to breach this hierarchy. hooks finds in this great-grandmother figure the identification necessary to take ownership of the word, spoken and written, attesting to her ability of talking back, a feature that was also explored by the protagonists of two other narratives that were examined in this study: Kincaid's Lucy and Morrison's Bride. The usage of lowercase is also an intentional move, to practically differentiate herself from the original person, and also to focus the significance of her production in her work, not her person. The author comments: "[t]he point of the pseudonym was not to mask, to hide my identity but rather to shift the focus, to make it less relevant" (*Talking Back* 276).

hooks comments on the need to write her coming-of-age story, describing it as an exercise on overcoming trauma, as a means of dealing with all the negative feelings that were associated with her childhood and adolescence, textually killing Gloria Jean in a cathartic way:

> To me, telling the story of my growing up years was intimately connected with the longing to kill the self I was without really having to die. I wanted to kill that self in writing. Once that self was gone—out of my life forever—I could more easily become the me of me. It was clearly the Gloria Jean of my tormented and anguished childhood that I wanted to be rid of, the girl who was always wrong, always punished, always subjected to some humiliation or other, always crying, the girl who was to end up in a mental institution because she could not be anything but crazy, or so they told her. [...] By writing the autobiography, it was not

just this Gloria I would be rid of, but the past that had a hold on me, that kept me from the present. I wanted not to forget the past but to break its hold. This death in writing was to be liberatory. (*Talking Back* 261)

Bone Black – Memories of Girlhood, published in 1996, may be seen as a complex example of autobiography[10], a coming-of-age story that is post-modern in its form and content, and that subverts many of the expectations that are assigned to the *Bildungsroman* genre. Sixty-one short vignettes comprise the text, which are not clearly bound by linear progressive time in their development, though the reader might grasp a few clues that help create a sense of direction and development, such as the few mentions of age, and the historical facts that are used as the background for some vignettes. There are shifts in the narration, moving from the first person to the third, demonstrating flexibility in the narrativization of her life story, which is told mostly from an outside perspective. There is also an intermingling of a realist discourse with dreams and projections, demonstrating another layer of the hybridity in this narrative. hooks comments on this specific aspect of her writing in the preface to the autobiography, stating:

> An unconventional memoir, it draws together the experiences, dreams, and fantasies that most preoccupied me as a girl. [...] This is autobiography as truth and myth – as poetic witness. [...] In *Bone Black* I gather together the dreams, fantasies, experiences that preoccupied me as a girl, that stay with me and appear and reappear in different shapes and forms in all my work. Without telling everything that happened, they document all that remains more vividly. They are the foundation on which I have built a life in writing, a life committed to intellectual pursuits. (*Bone Black* XIV)

Most importantly, the author highlights that: "[t]he events described are always less significant than the impressions they leave on the mind and the heart" (*Bone Black* XV). hooks also comments on the limits of autobiography, and the process of writing the vignettes that compose the text in *Talking Back*, stating:

> Each day I sat at the typewriter and different memories were written about in short vignettes. They came in a rush, as though they were a sudden thunderstorm. They came in a surreal, dreamlike style which made me cease to think of them as strictly autobiographical because it seemed that myth, dream, and reality had merged. There were many incidents that I would talk about with my siblings to see if they recalled them. Often we remembered together a general outline of an incident but the details were different for us. This fact was a constant reminder of the limitations of autobiography, of the extent to which auto-

---

10 The terminology memoir/autobiography is used interchangeably by hooks when commenting on *Bone Black* in other works, demonstrating flexibility in adhering to the conventions of each genre.

biography is a very personal story telling a unique recounting of events not so much as they have happened but as we remember and invent them. (*Talking Back* 264–265)

Regarding the canonical *Bildungsroman*, the open endedness of hooks' text is another feature that deviates from the conventional expectation, as the main character does achieve the realization of her formative years in the process of her identity construction, but by no means it may be considered that she becomes part of a larger whole in society. The question of belonging remains open at the end of the narrative, as the protagonist is able to find her self in the world of books, poetry and writing, but does not seem to find the same sense of belonging in the intimate universe of her family life, persisting as a stranger in the nest. The recognition of a position of autonomy is achieved in a sense of self-sufficient alienation from her family, as she finds her rightful calling in the world of writing and books, something she was continuously advised against by her parents and siblings, and therefore distances the protagonist further from her family.

Bone Black is a tale of self-discovery and self-making that is painfully punctuated by the limitations imposed by sex, race, and class on the experience of a black girl during the 1950s and 1960s. Most notably, the direct family members that figure in the narrative are the parents, grandparents, five sisters and one brother that comprise the domestic nucleus of the author's childhood. In addition to that, hooks is part of a large family, in which many characters who people her stories are part of an extended family, all contributing in different ways to the self-fashioning of her identity. Interestingly, hooks gives more relevance to some relatives such as her grandparents and parents, and her brother, contrastingly lumping the rest of the family into the pronoun "they", which seems to encompass all the vigilant eyes and ears that would promptly denounce every transgression that she made. This "they" most often comprises her sisters, who do not figure much in a direct form in the narrative, creating a silence regarding their relation with the protagonist, pointing to a distancing from them in the narrative[11]. Susana Vega-González describes hooks' autobiography in the following terms in "The Dialectics of Belonging in bell hooks' *Bone Black: Memories of Girl-*

---

[11] "Reading the completed manuscript, I felt as though I had an overview not so much of my childhood but of those experiences that were deeply imprinted in my consciousness. Significantly, that which was absent, left out, not included also was important. I was shocked to find at the end of my narrative that there were few incidents I recalled that involved my five sisters. Most of the incidents with siblings were with me and my brother. There was a sense of alienation from my sisters present in childhood, a sense of estrangement. This was reflected in the narrative" (*Talking Back* 267).

hood": "hooks' childhood is a story of loneliness and misunderstanding, fear and incomprehension, but above all it is a story about a rebellious spirit coupled with an eternal yearning for belonging" (237).

hooks turns to her tender years in *Bone Black*, taking advantage of this period of her life, stressing "[...] the significance of girlhood as a time when females feel free and powerful" (XII). This freedom and power, as the author explains, seems to come from the similarity between the sexes at a young age, when the bodies of females and males are not completely distinguishable, and the energy to live and explore that they have is also similar. Thus this period seems to be a privileged moment to investigate matters of identity construction, as well as matters related to race and class. hooks' account of her childhood is a narrative that explores the matters of talking back and silencing, as the protagonist strives to develop her own sense of identity. It traces the path from (being) silence(d) to speech, as the author slowly discovers the ways in which the word, and more specifically writing, is a core element of her identity. In the words of the author:

> Writing imagistically, I seek to conjure a rich magical world of southern black culture that was sometimes paradisiacal and at other times terrifying. While the narrative of family life I share can be easily labeled dysfunctional, significantly that fact will never alter the magic and mystery that was present – all that was deeply life sustaining and life affirming. The beauty lies in the way it all comes together exposing and revealing the inner life of a girl inventing herself – creating the foundation of selfhood and identity that will ultimately lead to the fulfillment of her true destiny – becoming a writer. (*Bone Black* XI)

The use of the word "dysfunctional" is revisited by the author years later in *All About Love – New Visions*, in which the author admits that using this word to describe her family might erase much of what was good during her upbringing, something that she does not intend to do. She states:

> Raised in a family in which aggressive shaming and verbal humiliation coexisted with lots of affection and care, I had difficulty embracing the term "dysfunctional." Since I felt and still feel attached to my parents and siblings, proud of all the positive dimensions of our family life, I did not want to describe us by using a term that implied our life together had been all negative or bad. I did not want my parents to think I was disparaging them; I was appreciative of all the good things that they had given the family. With therapeutic help I was able to see the term "dysfunctional" as a useful description and not as an absolute negative judgment. My family of origin provided, throughout my childhood, a dysfunctional setting and it remains one. This does not mean that it is not also a setting in which affection, delight, and care are present. (*All About Love* 6–7)

This positivity is also seen throughout *Bone Black,* as the author explores episodes in which her mother, her siblings and her grandparents act as constructive and loving influences in her life. Later on, as a full writer, hooks would comment on the difficulties of her childhood, and the subsequent learned lessons, developing a sense of self-care and protection that would finally be developed in her theories of equality. Commenting on the frequent punishments that she would receive, the author stresses the capacity of resistance that are a direct consequence of her reality:

> Certainly, when I reflect on the trials of my growing-up years, the many punishments, I can see now that in resistance I learned to be vigilant in the nourishment of my spirit, to be tough, to courageously protect that spirit from forces that would break it. While punishing me, my parents often spoke about the necessity of breaking my spirit. Now when I ponder the silences, the voices that are not heard, the voices of those wounded and/or oppressed individuals who do not speak or write, I contemplate the acts of persecution, torture—the terrorism that breaks spirits, that makes creativity impossible. I write these words to bear witness to the primacy of resistance struggle in any situation of domination (even within family life); to the strength and power that emerges from sustained resistance and the profound conviction that these forces can be healing, can protect us from dehumanization and despair. (*Talking Back* 26)

4.2.1 Silences

Not conforming to a patriarchal mindset of identity construction, hooks explores the different ways that she could be herself despite the violent and cruel reminders that her family unremittingly would give her every time she strayed from the path of the "normal" and the "expected". The author explores the ways in which silences regarding the black experience are pervasive, exposing publicly the private life of her childhood, demonstrating that her backchat was perceived clearly as a form of defiance to the patriarchal heteronormative order. Commenting on the division between private and public, the author writes:

> In reflection, I see how deeply connected that split [between private and public] is to ongoing practices of domination (especially thinking about intimate relationships ways racism, sexism, and class exploitation work in our daily lives, in those private spaces—that it is there that we are often most wounded, hurt, dehumanized; there that ourselves are most taken away, terrorized, and broken). The public reality and institutional structures of domination make the private space for oppression and exploitation concrete—real. (*Talking Back* 18)

hooks' memoir deals intimately with the matters that create dehumanization in the private space, questioning the structures of racism, sexism, and classism from within the African-American experience, demonstrating the ways in which these processes of discrimination and the creation of hierarchies happen in the microcosms of domestic life. The infancy that is described in this narrative show-cases the ways in which hooks' personality was deeply affected by these process-es, as the protagonist questions the decisions that are imposed upon her exis-tence, defying the authority of those who try to format her experience to fit the expected gender/racial/class roles upheld by the *status quo* of her generation. Ex-posing this reality in her narrative also raises questions that pertain to the studies of autobiography and memoirs, as the author claims that to be so open about her personal life was something that, although present in her previous work, did not figure so explicitly in her academic writing. Commenting on the difficulty of ap-proaching the matters of privacy and secrecy that are found in the way she was raised, hooks states:

> Secrecy and silence—these were central issues. Secrecy about family, about what went on in the domestic household was a bond between us—was part of what made us family. There was a dread one felt about breaking that bond. And yet I could not grow inside the atmos-phere of secrecy that had pervaded our lives and the lives of other families about us. (*Talk-ing Back* 262)

By approaching these matters in her memoir, hooks is making the realm of the private public for scrutiny, revealing not only the reality of her life from her perspective, but also the realities of all those who were also part of her com-ing of age. The reticence to write these stories reiterates the vows of silence that were imposed on her by her upbringing, revealing not only the silences that were created inside her family, but also that of the African-American community in general:

> The willingness to be open about personal stuff that has always been there for me in talk-ing has only recently worked its way fully into my writing. It has taken longer for me to be publicly private in writing because there was lurking in me the fear of punishment—the fear of saying something about loved ones that they would feel should not be said. The fear that the punishment will be loss, that I will be cut off from meaningful contacts. This is truly, on a deep level, a real race and class issue 'cause so many black folks have been raised to be-lieve that there is just so much that you should not talk about, not in private and not in public. So many poor and working-class people of all races have had the same stuff pushed down deep in them. (*Talking Back* 19)

The race and class issues that have been mentioned by hooks are dealt in depth in her autobiography, as she openly approaches questions of poverty, violence, and discrimination that take place inside the African-American community, turn-ing issues that have been considered to be private finally public. The resistance to expose such matters is also investigated in *Sisters of the Yam: Black Women and Self-Recovery* in which hooks explores issues related to the willingness of the African-American community to confront the difficult reality of their existence, as well as their role as the perpetrators of violence, oppression and discrimination inside the community, using the example of literature as an illustrative representation of this problematic:

> In Alice Walker's novel *The Color Purple*, Celie, the black heroine, only begins to recover from her traumatic experiences of incest/rape, domestic violence, and marital rape when she is able to tell her story, to be open and honest. Reading fictional narratives where black female characters break through silences to speak the truth of their lives, to give testimony, has helped individual black women take the risk to openly share painful experiences. We see examples of such courageous testimony in *The Black Women's Health Book*. Yet many black readers of Alice Walker's fiction were angered by Celie's story. They sought to "punish" Walker by denouncing the work, suggesting it represented a betrayal of blackness. If this is the way folks respond to fiction, we can imagine then how much harder it is for black women to actually speak honestly in daily life about their real traumatic experiences. And yet there is no healing in silence. Collective black healing can take place only when we face reality. (35)

In addition to that, hooks reports that one of her favorite novels is Toni Morrison's *The Bluest Eye*. Morrison's exemplary novel focuses on those who were always at the margin of representation, namely a little black ugly girl. hooks would be forever changed after reading the novel, as she deeply identified with the narrative. hooks states: "[...] she [Morrison] gave us girls confronting issues of class, race, identity, girls who were struggling to confront and cope with pain. And most of all she gave us black girls who were critical thinkers, theorizing their lives, telling the story, and by so doing making themselves subjects of history" (*Bone Black* XII).

By exploring questions of poverty, violence, and discrimination through her autobiography, it may be said that hooks is once again talking back to her community, challenging the imposed silences of her childhood, in addition to the ones she has carried to her adult life. The author ponders on the ability to talk back as a form of resistance that may not be confined to the personal and the particular, but to all the oppressed, that engage in resistance through backchatting the oppressive powers that try to maintain the power hierarchies to their benefit:

> In retrospect, "talking back" became for me a rite of initiation, testing my courage, strengthening my commitment, preparing me for the days ahead—the days when writing, rejection notices, periods of silence, publication, ongoing development seem impossible but necessary. Moving from silence into speech is for the oppressed, the colonized, the exploited, and those who stand and struggle side by side a gesture of defiance that heals, that makes new life and new growth possible. It is that act of speech, of "talking back," that is

no mere gesture of empty words, that is the expression of our movement from object to sub-ject—the liberated voice. (*Talking Back* 29)

Breaking these silences is thus not only a matter of healing and overcoming, but also a political act that intends to foster a better future for all those who partic-ipate in the making and the maintaining of these silences. The author claims that daring to speak about much of what has been silenced within/by the African-American community has become a political struggle, as she deems necessary to publicly share this information if the intent of this community is healing, re-covering and realizing themselves (19). hooks' own writing attests to the unmak-ing of silences, private and public, stressing the importance of voicing the con-cerns of the self, as well as the concerns of the collective, demonstrating how the ownership of the word, spoken and written, may be used as both a form of pro-tection, as a way of denouncing the creation of circumstances and structures of dehumanization, becoming eventually a form of resistance (27).

For hooks, to break these silences, to produce speech that challenges these narratives, is a form of unmaking the colonial oppression that has shaped much of the existence of racialized groups, as well as discrimination based on gender and on class, since the discourses that comprise the narratives of these domina-tion processes are to blame in the silencing of oppressed subjects. To overcome the silences by creating discourses departing from the experience of the op-pressed is a way of healing the wounds that have been created by these systems of domination and dehumanization, functioning as a form of reparation that symbolically restores the humanity of these subjects.

hooks' inclination to challenge narratives and authorities was something present since her tender years. During hooks' infancy simple acts of expression as a child would be interpreted as a form of defiance for those in her communi-ty. This characteristic in her personality, something she shared with her great-grandmother, is something that was often suppressed, as it was not expected of a child, to participate in conversation with adults, unless they were spoken to first. By challenging the hierarchies that were already in place all around her, the author often was the receiver of violence, employed here as a corrective force:

In the world of the southern black community I grew up in, "back talk" and "talking back" meant speaking as an equal to an authority figure. It meant daring to disagree and some-times it just meant having an opinion. In the "old school," children were meant to be seen and not heard. My great-grandparents, grandparents, and parents were all from the old school. To make yourself heard if you were a child was to invite punishment, the back-hand lick, the slap across the face that would catch you unaware, or the feel of switches stinging your arms and legs. (*Talking Back* 21) hooks makes explicit that a traditional upbringing in the Southern part of the United States was typically harsh and violent, not recognizing children as full subjects who were expected to participate in the social life of the family the same way adults were. Talking back elicited responses that were violent, and that served the purpose of re-ascertaining the place that the child occupied in the social hierarchy, both in the familial sphere and in the public world. Sex was a determining factor that should be accounted for in this process of socialization and reinforcing of hierarchies, as the word was granted more often to boys than to girls in this Southern reality. The author comments on the discrepancies of treatment that existed between the two sexes. Therefore, the will (and the freedom), to speak was something that would have positive consequences in the lives of boys and not in the lives of girls:

> Needless to say, the punishments for these acts of speech seemed endless. They were intended to silence me—the child—and more particularly the girl child. Had I been a boy, they might have encouraged me to speak believing that I might someday be called to preach. There was no "calling" for talking girls, no legitimized rewarded speech. The punishments I received for "talking back" were intended to suppress all possibility that I would create my own speech. (23)

The disparity between the education that was given to the different sexes would be a force that ultimately would shape these subjects into the expected social roles of society later on. Boys that defied the silence rule would be perceived in a positive light, confirming once again the double standard that would certainly favor males in a patriarchal heteronormative society. Religion was also a force that would contribute to this disparity, as the teachings of a Christian belief would profess the subordination of the female to the male, and the reference of boys who broke the silent children rule as a possible preacher, one who was worthy of spreading the word, would confirm men's "monopoly" of speech. There is, however, a contrasting idea related to the education of black girls in the South, one that stipulated that, though girls were not expected to participle in conversation with adults unsolicitedly, they should know how to speak once they were spoken to, as hooks explains: " [...] it was and is expected of girls to be articulate, to hold ourselves with dignity. Our parents and teachers were always urging us to stand up and speak clearly. These were traits that were meant to uplift the race" (*Bone Black* XIII). hooks also compares this education to the one that was given to the white counterparts, in which silence in all occasions was encouraged. Adult black women, however, would enjoy the right to speech in the realm of the home, and mostly in the company of their peers. hooks describes the world of female speech as something that deeply urged her to participate:

> Black men may have excelled in the art of poetic preaching in the male-dominated church, but in the church of the home, where the everyday rules of how to live and how to act were established, it was black women who preached. There, black women spoke in a language so rich, so poetic, that it felt to me like being shut off from life, smothered to death if one were not allowed to participate. It was in that world of woman talk (the men were often silent, often absent) that was born in me the craving to speak, to have a voice, and not just any voice but one that could be identified as belonging to me. To make my voice, I had to speak, to hear myself talk—and talk I did—darting in and out of grown questions that were folks' conversations and dialogues, not directed at me, endlessly asking answering questions, making speeches. (*Talking Back* 22–23)

hooks asserted the ownership of her voice through a constant exercise of resistance against everything that attempted to place her in the silent reality of womanhood under patriarchal heteronormative discourses. By participating in the conversations of adults, even if unsolicitedly, and often being ignored or punished by doing so, hooks practiced this exercise of resistance at every chance she got, fostering her own sense of identity. hooks also comments on the racial differences found in the speaking environment between black and white homes, stating that though silence was imposed upon females everywhere, describing the experience of black women as completely relegated to a position of passivity and silence would not be precisely true, as the author states:

> Within feminist circles, silence is often seen as the sexist "right speech of womanhood"—the sign of woman's submission to patriarchal authority. This emphasis on woman's silence may be an accurate remembering of what has taken place in the households of women from WASP backgrounds in the United States, but in black communities (and diverse ethnic communities), women have not been silent. Their voices can be heard. Certainly for black women, our struggle has not been to emerge from silence into speech but to change the nature and direction of our speech, to make a speech that compels listeners, one that is heard. (*Bone Black* 23)

hooks' description could be compared to the struggle explored by Spivak in "Can the Subaltern Speak?", concluding that what was necessary in this context was not for these subaltern subjects to acquire the language that would make possible the challenging of patriarchal domination, but the creation of the conditions for this speech to be heard. hooks describes the speech of black women as something that was always present but that was not minded, specifically by male peers, something more like a background music than like a sound:

> Our speech, "the right speech of womanhood," was often the soliloquy, the talking into thin air, the talking to ears that do not hear you—the talk that is simply not listened to. Unlike the black male preacher whose speech was to be heard, who was to be listened to, whose words were to be remembered, the voices of black women—giving orders, making

threats, fussing—could be tuned out, could become a kind of background music, audible but not acknowledged as significant speech. Dialogue—the sharing of speech and recognition—took place not between mother and child or mother and male authority figure but among black women. (*Bone Black* 23–24)

The hierarchy represented here demonstrates that the meaningful exchange of speech was something that really existed among black females, as their soliloquy would not be ignored among peers. It is also interesting to point out that the verbs used by hooks to exemplify the speech produced by black women in their households ("giving orders, making threats, fussing") are all somewhat related to the world of discipline, as actions connected to controlling the life of a child, restricting once again the soliloquy to the domestic and to the sanctioned role of authority a female was allowed to occupy: motherhood. To the young author, writing became the way she could more effectively deal with language in her own terms, as she reveals in retrospect:

Writing was a way to capture speech, to hold onto it, keep it close. And so I wrote down bits and pieces of conversations, confessing in cheap diaries that soon fell apart from too much handling, expressing the intensity of my sorrow, the anguish of speech—for I was always saying the wrong thing, asking the wrong questions. I could not confine my speech to the necessary corners and concerns of life. I hid these writings under my bed, in pillow stuffings, among faded underwear. When my sisters found and read them, they ridiculed and mocked me—poking fun. I felt violated, ashamed, as if the secret parts of my self had been exposed, brought into the open, and hung like newly clean laundry, out in the air for everyone to see. (*Bone Black* 24)

Shame is seen here as the violation of her privacy, since the mockery from her sisters may be understood as the rebuttal of her identity construction process. Her first attempts of writing functioned as a means to play with language in an apparent safe space, one that was breeched and attacked by her sisters. The symbolic stance of the places hooks used to keep her secret writing is also telling, as it aligns the writing exercise with objects that either are supposed to provide comfort, or with markers of the personal and intimate.

Closing the first vignette, hooks explores a dream, one which starts with the protagonist going away, only to return and find that her house has burned down. Her loud cries make her family appear with candles, coming for her comfort, and together the family rummages through the remains for fragments of her life that might have survived the fire. The only not completely burned piece is the hope chest, and the family weeps together as they find bits and pieces of their past. They are interrupted by a familiar voice:

> Louder than our weeping is a voice commanding us to stop our tears. We cannot see who is speaking but we are reminded of the stern sound of our mother's mother's voice. We listen. She tells us to sit close in the night, to make a circle of our bodies, to place the candles at the center of the circle. The candles burn like another fire only this time she says the fire burns to warm our hearts. She says Listen, let me tell you a story. She begins to put together in words all that has been destroyed in the fire. We are all rejoicing when the dream ends. (*Bone Black* 3)

The inclusion of this dream in the opening chapter of her autobiography attests to the importance of this kind of discourse in the work as a whole, and more importantly, it showcases the relevance of this specific family member in hooks' life story. hooks identifies the voice as belonging to Sarah, whom hooks dearly calls Saru. This storyteller is the one who is capable of creating meaning even when all seems to be lost, all through the power of words, teaching the protagonist the importance of protecting memory. This moment of communion, which takes place only in dreams, also reveals the desire of the protagonist of fully being embraced by her family. The next day, hooks asks for the help of Saru to understand her dream:

> Saru, mama's mother, is the interpreter of dreams. She tells me that I should know the storyteller, that I and she are one, that they are my sisters, family. She says that a part of me is making the story, making the words, making a new fire, that it is my heart burning in the center of the flames. (*Bone Black* 3)

Saru's remarks display much of what hooks autobiography is comprised of, stating that she is also a maker of words, and thus is capable of directing her own life story. Her heart symbolically stands for her yearnings in the fire, which are to become an intellectual who lives in the world of books and language, and the making of her narrative, the telling of her own story, is the action that will finally save her through the darker days of her childhood.

Silence takes many forms in the narrative, not only relating to the reality of having her point of view being curbed in a heteronormative patriarchal culture, but also in the way certain topics are not discussed with children. One significant form of silence during her childhood regards race. As the protagonist was raised partly in the countryside in the segregated South white people were not characters that peopled much of her reality, but their presence was surely felt. The narrator asserts that "[w]hite folks mean little to them. They pay them no mind. It is black people of all colors who are at the center of their world" (32). Most of her confusion regarding white people derives from the lack of communication regarding this subject in her house, as the protagonist senses that they are associated with negative feelings, but is never directly told why:

SHE HAS LEARNED to fear white folks without understanding what it is that she fears. There is always an edge of bitterness, sometimes hatred, in the grown-ups' voice when they speak about them but never any explanation. When she learns of slavery in school or hears the laughter in geography when they see pictures of naked Africans – the word *savage* underneath the pictures – she does not connect it to herself, her family. She and the other children want to understand Race but no one explains it. They learn without understanding that the world is more home for white folks than it is for anyone else, that black people who most resemble white folks will live better in that world. (*Bone Black* 31)

The disconnection between the reality of slavery and their own lives, the Africans depicted in the geography book and their characterization as savages are symptomatic of the lack of communication that pervades her childhood, and even her adolescence. As race is not explained but only lived the protagonist derives her conclusions from the appreciation of the reality around her, trying to figure out her own place in the world. The cruel understanding that this world belongs more to white people than to anyone else condensates in simple child-like language the consequences of colonialism, and more importantly, the dynamics that rule their existence. The lack of understanding of how this situation was constructed, and how much it is related to their own poverty and disenfranchisement, their only partial citizenship and the violence and fear that inherently shape their lives demonstrates the difficulties in the protagonist's experience and development. The white people that they do know are mostly the poor folk who live near their house in the country, a description that shows how much poverty is a social reality that also disenfranchises subjects, but that does not unite these different groups in solidarity:

Where they live, on hill land, they have white neighbors. They are not the good white folks, the ones that look at you with sugary smiles. They are the peckerwoods, the crackers, the ones who look at black folk with contempt and hate. Children are warned to stay away from them, to pretend not to hear when they talk, to stay away from their yards, and not to enter their houses, not for anything. All such warnings make them curious, make them desire to know more. They go to the poor white folk's houses every chance they get. They see that the food they eat looks better than the food they see over there. They see that all these white folk have hardly any furniture. (*Bone Black* 33)

The racial slurs used to describe the white neighbors make evident the animosity between blacks and whites, demonstrating that the mistrust mas mutual. Poor white people, even having less conditions of living, as the narrator describes, still maintained a color privilege over the blacks, something that granted them some power over black subjects in the perspective of a white supremacist society. Their poverty, however, forced them to share the space with subjects that are clearly considered inferior. The children, not being told what to fear and even the reason for that, feel curious about these subjects, exploring the boundaries of this separation, trying to understand the divide in their own terms.

Colorist discrimination as a reality is also present in the protagonist's reality, as she is already aware that the closer to looking white, the more chances of a better life black subjects would have access to. This understanding is directly connected to a practical example in her life, as the protagonist describes a grandmother that passes for white, also recounting the prejudices that this grandmother displays toward darker-skinned relatives, and even the children:

> They have a grandmother who looks white who lives in a street where all the other people are white. She tells things like Black nigger is a no-good nigger, that Papa looked like a white man but was a nigger. She never explains to them why she is married with a man whose skin is the color of soot and other wonderful black things, things they love – shoe polish, coal, women in black, slips. [...] They cannot wait to get away from this grandmother's house when she calls one of them blackie in a hateful voice, in a voice that seems to say I cannot stand the sight of you. They want to protect each other from forms of humiliation but cannot. They stand cringing and weeping inside saying nothing. They do not want to be whipped with the black leather strap with holes in the hanging on the wall. They know their place. They are children. They are black. They are next to nothing. (*Bone Black* 31–32)

The privileges granted by passability are seen here, as this grandmother seemingly is able to live beyond the color line, inhabiting white spaces. The colorist discrimination is ironically denounced as the narrator reveals that despite her clear dislike for dark-skinned African-Americans, she is married to an undeniably black man. The maxim "Black nigger is a no-good nigger" summarizes her attitude in relation to the creation of hierarchies, as she appropriates the derogatory language of dehumanization, and continues to exert a racist classification of subjects based in their skin color.

The threat of violence resulting from any possible misbehaving is a theme that is pervasively present during the protagonist's childhood, a violence that would be perpetrated by the father figure and that was meant as a form of education, disciplining the body into compliance. The conclusion that as black children they amount to very little describes accurately their precarious position in the hierarchies of the family and of society at large, and as their subjectivity is not recognized, the creation of silences, the lack of explaining and understanding, and finally of support in their process of identity construction will create the conditions for a difficult and painful coming-of-age path.

4.2.2 Heirlooms, toys and books – education through Objects

The opening episode in *Bone Black* sets the tone of the narrative in an environment that investigates womanhood from the first page. Without giving much detail about the age of the protagonist, hooks reminisces about a private moment of sharing between mother and daughters. Just like in Kincaid's *Annie John*, a special memory is constructed as the mother opens a chest full of objects that tell stories about themselves and generations past. Here, the container is called a "hope chest", a wooden box into which girls deposit things they would take with them once they were married. As the mother carefully examines different items, hooks is given a quilt, a heirloom from her grandmother, a person with whom she develops an important relationship during her childhood. She is also given a purse, from a great-aunt who did not specially care for her, generating conflict among the sisters:

> MAMA HAS GIVEN me a quilt from her hope chest. It is one her mother's mother made. It is a quilt of stars – each piece taken from faded-cotton summer dresses – each piece stitched by hand. She has given me a beaded purse that belonged to my father's mother Sister Ray. They want to know why she has given it to me since I was not Sister Ray's favorite. They say that she is probably turning over in her grave angry that I have something of hers. (*Bone Black* 1)

Hooks' autobiography may be compared with the quilt she has just received, as it is comprised of small pieces of memory that, though faded like the former summer dresses, are able to make up a whole new object of remembrance. Each piece stitched together will tell a complex story, just like the sixty-one vignettes that form the narrative. The protests from her sisters are also telling, since they serve the purpose of exemplifying the constant dichotomy of this family: she and them.

The tradition of filling the chest with things that they would carry into their adulthood is also seen from the perspective of the mother, who is able to retrieve something from her girlhood at this moment. These memories of girlhood stand for much this woman has given up when becoming a wife, and not much later a mother. "I see her remembering, clutching tightly in her hand some object, some bit of herself that she has had to part with in order to live in the present. I see her examining each hope to see if it has been fulfilled, if the promises have been kept" (*Bone Black* 2). The mother, with tears in her eyes, reexamines the circumstances and consequences of becoming a woman, leaving the powerful state of girlhood behind. Marriage stands as the compulsory rite of passage in which girls renounce their own plans, and assume their position in the patriarchal order. hooks, however, is aware that she does not want to participate in this logic, which places her in a limbo in regards to her development as a woman in this normative perspective. Thus, the narrator comes to a conclusion:

> I am clutching the gifts she hands to me, the quilt, the beaded purse. She knows that I am often hopeless. She stores no treasure for my coming marriage. I do not want to be given away. I cannot contain my dreams until tomorrow. I cannot wait for someone else, a stranger, to take my hand. (*Bone Black* 2)

In this episode, hooks is claiming her place as the outsider in a heteronormative culture, in which the destiny of women was already determined by a patriarchal order. hooks relies on herself to achieve her future goals, not conceiving the possibility of a stranger to direct her future. Her aspirations and dreams are her own and are already present, not depending on the existence of someone else to be achieved, straying farther away from the prescribed norm of womanhood. Though her mother recognizes that this particular daughter apparently will not follow the lead to standard womanhood, her gifts are a sign of positive nurturing. Later hooks would investigate the role of brides in a school play, indicating her unease with this particular female fate, stating "[w]e are practicing to be brides, to be girls who will grow up to be given away. My legs would rather be running, itching to go outdoors. My legs are dreaming. Adventurous legs. They cannot walk down the isle without protest" (9). Her legs stand here for her will to escape the expected and to enjoy life like boys would usually do in their tender years. Her legs as symbols of her restlessness are also mentioned in regards to a school play, in which hooks wears a red crape paper dress that is torn during rehearsal, since she cannot figure the perfect pace for her legs to walk down the isle, providing the author with a comparison about the female condition: "[t]he tear must be mended. The red dress like a woman's heart must break silently and in secret" (9).

One of the clear markers of difficulty in hooks infancy was poverty. Her first years were spent in the countryside of Kentucky, and the segregated kind of poverty experienced by the protagonist is described in many examples of everyday life, showcasing a life that is impacted by the structural poverty found in underprivileged low-income households. All these experiences are framed through the eyes of young children, as hooks states:

> WE LIVE IN the country. We children do not understand that that means we are among the poor. We do not understand that the outhouses behind many of the houses are still there because running water came here long after they had it in the city. We do not understand that our playmates who are eating laundry starch do so not because the white powder tastes so good but because they are sometimes without necessary food. We do not understand that we wash with heavy, unsmelling, oddly shaped pieces of homemade lye soap

because real soap costs money. We never think about where lye soap comes from. We only know we want to make our skin itch less – that we do not want our mouths to be washed out with it. (4)

The different aspects of poverty enumerated in this passage illustrate the kind of environment that the protagonist experienced during her first years, before moving to the city. Though hunger is never explicitly addressed in the narrative as something hooks personally experienced, here it is seen through the experience of her playmates, stressing that it was a reality for many around her. The mention of the lye soap also illustrates the extent to which expenses were justified in a strictly practical sense, inferring that regular perfumed soap, real soap as the protagonist calls it, was a kind of spending that was not allowed. As a child, hooks does not understand the role of money and its influence in their lives, something that is never explained to children. However, what is clearly understood is punishment, as exemplified in the case of lye soap, hinting that communication inside the family, and more specifically between adults and children, is more effective through acts of discipline than anything else.

School life is also telling of the poverty that characterized hooks' infancy, which is seen from the perspective of a segregated busing system, to the actual classroom and its activities. The different schools that white and black children are allowed to attend also demonstrate the hardships associated with this separation of the races, which forced many black children to leave their homes hours before the classes would start, due to the distance between their homes and the education facilities:

Because we are poor, because we live in the country, we go to the country school – the little white wood-frame building where all the country kids come. They come from miles away. They come so far because they are black. As they are riding the school buses they pass school after school where children who are white can attend without being bused, without getting up in the wee hours of the morning, sometimes leaving home in the dark. (*Bone Black* 4–5)

hooks and her siblings, however, are not bused to school, since their home is within walking distance. Their father tells them that he too had to walk to school, even in snow, without boots or gloves. "We are not comforted by the image of the small boy trudging along many miles to school so he can learn to read and be somebody" (5) hooks claims, stressing the historical forms of inequality that the past generation had gone through, and the different ways in which this inequality continues to be perpetuated. This interference of the father also serves the purpose of illustrating the frustration he felt, giving the readership some background about this character, as the protagonist claims: "[w]hen we close our eyes he becomes real to us. He looks very sad. Sometimes he cries. We are not at all comforted. And there are still days when we complain about the walk, especially when it is wet and stormy" (5).

hooks also reports that there are "tasting parties", a moment in which the teacher would bring different kinds of food for the children to taste, since they would not have access to them in their homes.

> Mama tells us that most of that food we taste isn't good to eat all the time, that it is a waste of money. We do not understand money. We do not know that we are all poor. We cannot visit many of the friends we make because they live miles and miles away. We have each other after school. (6)

The tasting parties function in a bittersweet mode, as they show these children a universe they might not have access to, only to be confronted with the reality of poverty back in their homes. Their economic condition as a fact that might be explained is still not clear to these children, as evidently stated by the protagonist, but the hard conditions of living that are experienced in their homes and in their friends' homes are surely perceived. The impossibility to visit their colleagues is also relevant, since it clarifies the reason why hooks is so involved in her familiar universe, living a kind of isolated life.

The impoverishment determined by race emerges in the narrative as a determining factor during her childhood, as many school activities are devised to raise money so the school may carry on functioning. hooks confesses that often the family is shamed into contributing, and that often a color line appears as a clear demarcation of wealth in their imaginary, as subjects that are related to money seem always to involve white people. This imaginary is portrayed by the protagonist through objects and activities that clearly pertain to a child's universe, such as the characters in storybooks:

> The people with lots of money can buy many tickets – can show that they are "big time." Their flesh is often the color of pigs in the storybook. Somehow they have more money because they are lighter, because their flesh turns pink and pinker, because they die their hair blond, red, to emphasize the light, lightness of their skin. We children think of them as white. We are so confused by this thing called race. (7)

From the description supplied by the author, it seems that these children only indirectly know of the existence of these other colored people, through pictures in books, implying that the segregation they experience is possibly absolute at this age. Her first contact with a "white" man comes when she wanders far from her house, and is found crying with scraped knees. hooks describes her house in this episode, detailing that it is made out of concrete and stone and that it was left behind by white men who tried to explore oil in those lands without success. The narrator also indicates that all that land is owned by a black man, who also lives in a house without other houses around it. hooks tells that she does not mind the coldness and dampness of her house, as she likes to spend her time outside connecting with nature. At this point, hooks introduces a fable-like discourse, in which she befriends a snake, telling it all the troubles she cannot tell adults. When she is rescued by a man, who is described as having pink skin like the pig's skin, with black curly hair. He introduces himself as the son of the landowner, telling the little girl that he would be back to marry her when she was older, something that seems to please the family. hooks confesses to the man that she would not be able to do so, since she promised the snake she would be back to spend all afternoons with it. Finally, this episode veiledly informs the reader about the direct relation between social mobility and whiteness. If the reader is to trust the first description that the land is black-owned, and that this pink man is the son of this landowner, one can imply that the accumulation of wealth (owning of the land, accruing capital over generations) is connected to a pattern of whitening.

The approval of the family is also relevant, as they see the possibility of whitening in the horizon of their daughter's offspring, and the subsequent economic uplifting as something positive, exercising a form of colorism, even if only hypothetically. Interestingly, hooks is considered to be "too white" by some of her relatives and effective colorism is seen as one of the aunts, Sister Ray (the owner of the beaded purse hooks would inherit), despised the protagonist because of her light skin, something hooks remembers as her first experience of rejection (*Bone Black* 14). Sister Ray's deep blackness is contrasted with hooks' light skin, and here colorism is enacted in a different manner, as her light skin assigns her to a lesser category in a scale of black purity. Later, hooks and her siblings would prank this aunt by hiding inside a no longer working old car that would sit in their yard. When the distraught aunt looked for the children, they would call her from inside the car, locking all doors and windows. While her siblings wanted to leave the car upon seeing their aunt's angry face, hooks defiantly spit in her direction, with all the saliva getting caught midway by the glass, in a clear provocation of her authority, which might have stemmed from the rejection she felt by this relative. By the look in Sister Ray's face hooks knew she would be severely punished by such action.

> Defiant, determined, they refused to budge even though they were beginning to feel afraid. Their resolve was weakening. They knew that the longer they remained in the car, the more intense the punishment. They wanted to relent but the particular little one, the one who was not her favorite, couldn't resist a last rebellious display. She spit right in the direction

of that stern face. Only the rolled up window kept it from reaching the target. Shocked, they opened the doors and ran, leaving her sitting alone contemplating the coming punishment. (*Bone Black* 15)

The author explores the universe of games and play in many different moments in the narrative, as it often offers the context for the showcasing of the learning of social norms and their respective defiance. One example is how the protagonist describes the fraught relation that she developed with her older brother, who was close to her at first, but who drifted apart due to matters related to masculinity and gender roles, differences that are focally constructed around another toy. The narrator reports the identification of the protagonist with her brother starting with their physical resemblance:

> SHE WAS CLOSEST to her brother. Not only were they months apart in age but they looked alike. They looked like twins even though he was older. Like twins they shared the same dreams and longings, the same devotion to one another strangely enough it was a toy that separated them, that forced upon them different roles, different identities. (19)

The toy in question was a red wagon that was supposedly shared between them. The gender roles that are mentioned by the author are related with how each child was supposed to enjoy the toy. hooks was supposed to be pulled, while her brother was the one who should pull the wagon around, roles that would correspond with traditional views regarding masculinity and femininity. Through the imagery of Europeanized relations in fables, hooks depicts the gender roles associated with these traditional views relating them to the expected roles of princes and princesses. These characters, very much present in the universe of children, represent the ideals that boys and girls should strive for, and the red wagon was simply another object that served to display this logic:

> She was to ride in the red wagon and he was to pull it. She was to ride because she was a girl – a would-be princess whom some rich prince would come seeking, take away to his palace, and keep her there in splendor forever. He was to pull it because he was a boy – a would-be prince, who should do all the hard work, slay the dragons, fight the slimy creatures, challenge the fat ugly men so that he could carry away the beautiful princess. (19–20)

The discordance, however, takes place as the brother prefers to be pulled, and makes the sister be the one pulling the toy, something that was not approved by the adult family members (*Bone Black* 20). Every time adults would find the brother being pulled, they would chastize him, asserting that he should not challenge the expectations that were assigned to his sex, in addition to telling him that if he did not do otherwise the toy would be taken away and given only to the girl. hooks, would also reinforce this idea when the brother refused to pull her, asserting what she calls her "girl rights", threatening him of telling the adults that he was not fulfilling his masculine duty. This dispute would lead to the boy claiming that he hated her because she was a girl, finally distancing from her. Another layer that is interesting in this particular story is related to the ways in which hooks remembers the toy. Upon seeing it in photographs, the red wagon turned out to be a wheelbarrow, demonstrating how much the affectionate memories of the author are dressed up in remembrance. This revelation is also telling of the poverty experienced by these children, turning a wheelbarrow into a toy, fashioning it in their imagination as something better than it actually was. Her relationship with her brother, and the reasons that would set them apart are revisited when she is older, as the narrator states: "[g]rowing up divides them. He is shamed to love a girl. He must show in every way that there is nothing about her that he can stand. He must not be on the side of an outcast" (166).

The motive of marriage is retrieved later in the text, as the protagonist writes about her dolls, and particularly about the life of Barbie. One of the most common girl child activities is playing with dolls, as they help them invent social worlds that are enacted and reenacted continuously, mirroring social relations, as well as serving as a means to project alternative realities. One of her sisters receives a Barbie doll as present, which has tanned skin, but its blond hair confirms her racial identity as white. The protagonist comments on the discrepancies between them and the idealized Barbie life:

> Barbie is anything but real, that is why we like her. She never does housework, washes dishes, or has children to care for. She is free to spend all day dreaming about the Kens of the world. Mama laughs when we tell her there should be more than one Ken for Barbie, there should be Joe, Sam, Charlie, men in all shapes and sizes. We do not think that Barbie should have a girlfriend. We know that Barbie was born to be alone – that the fantasy woman, the soap opera girl of *True Confessions*, the Miss America girl was born to be alone. We know that she is not us. (*Bone Black* 22–23)

The appreciation for the doll stems from the dream of a life with no limits, such as the one the doll enjoys in its fictional world. The lack of identification comes not only from the physical attributes that characterize these fantasy women, but also from the reality of no responsibility that they enjoy. Furthermore, it seems that the protagonist is able to connect the reality of enjoyment of the inherent independence of these fictional women with a life of solitude, as she states that "she is not us", not explicitly commenting on her whiteness, but making the point that womanhood for them is connected with traditional female social roles of marriage and motherhood. Other toys that would reveal some of the hardships in hooks' childhood are her dolls, and more specifically hooks' favorite doll: a brown baby given to her by her mother. Differently from the Barbie dolls she would play with her sisters, hooks would develop a better relation of identification with this particular toy, which she called simply "Baby". The protagonist reports the story of the doll, already hinting at her complicated relation with the rest of the family, and more importantly, the relation of the world with brown babies such as herself:

> She tells us that I, her problem child, decided out of nowhere that I did not want a white doll to play with, I demanded a brown doll, one that would look like me. [...] Deep within myself I had begun to worry that all this loving and care we gave to the pink and flesh-colored dolls meant that somewhere left high on the shelves were boxes of unwanted, unloved brown dolls covered in dust. I though that they would remain there forever, orphaned and alone, unless someone began to want them, to want to give them love and care, to want them more than anything. (*Bone Black* 24)

Just like the unwanted and forgotten brown babies, hooks wishes she was cared for, being rescued from the abandonment she feels inside her family. When her mother characterizes her as her "problem child", singling hooks out from the other siblings in a negative manner, she reinforces the idea that hooks already perceives that she is not really part of this whole, that her experience is challenging not only to her, but to those around her, cementing the idea of negative differences.

Another toy-related episode that is relevant in hooks' story are marbles, which she was not supposed to play since they were the property of her brother. In a Saturday afternoon, while the father would watch sports on television, a sacred activity for the patriarch, hooks demands that she would like to play marbles with her brother, something he did not want to share, asserting his "boy rights" over the small glass spheres. She threatened to scatter all the marbles if she woudn't be allowed to play with them, something her brother dared her to do, a situation that would quickly escalate, culminating in violence:

> Mama had already begun to encourage her to leave the boy alone. Several times the father had interrupted his game to tell her to leave him alone. Again the boy dared her. She hesitated only for a few seconds before stomping her feet onto his marbles. Jumping from his chair the father began to hit her – not wanting to damage his hands since he needed them for work, he tore a piece of wood from the screen door that kept flies out. As he hit her with the wood he kept saying Didn't I tell you to leave those marbles alone? Didn't I tell you? The mama stood watching, afraid of his anger, afraid of what he might do, but too afraid to stop it. The spectators knew not to cheer this punishment or they might be next. They would cheer afterward, they would tease her afterward when he could not hear, when they did not need to fear being next. (*Bone Black* 29 – 30)

This episode is illustrative of many different aspects of violence that hooks would have to navigate during her childhood. Firstly, the role of this violent father figure, who would not demonstrate love for his children, showing more specifically his distaste for this specific child because of her frequent non-compliant attitude, straying from the expected behavior of a girl. In addition to that, her mother's inability to interfere and protect her daughter, possibly fearing that that violence would turn on herself, is also an important factor to be considered. This vulnerability aspect would shape hooks' opinion of her mother, and of marriage as a whole, as her defenseless body would be under attack and would not be protected by those who were suppose to care for it. The latest aspect is related to the cruelty of children, who would not stand in solidarity with her later on, but who would continue to harass her in a different manner. These kinds of interactions would shape her family life, creating an even larger distance between daughter and parents, as well as sister and siblings. Concluding the episode, the narrator tells the readership what ensued, delivering the perspective of the father from the indirect retelling of the protagonist, stating that hooks had "too much spirit", that this spirit needed to be broken, and violence seemed to be the only way this father figure could effect this change:

> She was sent to bed without dinner. She was told to stop crying, to make no sound or she would be whipped more. No one could talk to her and she could talk to no one. She could hear him telling the mama that the girl had too much spirit, that she had to learn to mind, that that spirit had to be broken. (*Bone Black* 30)

hooks' disposition, the "too much spirit" as perceived by this patriarchal figure, in addition to his wish to domesticate her impulses, are telling aspects of the position that the father wished his daughter occupied in the social hierarchy. The disciplining of this body would happen in many different ways, as the narrative develops, and would also be resisted, as the protagonist asserts her sense of self. Her relationship with her father would deteriorate as she grew up, when she felt ever more distant from this masculine figure. hooks would comment on the strictness of black parenting in *Sisters of the Yam*, relating the severity of their actions to the need of protecting them from a racist environment (similarly to Sweetness in *God Help the Child*):

> Black parents' obsession with exercising control over children, making certain that they are "obedient" is an expression of this distorted view of family relations. The parents' desire to "care" for the child is placed in competition with the perceived need to exercise control. This is graphically illustrated in Audre Lorde's autobiographical work *Zami*. Descriptions of her childhood here offer glimpses of that type of strict parenting many black parents

felt was needed to prepare black children for life in a hostile white society. (*Sisters of the Yam* 47)

There was at first a good relation between father and daughter when she was a baby girl, possibly because she did not at the time display signs of the rebellious traits that would later characterize her. As she develops, however, the father makes his affection unavailable to her, creating at first a desperate need to reclaim the love she first experienced from him and later, as she feels that he is never going to accept her as she is, and that the previous love they shared would not exist the same way, the narrator gives up trying to develop any relation to him. The narrator confesses:

> Growing up she stopped trying. He mainly ignored her. She mainly tried to stay out of his way. In her own way she grew to hate wanting his love and not being able to get it. She hated that part of herself that kept wanting his love or even just his approval long after she could see that he was never, never going to give it. (*Bone Black* 146)

The protagonist's sour mood in the eyes of the family contrasts with cultural expectations regarding childhood, and according to the protagonist: "I am a child who is sad all the time. They tell us children should be happy, should love to go outside and play. I would rather read books in our house even when we were very young" (*Bone Black* 76). And later: "[w]hen I become the problem child they blame it all on the books" (*Bone Black* 77). The habit of reading is not something that is completely out of the reality of this family, as the protagonist describes the presence of a reader in the house:

> I remember daddy reading paperback novels, detective stories. I know he reads dirty books because I read them too. I know that there are black writers like James Baldwin, Frank Yerby, and Ann Petry because their books are on the shelf. They tell me that these books are for when I am older (*Bone Black* 76)

It is possible to state that the protagonist is part of a family or readers, that they might even have encouraged the girl to develop a reading habit. While her father believes it is a waste of money to buy a child books, her mother is the one who gifts her with a hardback book of religious stories, which she would later use in her missionary work: reading to sick people who were not able to leave their houses as a form of charity. What they did not expect and cannot fathom is how important books are in the life of this lonely child. The reference to black writers is also relevant, as the protagonist knows from an early age the world of literature and that the occupation of writer was available to all, black or white, men or women. Her reading opportunities expand once she has access to a library, not depending only on the books she could find at home.

> When we leave the country and move to the city we have a library to go to. We have library period. This is my favorite time. I love biographies. I read about George Washington Carver, Mary McLeod Bethune, Booker T. Washington, Louise May Alcott, Amelia Erheart, Abraham Lincoln. We are not allowed to take books home. There is no money to replace them if they are lost. I am a good reader, careful with books, a library helper. I take as many books home as I like. I read Laura Ingalls Wilder's *Litte House on the Prairie* and Alcott's *Little Women* – and even other Alcott book. I find remnants of myself in Jo, the serious sister, the one who is punished. I am a little less alone in the world. (*Bone Black* 76–77)

The identification that the protagonist feels with characters in the books that she reads demonstrates the power of literature in diminishing the feeling of alienation that she experiences at home. She is able to find in the pages of books the referents that she cannot find in her real life, providing her with a strategy to counterattack the isolation from her siblings and even from her parents. When a neighbor who knows about her reading habits calls to say she has a collection of books she is discarding and that if she wants she could go there and fetch them, the protagonist must wait for her sister to be able to accompany her. As they get there, the books have already been tossed in the trash. The protagonist would salvage them and take them home, happy with her new friends:

> Inside the trash are cartons of tiny books called *A Little Leather Library*. They can be carried in the pocket. The green leather covers have become dry and brittle. The works of Shakespeare, Homer, Dickens, and All the Romantics are there. The novels of George Elliot, the Brontës, the poetry of Poe and Emily Dickinson. We have a hard time carrying the cartons. They smell of mold and decay, but to me they are a treasure. The print is so tiny I am sure it will take hours to read each little line but I am ready. When we get home they say More trash but they are happy because I am happy. The books are a new world. I am even less alone. (*Bone Black* 78)

The parents are able to recognize the joy that the protagonist derives from these books, despite the bad conditions in which they are found. Her loneliness is made lesser because she has now more stories to spend her time with, in her own company. Books, however, become objects that signify to the family that the protagonist does not conform to what is perceived as a "normal" child. They also become the way in which her parents are able to control her, becoming a marker of difference:

> When I become the problem child they blame it all on the books. They make me stop reading unless all my chores are done. They make me stop reading to go outside and play. They

snatch the book out of my hand and throw it away because I am not listening when someone is talking to me. (*Bone Black* 77)

The constant punishments for her backchat make her feel like the scapegoat of the family, a position she states when she affirms: "[e]ach time she opens her mouth she risks punishment. They punish her so often she feels they persecute her" and "[e]ven though she is young she comes to understand the meaning of exile and loss" (130). The sensation of not belonging to the world of the family, the only world she knows, makes her feel like she does not belong to any place at all, except the world of books.

> She finds another world on books. Escaping into the world of novels is one way she learns to enjoy life. But novels only ease the pain momentarily, as she holds the book in hand, as she reads. It is poetry that changes everything. When she discovers the Romantics it is like losing a part of herself and recovering it. She reads them night and day, all the time. She memorizes poems. She recites them ironing or washing dishes. Reading Emily Dickinson she sense that the spirit can grow in the solitary life. She reads Edna St. Vincent Millay's "Renascence,", feels with the lines the suppression of spirit, the spiritual death, and the longing to live again. She reads Whitman, Wordsworth, Coleridge. Whitman shows her that her language, like the human spirit, need not be trapped in conventional form or traditions. For school she recites "O Captain, My Captain." She would rather recite from *Song of Myself* but they do not read it in school. (*Bone Black* 131–132)

The protagonist extends the identification she feels toward the characters to the authors she reads. The wish not to conform mirrored in Whitman, and the joy found in solitude in Dickinson are representative of the values that the protagonist shares with these authors, finding in them a possibility for a better life that she cannot find in her actual life at this moment. The next step would be to fulfill her dream of becoming a writer, something she attempts from an early age, expressing her internal anguish: "[s]he writes her own poetry in secret. She does not want to explain. Her poems are about love and never about death. She is always thinking about death and never about love" (132). Finally, she uses her own writing as a form of resisting the miserable live she experiences in her house: "[t]he punishments continue. She eases her pain in poetry, using it to make the poems live, using poems to keep on living" (132).

The protagonist tells of an episode in which she was prevented from reading as a form of punishment, having to iron sheets instead. This is also an episode in which the protagonist expresses her frustration in her house, in addition to showing the relentless psychological abuse she suffered in the hands of her family.

It is my turn to iron. I can do nothing right. Before I begin I am yelled at, I hear again and again that I am crazy, that I am going to end up in a mental institution. This is my punishment for wanting to finish reading before doing my work, for taking too long to walk down the stairs. Mama is already threatening to smack me if I don't stop rolling my eyes and wipe that frown off my face. It is times like these that I am sorry to be alive. That I want to die. In the kitchen with my sisters, she talks on and on about how she cannot stand me, about how I will go crazy. I am warned that if I begin to cry I will be given something to cry about. The tears do not fall. They stand in my eyes like puddles. They keep me from seeing where the ironing is going. I want them to shut up. I want them to leave me alone. I shout at them Leave Me Alone! I sit the hot iron on my arm. Already someone is laughing and yelling about what the crazy fool has done to herself. Already I have begun to feel the pain of the burning flesh. They do not stop talking. They say no one will visit me in the mental hospital. Mama says it does not matter about the pain. I must finish ironing the clothes in my basket. (*Bone Black* 101–102)

In this passage the protagonist explores the extent to which the psychological abuse and the lack of empathy from her family would figure as factors that induced her to suicidal ideation. It is clear that the protagonist feels she does not belong on this family, and that the ones who should provide her with understanding are responsible for the creation of a hostile environment. Her mother's remarks describe the dynamics of violence, physical and psychological, that are used to "educate" the protagonist, different ways of "breaking her spirit", in the words of her father. Her rebellion is not seen as a typical phase of adolescence, the frowning and the rolling of the eyes, as coded representations of the questioning of authority. When she is timelessly called "crazy", the family is reinforcing the idea that those who do not belong to the order, who stand against the values and expectations preconized by a normative idea of personhood, are deemed not to have the same mental faculties than the dominant group, becoming less of a human. The association of reading too much with mental illness is determinant mainly because the protagonist is a woman. By deviating from the norm that defines that positive feminine attributes are mainly related to physical appearance and politeness; and not with mental capabilities, a woman who prioritizes her intellect is seen as not belonging to the social order. The lack of empathy regarding the accidental burn from the ironing is telling of how difficult it is for this protagonist to achieve a sense of self and of community inside her own house, as no one, adult or otherwise, seems to demonstrate that they care much for her needs.

4.2.3 We are women together. Womanhood, sexuality and race.

The difficult relationship between the protagonist and her family is illustrated by the many instances in which she is teased, provoked, ignored, subjected to violence, and ultimately excluded, instances that happen during all her childhood and adolescence. Despite this, the protagonist confesses her love for her mother as the one person in the house who is not exclusively negative towards her. The mother becomes the locus of a conflicting reality: she is both the caretaker, who does show compassion to some extent, but she is also complicit in the creation of oppression in the life of her daughter, siding with the father and the other children in the exiling of the protagonist (*Bone Black* 139).

Her mother figures as both caring and disciplining, ultimately becoming the person who accepts the protagonist's needs inside their home, and who tries to meet them. The protagonist suffers most for not having her mother's full acceptance, something that she believes she offers her mother, since she is able to recognize her flaws, but does not use them to recriminate her:

> I am a pain to her. She says that she is not sure where I come from, that she would like to send me back. I want so much to please her and yet keep some part of me that is myself, my own, not just a thing I have been turned into that she can desire, like, or do with as she will. I want her to love me totally as I am. I love her totally without wanting that she change anything, not even the things about her I cannot stand. (*Bone Black* 140)

The kind of acceptance that the protagonist wishes would only be available in the older generation of her family, in the relationships she would develop with her grandparents. Her mother, however, does not seem to be emotionally able to deal with all the matters that might be considered stressors in the life of the protagonist, avoiding discussions about her feelings and fears. This dynamic might be related to the mother's own adolescence, as the narrator implies: "[s]he does not want to hear the word *loneliness*. She does not want to remember" (*Bone Black* 140). The mother, however, does seem to be a positive influence, as the narrator lovingly describes her care:

> We can see that she is working hard to give us more than food, shelter, and clothes to wear, that she wants to give us a taste of the delicious, a vision of beauty, a bit of ecstasy. [...] Even so she is moving away from her awareness of the deeper inner things of life and worrying more about money. I watch these changes in her and worry. I want her to never lose what she has given to me – a sense that there is something deeper, something more to this life than the everyday. (*Bone Black* 141)

At one point in the narrative, her mother is taken to the hospital, after not telling for a long time to all her children that she had been diagnosed with cancer, believing it to be a way of protecting them. When they are all instructed to visit her in the hospital because it might be the last chance to say goodbye, the protagonist refuses to go, not wanting to carry the image of a sick mother as the last memory she had of her, instead preferring to remember her mother as the beautiful and active woman she knew and loved. The mother, however, recovers fully and returns home, resenting the "uncaring child" who did not even want to see her in her deathbed. The protagonist is never able to explain to her mother her reasoning for not going. "When I go to see her, sitting on the bed, with my longing and my tears she knows that she breaks my heart a little. She thinks I break her heart a little. She cannot know the joy we feel that she is home, alive" (*Bone Black* 144). Commenting on the issue of controlling mothers, hooks writes in *Sisters of the Yam:*

> It troubled me that it was difficult to find autobiographical narratives where black daughters describe loving interactions with black mothers. Overall, in fiction and autobiography, black mothers are more likely to be depicted as controlling, manipulative, and dominating, withholding love to maintain power over. (*Sisters of the Yam* 153)

Her mother's desire to love and to control is seen through the education on gender that she provides her daughters, crafting their image so they conform to social expectations. One of the desires of the protagonist is wearing the color black, something her mother never allows her to do, since black was a color for grown women and not for children. Another episode that illustrates this principle is concerned with the protagonist's poor eyesight, as she needs glasses to correct her vision. Upon visiting the doctor, hooks states that she prefers to have black frames, "[...] shaped almost like dominoes" (*Bone Black* 38), but her mother buys pink glasses, which would be more feminine, though a color her daughter hates. "She buys the ones that are pink and shaped like the end of wings" (*Bone Black* 38). While developing a friendship with another black girl in school, Rena, she mentions once again her rejection of the color pink, stating: "Rena lives at the very end of Younglove Street. She walks to meet me at the corner of Younglove and Vine. She lives in a house that is painted pink. I do not tell her how much I hate the color pink" (*Bone Black* 38). In another passage the describes pink and what is associated to it: "[g]rown-ups think it should be her favorite color. Pink innocence, pink dreams, pink the color of something alive but not quite allowed to be fully living" (*Bone Black* 110). This comment stresses once again how much the protagonist wishes to distance herself from the indexes of compulsory heteronormative femininity, as it feels is restricts her full potential of enjoying life and feeling really alive. However, by withholding her opinions, she also shows that she is not completely confident in displaying this position to the social world. While discussing their future, the two girls talk about their plans, and once again hooks demonstrates that she does not wish to participate in the traditional role imposed on women:

> She is going to be a doctor and make sick people well, or a teacher, or a housewife with two children. She is an only child. She says that it is sometimes lonely. I do not tell her how lonely it can be to be one of many, especially if you do not fit in. Instead I tell her that I will be a librarian, a writer, and will never marry. She laughs at me when I say that I will never marry, and tells me of course I will. (*Bone Black* 39)

Rena's disbelief plays with the notion of the hegemonic aspect of this compulsory femininity that inherently would involve marriage. Another reveling aspect of this passage is related to the feeling of loneliness experienced by the protagonist, which is hidden from her interlocutor and confessed to the reader, as her difficulties to assimilate to the patriarchal order exile her from siblings and friends.

Hair is another aspect that is explored by the author in this narrative, as it plays in an important part as a signifier of womanhood. In the context of the African-American experience the hierarchies created based on racialized beauty standards come to the fore, as exemplified in the grading of "good hair" and "bad hair", characterizations based on white definitions of desirability which are imposed, creating hierarchies outside and inside the black experience. The protagonist states:

> GOOD HAIR – THAT'S the expression. We all know it, begin to hear it when we are small children. When we are sitting between the legs of mothers and sisters getting out hair combed. Good hair is hair that is not kinky, hair that does not feel like balls of steel wool, hair that does not need tons of grease to untangle, hair that is long. Real good hair is straight hair, hair like white folk's hair. Yet, no one says so. No one says Your hair is so nice, so beautiful because it is like white folk's hair. We pretend that the standards we measure our beauty by are our own invention – that it is questions of time and money that lead us to make distinctions between good hair and bad hair. I know from birth that I am lucky, lucky to have hair at all for I was bald for two years, then lucky finally to have thin, almost straight hair, hair that does not need to be hot-combed. (*Bone Black* 91)

Commenting on her family's hair texture, hooks states that all six girls have different types of hair, a difference that is not celebrated, but instead yearns to be homogenized in accordance to a white standard of "good hair". The straightening of hair is seen not as a way of emulating whiteness, but as a rite of passage to womanhood, as plaits and braids were worn mostly during girlhood. Hot combing, a process in which heat and grease are applied to the hair with the intent of straightening it, demarcates the difference between these stages:

> For each of us getting our hair pressed is an important ritual. It is not a sign of our longing to be white. It is not a sign of our quest to be beautiful. We are girls. It is a sign of our desire to be women. It is a gesture that says we are approaching womanhood – a rite of passage. Before we reach the appropriate age we wear braids and plaits that are symbols of our innocence, our youth, our childhood. Then we are comforted by the parting hands that comb and braid, comforted by the intimacy and bliss. (*Bone Black* 92)

This time of infancy is left behind as the growing girls and their hair are tended in different ways, creating the opportunity of communion. hooks describes the ritual of pressing hair as a moment in which girls-becoming-women have the opportunity to develop a sense of sisterhood outside the gaze of men, focusing their attention and effort on each other. hooks comments in this special moment of care in *Sister of the Yam*, stating:

> As grown-ups, many of us look back at childhood years of having our hair combed and braided by other black women as a moment of tenderness and care that was peace-giving and relaxing. This dimension of sharing in care of the black female self is necessary in our life and we should seize all opportunities to feel caring hands tending our hair.(98–99)

This ritual would take place in a quarter of the house that would symbolically belong to women, the kitchen, which would turn into a beauty parlor every Saturday. In resemblance of the Greek gynaeceum, this space would become the place in which the women gathered to attend their own needs, to display their vulnerabilities, and to create a sense of belonging:

> There is a deeper intimacy in the kitchen on Saturday when hair is pressed, when fish is fried, when the sodas are passed around, when soul music drifts over the talk. We are women together. This is our ritual and our time. It is a time without men. It is a time when we work to meet each other's needs, to make each other beautiful in whatever way we can. It is a time of laughter and mellow talk. Sometimes it is an occasion for tears and sorrow. Mama is angry, sick of it all, pulling the hair too tight, using too much grease, burning one ear and then the next. (*Bone Black* 92)

This moment of care amongst the flourishing women is also seen in a more negative light, as hooks' mother does not seem to enjoy the task of pressing the hair of her many daughters, since it adds up to her house duties as a mother. hooks does not account in this passage for the humongous workload of her mother, seeing simply from the perspective of the child who wished to be a part of this moment. hooks is excluded from this moment not because of her age, something that is never stated clearly during the episodes in the narrative, but because of the texture of her hair, which is straighter than her sisters' and does not need to be pressed in order to be styled. She confides:

> At first I cannot participate in the ritual. I have good hair that does not need pressing. With-out the hot comb I remain a child, one of the unitiated. I plead, I beg, I cry for my turn. They tell me once you start you will be sorry. You will wish you had never straightened your hair. They do not understand that it is not the straightening I seek, but the chance to belong, to be one in this world of women. (*Bone Black* 92–93)

By remaining outside the ritual of pressing hair, hooks is further exiled from her own family, being kept in the realm of childhood for longer than she had expected. Her hair, which is considered to be good, works against her development as a more complete social being within the family structure. The sisters advise her that regret is expected once she crosses the barrier and starts participating in the ritual, an admonishment that might even be seen as a caution tale about leaving childhood to become a woman. When the protagonist finally takes part in this rite of passage, she is happy to finally be a part of the group of "women" in the house, although the result of the procedure deeply disappoints her.

> It is finally my turn. I am happy. Happy even though my thin hair straightened looks like black thread, has no body, stands in the air like ends of a barbed wire; happy even though that sweet smell of unpressed hair is gone forever. Secretly I had hoped that the hot comb would turn the thin good hair into thick nappy hair, the kind of hair I like and long for, the kind of hair you can do anything with, wear all kinds of styles. I am bitterly disappointed in the new look. (*Bone Black* 93)

In a reversal of the white centered image that was cultivated, hooks wishes that her thin "good" hair would become thick, giving her more options to style it as she wanted, something the hot comb could not do. Her desire for "thick nappy hair" demonstrates her Afrocentric worldview, showcasing a shift from the perspective that dominated the standards in her family. This desire would resurface later on, as hooks would grow a natural Afro style in high school.

> I want to wear a natural, an Afro. I want to never get my hair pressed again. It is no longer a rite of passage, a chance to be intimate in the world of women. The intimacy masks betray-al. Together we change ourselves. The closeness in an embrace before parting. A gesture of farewell to love and one another. (*Bone Black* 93)

Here, the ritual of pressing hair, of entering womanhood, is seen from a different perspective, not as a moment of nurturing and communion, but as the moment before the departure from the previously established sisterhood. Her wish to sport a natural hair style confirms the development of her awareness of the social role of women in a patriarchal society, functioning as a move away from adapting to the norm. The ritual is then reevaluated, and seen as a betrayal of the sisterhood, who would be grooming themselves to become full women in the patriarchal order, seen only as wives and mothers.

Her physical appearance would also be criticized later in the narrative, as her slim figure would come under scrutiny. As a teenager, her family continues to criticize her, telling her she is too skinny, looking like pictures of starving children. Her self image is very negative at this point: "[s]he looks at her eyes in the mirror and sees that they are red with crying, red with fear of being too skinny, ugly, unwanted" (*Bone Black* 166). Many remedies are given to the protagonist, to stimulate her body to gain some weight, such as pills, vitamins and tonics, and the addition of raw eggs to her diet, which do not stop her family members from teasing her. She is given larger clothes, so that she would grow into them, but end up accentuating her slimness instead. She is compared to Saru, the grandmother, because of her fragility, something that does not upset her, since she is very much fond of this grandmother. They say she eats like a bird, and that she plays with her food (*Bone Black* 167).

She is constantly surveilled by her family members, who either punish her or denounce her to their parents, feeding a vicious cycle in which the protagonist is continuously in an uneasy position. She finds solace in literature and in the occasional gesture of solidarity of others, as she uses these moments of loneliness to consider ways of escaping this reality of violence and isolation.

> In the cold kitchen, staring at the window, she thinks about Wordsworth and Shelley, about Dickinson, Whitman and Frost. She thinks about ways to escape her punishment. Every now and then, if the food is still hot, someone will help her eat it because they feel sorry, because they cannot stand to see her sitting alone crying. If they are caught she will be punished. She knows that no matter what they do she will not get bigger. They cannot give up trying. They believe they are saving her life. (*Bone Black* 168)

hooks is able to perceive that this treatment is a form of care, even if misguided, and even reports a little glimpse of support from her family. It is not clear who are the agents that help her, possibly one of her siblings, or even her mother.

As her body develops, the protagonist experiences an early episode of urinary tract infection, causing a white discharge to appear in her underwear. Her mother notices this event, and the shame associated with matters related to the body are explored in this intimate perspective. Though the protagonist is afraid to reveal that she is experiencing this issue, ashamed of having to explain her unexpected bodily fluids, the mother is attentive and supportive:

> When mama finally asks me if it is me that has the panties with the discharge, with the sometimes funny smell, I do not ask her how she knows – she finds out everything. Yet she is mostly gentle when she comes across a secret that might hurt in the telling. I tell

> her I suppose they are mine. She wants to know have I been doing anything with boys. I do not know what this anything is. When I say no, she asks again and again. I always answer no. When I become tired of answering this same old question I ask her a question. I ask her What is this anything that one can do with boys. I am so angry at boys – the ones I do not know, who are capable of this anything that makes me be questioned in a way that feels like I have done something wrong, like I'm on trial. She does not want to tell me what this anything is. She believes me. (*Bone Black* 94 – 95)

The protagonist's mother suspicion that her daughter, though still young, could have already been involved in sexual activity, is similar to the preoccupation demonstrated by Sophie's mother in *Breath, Eyes, Memory,* and is telling of the chastising of premature sexualization of these vulnerable subjects. The protagonist's innocence is shown by her lack of understanding of her mother's question, as the vague "anything", a coded word for sexual behavior, is not grasped. There is also a sense of culpability, denoted by the trial characterization of the questions, as the girl seems to be defending her honor without ever fully understanding what her mother is judging. The protagonist is taken to the doctor's office, in which he assures the mother that the discharge occurs naturally and is not related to anything linked to sex. The remedy for her condition is having a water and vinegar douche, something that the protagonist does not understand, and the impatience of her mother and that of the doctor is shown, as it seems to be an embarrassing matter even to the adults:

> They do not try and explain. They are annoyed that I am so ignorant when it comes to matters of the body. Yet they have always made us ashamed of the body, made us tuck it away under our pillows like some missing tooth for which the fairy will reward. They reward our silence about the body. (*Bone Black* 95)

The narrator's last remark in this quotation summarizes her education regarding the body, as the less it was discussed, the better. When time comes for the treatment, the protagonist is faced with her own nakedness in front of her mother, something that she is no longer used to, implying that somehow shame regarding the body, and specifically about genitals, has already been instilled in her. She confesses: "[s]tanding naked before her I pretend I am wearing clothes, that she cannot see the parts of me that I have chosen to show to no one, the parts I no longer see myself even as I undress them, wash them" (*Bone Black* 95). This confession illustrates the ways in which the protagonist denies, even to herself, the existence of a sexual being, by suppressing from thought any reference of its sexual existence. One may conclude that according to the logics of her body disciplining, if it is not mentioned, it therefore does not exist. Following the doctor's orders, the mother prepares the douche, pointing the apparatus to the child, who does not understand exactly what is soon going to happen. She recognizes the douche as a "red balloon that could never be blown up", instating her lack of familiarity with the object, its referent being only a toy in her imagination. When the douche is applied, the protagonist describes the act as an intrusion, a break in the barriers created to protect her body:

> When she tries to place the nozzle inside me, I know that I am naked, I know that this is my body, that she has no right to touch or enter. I begin to scream and scream – cries that sound as if a terrible crime is being committed. Worried that the neighbors will hear, she demands that I shut up before she kills me. (*Bone Black* 96)

The threat of violence makes more evident the no-talking-about-the-body politics, as the mother silences the protagonist even as she feels her body is in danger. Though the mother is trying to administer a medical treatment, the silences regarding the body create an environment that feels dangerous to the protagonist, not one that might be understood as healing. Had the mother explained clearly what the procedure was, and its desired result, the child might have felt secure enough to understand this violation of barriers as a necessary step towards a healthier body. The situation becomes even more traumatizing as her older sister joins the mother so as to perform the task at hand.

> My oldest sister enters the bathroom with a smirk on her face that tells me right away that she sees that I am naked, afraid, ashamed; that she enjoys witnessing this humiliation. Together they struggle to perform the task. Mama asks angrily What are you going to do when some boy sticks his thing up you? I am shocked that she could think that I would ever be naked with a boy, that I would ever let anyone touch my body, or let them stick things in me. When I say this will never happen to me they stop their task and laugh, long and loud. I weep at their refusal to believe I can protect myself from further humiliation. (*Bone Black* 96)

The presence of her sister and the further shame derived from another person invading her privacy creates another layer of cruelty to this moment as her sisters are represented often in the role of teasers in her life. Her mother's comment is an example of the oblique education the protagonist receives in matters related to the body, since no one directly explains what is to be expected in her future relations with men, not clarifying even the basic facts about sex and the body. It seems that this knowledge is something that the girl should have already obtained somehow, but not via her mother. Therefore, her lack of understanding and her childish refusal to accept that she will be sexual someday in the future are met with ridicule further instigating the sense of humiliation that she experiences during this episode. Finally, it seems that at this point the protagonist strictly associates her sexuality to humiliation.

Another issue that is involved in the matters of the developing body is masturbation, a topic that was never discussed openly in her family, but that has shaped her early years as a sexual being. The difference between the approaches related to masturbation in relation to boys and girls is a matter that is also explored by the narrator, as manhood, even in its early stages is celebrated in a patriarchal culture, in contrast with the initiation of girls, which takes place in silent denial, and only figures as a means for chastising female subjects:

> MASTURBATION IS SOMETHING she has never heard anyone talk about girls doing. Like so many spaces of fun and privilege in their world, it is reserved for the boy child – the one whose growing passion for sexuality can be celebrated, talked about with smiles of triumph and pleasure. A boy coming into awareness of his sexuality is on his way to manhood – and it is an important moment. The stained sheets that show signs of his having touched his body are flags of victory. They – the girls – have no such moments. Sexuality is something that will be done to them, something they have to fear. It can bring unwanted pregnancy. It can turn one into a whore. It is a curse. It will ruin a young girl's life, pull her into pain again and again, into childbirth, into welfare, into all sorts of longings that will never be satisfied. (*Bone Black* 112)

The topic of masturbation is approached with caution, since female pleasure is something that is not discussed and seems to be almost inexistent around her. While the male experience of a developing body is celebrated, there is a silence regarding the female experience of self-pleasure, and any idea related to the inception of a sexualized female self is automatically connected to negative consequences. The social stigma that is placed over the sexuality of women, especially developing women, is described as a path for inherent misfortune, and more importantly, female sexuality exists in a context of passivity, as their bodies are used in sexual activity without their agency.

The control over females regarding their sexuality takes place from a very early age, however, it functions silently, as information regarding their own bodies might instigate this process, and it must be deferred as much as possible inside a patriarchal logic. By associating sexuality to inherent pregnancy, the young females are advised against getting to know their own bodies, as they understand that their bodies are capable of reproducing, yet they are not completely aware of how this happens. The veiled association of sexuality with shameful images of femininity also feeds this silence, creating an understanding of the body that is clearly stigmatized. As sexuality is not discussed, or even mentioned, the longing for sensual pleasure is another source of stress and confusion for the protagonist, since there is no guidance or instruction regarding this specific aspect of development. Upon discovering the pleasures of her own body, the protagonist's first feeling is shame:

> When she finds pleasure touching her body, she knows that they will think it wrong, that it is something to keep hidden, to do in secret. She is ashamed, ashamed that she comes home from school wanting to lie in bed touching the wet dark hidden parts of her body, ashamed that she lies awake nights touching herself, moving her hands, her fingers deeper and deeper inside, inside the place of woman's pain and misery, the place men want to enter, the place babies come through – ashamed of the pleasure. (*Bone Black* 113)

This sense of shame worsens as one of her siblings walks into her during a moment of self-pleasure and threatens to tell it to their parents. At this point, the act that would give her pleasure becomes even more a burden of guilt and fear, as the repercussions of these acts would surely be violent. "Like a party ending because the lights are suddenly turned on she knows the secret moments are gone, the dark, the pleasure, the deep cool ecstasy" (*Bone Black* 114). After this moment the protagonist no longer touches herself, substituting once again the fear of punishment with reading, finding an escape in the stories of different realities told by others. The pleasure derived from reading is compared to that of the masturbation, as the narrator claims: "[s]he reads with passion and intensity. When she has read everything in sight she goes searching for something new, something undiscovered. Books, like hands in the dark place, are a source of pleasure" (*Bone Black* 115). These expeditions would take her to the discovery of her father's pornography books, which were hidden in her parents' bedroom. Not knowing what they were about, and not even knowing the meaning of the word "pornography", she discovers in these books the description of different kinds of sex, "[...] not the sex married, religious people have, but the dirty kind, the kind people have for pleasure" (*Bone Black* 116). It is possible to see here a better understanding of what sexuality was in the protagonist's perspective, as she can recognize that sex exists, and that there are different kinds of sexuality, as well as that sex exists in different contexts and with different purposes, a reality that is evidenced in her comments regarding the religious aspect of sex. These books also reveal for the character that there are different possibilities regarding the people involved in the act of sex, as she is confronted with lesbian and gay sex, as well as other practices involving simply visual pleasure and masturbation, or in groups, discovering that reading these books have a similar effect to touching herself. However, sex is presented differently in this kind of fiction than the one that was not discussed in her reality.

> Sex in these new books fascinated her. There are no babies to be had through the excitement these pages arouse, no pain, no male abuse, no abandonment. She never thinks much about the role of women and men play in the books. They have no relationship to real people. The men do not work, the women do not have children, clean up houses, go shopping. Sometimes the men make the women do sexual acts. She could never under-

> stand how the women did what they didn't want to do, yet felt pleasure in doing it. She
> never felt pleasure doing what she did not want to do. (*Bone Black* 116)

Sex pictured in these books is devoid of the harsh reality women face, and that
she sees and experiences in her own home. It is also devoid of shame, and of the
negative consequences that are preconized by the patriarchal order. Her final
comment on the not being able to derive pleasure from doing things she does
not want to do is testament to her own growing sense of identity which is mainly
concerned with developing a positive sense of self that is not dictated by the desire of others.

Later on the protagonist is caught reading these sex-filled books by her
mother, who simply orders her to put the book back in place without asking
any questions. This lack of reprimand may be seen as an attempt of the mother
to protect her daughter from the possible anger that this transgression would
elicit in the father. Those books eventually disappear from the house and the
protagonist once again turns to literature as a substitute. Having access to the
public library, a space in which she often feels like an intruder given the racist
treatment she receives from one of the librarians. Her presence in the library was
met with distrust:

> Like may places in the white folk's world she knew they considered her presence at the li-
> brary an intrusion. They watched her suspiciously. When she checked out books they
> turned them over and over in their hands, as if the books were hiding some secret, as if
> they needed to understand why this black girl was reading this or that. (*Bone Black* 118)

There was, however, a nicer librarian who in the words of the narrator was "[...]
the one who did not treat her like dirt, who did not ask her Are you sure you can
read all these books" (*Bone Black* 119), expanding the horizon of the young reader by suggesting she tried reading writers such as George Elliot, Henry James,
Emily Brontë, and Charlotte Brontë. After reading all she could on the early
1900s romance, the protagonist would turn to paperback romances. This kind
of narrative would help her escape her reality once again, offering identification
and hope in its often used plot:

> The woman was almost always poor, a working woman, almost always missing physical
> beauty but by some slight flaw, hair that was too red, too short, too thin, long nose,
> poor eyesight. What beauty they possessed was always the inner quality, unfolding, blos-
> soming, when the right man came along. These women were always in need of rescue de-
> spite their independence, their work ability. They always loved their children. They would
> abandon work, travel thousands of miles to strange countries, to care for children. They

were always virgins. They always married. Their stories always had happy endings. (*Bone Black* 119)

The happy endings provided by these novels were exactly the kind of assurance that the protagonist wished she could find in her life. They provided the comfort of a loving family, as well as the resolution of flaws that she desired she could apply to her own life. The narrator concludes that "[r]omance fiction gave her escape, release, a feeling of satisfaction, a belief in the possibility of self-recovery" (*Bone Black* 120). Another positive effect brought by the reading of these romances was the normalization of reading in the eyes of the family, as they perceived her as less strange when reading popular novels instead of the overly complex classics of literature.

Regarding her own sexuality, as she develops into a teenage girl, the protagonist's rejection of any kind of form of domination shuns her away from pursuing any kind of relationship with boys at that time. Her family, once again, believes that she must have a problem in this regard, and since matters related to the body are not discussed, her desires and inclinations are never made clear. The narrator describes the reasons why she does not feel comfortable when dealing with boys her age:

THEY ARE CONCERNED because she has not shown the right interest in boys. They do not talk to her about what it is about boys that she finds boring, uninteresting. She cannot talk to them. She cannot tell them how much she hates anyone to lord it over her. She cannot tell them that this is what boys often want to do. She cannot explain that she does not like to be touched, grabbed at, without agreeing to such touching. She is disgusted by the grabbing, the pleading that she let them do this and do that. Even when she is aroused, the feeling goes away when boys behave as though there is only something in this moment for them, something they are seeking that she must give. She is not ashamed to say no. She does not care that the word gets around that she will say no. She cares that she is left alone. (*Bone Black* 157)

The protagonist demonstrates that what she loathes are not boys specifically, but their attitudes, the scipt of heterosexual masculinity that must be followed in order to conform to the strict notions of gender identity construction, in addition to spelling out that both the constant pursuit of her body as a sexual object and the lack of consent to have her body touched distance her from getting involved in the rites of socialization that are expected at this stage of her life. Her admittance of feeling aroused, and later how this feeling is deflated, showcases that she identifies indeed as a heterosexual woman. Yet, the dynamics of courtship do not interest her, preferring to abstain from them at all. The concerns demonstrated by her parents are that she might be a lesbian, or as they describe it without ever explicitly saying it, that she might be "funny", as the narrator reports: "[t]hey are concerned that she might be growing up funny. They watch her behavior. They think about the way a certain funny grown-up woman showed intense interest in her. [...] They are sure that she is not showing enough willingness to seek out boys and do what girls do" (*Bone Black* 158). The first definition of "funny" comes when the protagonist talks about a group of homosexual men in their community, providing a characterization that starts with fear and shame, but that ultimately demonstrates respect and tolerance of this specific minority group. The narrator provides an account of their perception, stating:

> WHEN THEY TALK about same-sex love they use the word *funny*. They never say the word *homosexual*. As small children we think to be called funny is a nice way of talking about something grown-ups are uncertain about, ashamed and even a bit afraid of. Growing older we learn to be afraid of being called funny because it can change everything. Mostly men we know are funny. Everyone knows who they are and everyone watches and talks about their business. They are good men, kind men, respected men in the community and it is not their fault, not their choice that they are funny – they are just that way. They had to accept themselves and we had to accept them. We do not make fun of them. We know better. (*Bone Black* 136)

Regarding the understanding of the family in relation to sexuality, it is clear once again that there is a gender difference that tolerates male homosexuals but does not offer the same tolerance towards lesbians. This logic of respect and tolerance does not exist for women who display their homosexuality, as accounted by the narrator:

> When grown-ups talk about women who are funny, they are not accepting. Their voices are harsh and unforgiving. They do not see them as kind, respected, *good* women. They talk about them as unnatural, strange, going against god. I want so badly to know why these women must live secretly, must sometimes be married. It is hard to ask questions. When I do they let me know quickly that men have the right to do whatever they want to do and that women must always follow the rules. [...] Women who do not want to be with men must be made to feel bad, ashamed, must be excluded from all community of feeling so that they will come to do what is expected of them – if not, they will be punished, they will be alone – they will not be loved. (*Bone Black* 138)

Women are not granted the same acceptance as men as they exist in this patriarchal order, as men might enjoy the privileges of the dominating position even in situations of discrimination, maintaining their symbolic position of superiority over all other female subjects. As women stray from the heteronormative patriarchal domination, they become pariahs who must be brought to heel, so they can perform the few roles necessary for the mainlining of the order under these symbolic institutions. Female homosexuals occupy then a position of exile, and must exist in secrecy, so as to preserve a little more of their humanity in this discriminatory context. When the family suspects that she might be a homosexual, the ominous "they" constantly monitors her activities to assess any traces of deviance. The protagonist has a white friend who she occasionally can visit with. As their friendship develops she is watched closely, her phone calls are heard in secret, hoping some clue regarding her sexuality will come up. What the protagonist does share with this minor character is the feeling of being an outcast, the loneliness experienced by those who do not fit in. The narrator reports that the white girl had already attempted suicide as a consequence of this feeling of inadequacy, and thus what the protagonist offers her is her support and friendship. Once, after one of their visits, the white girl takes the protagonist home, and they embrace while they say their goodbyes, an action that is watched by the vigilant eyes of the parents, who tell her she must come in immediately.

> When she enters the door her mother and father say nothing, even though they have been watching. Later in the night they keep her downstairs and want to know what is going on between her and the friend. She tells them they are friends, nothing is going on. Her daddy says, Don't lie to me. She looks at him with anger and contempt. She has no answers for them. They tell her that they will have none of this in their house, that she will have to go. She is not sure why they are so upset. She does not understand. Shaken by the fear of being told to leave, by threats of punishment, she agrees to stop seeing her friend. She does not understand why they want to take this friend away from her. She does not know that they are worried that she may grow funny. (*Bone Black* 159)

Just like she had previously been warned, her possible deviance from the patriarchal order is threatened with exile, and like the women who are abused for their transgression, she is forced to abide and distance herself from her friend. When she does finally demonstrate interest for a specific boy, the family seems to feel more at ease, as she is definitely an outcast, but she is not an "unnatural" being. Her exploration of sex happens with a suitor that is inadequate in the perspective of the family, since he is younger than she, a fault that is overlooked since that is better than being considered "funny". The preoccupations of the family are different now, as they warn her not to become pregnant. The interest in this specific boy comes from his behavior that is different from the pushy teenagers she had met before. "He does not plead with her to give him some. They are content to touch each other, to explore" (*Bone Black* 160). These affirmations demonstrate that what she is interested in is finding a partner that would see her as an equal, in which their desires would meet, and not superimpose on each other. They break up later because he does not share much of her interests. She is then interested in another boy, a basketball player, and this time his physical attributes are the reason she mostly develops an attraction for him. As he demonstrates that the interest is reciprocal, she decides to move a step further in her sexual exploration, but her intentions are frustrated:

> He moves into the backseat as if he is entering a cage, a trap in which he and not she will be imprisoned. He feels her innocence is too much. She is beginning to feel afraid, afraid because she is innocent. He has never let innocence stand in his way. It is her trust that catches him. He is not to be trusted. Perhaps he, too, has heard the words A black nigger is no-good nigger. He wants to be trusted. His cold hands around her neck do not make her afraid. He says no to her in caresses and kisses. He says no, the night is fleeting – it must be late – he must take her home. (*Bone Black* 161–162)

In this episode hooks complicates notions regarding heteronormative archetypical figurations of masculinity through the insertion of the category of race in her description. Firstly, her love interest does invest in an image of masculinity that is standardized, exulting the male attributes that are expected in his manhood. This apparatus is challenged by her positioning as unafraid, as an assertive interlocutor that is playing along with his script of dominant man, but that certainly demonstrates that her attitudes derive from her own desire. Her innocence, possibly a veiled reference to her virginity and inexperience, is brought as one of the obstacles he must overcome, but it is one he would have no trouble navigating. The setting, an abandoned road, and the description of his cold hands around her neck are also triggers of a scene of violence, creating an atmosphere that is ambiguous, which would better contribute to represent the dispute that was taking place in him. The marker of race is the one which complicates their exchange, as he does not want to conform to the preconception that he is taking advantage of her, that he is indeed someone to be trusted, and to prove so he must withdraw from this arrangement, negating his first impulse to assert his dominant position as male.

Race also comes at play when a white classmate demonstrates interest in her. Her mistrust in white people is rooted in the common belief that white men and black women can only bring misfortune to each other. As a child, the protagonist experiences an exchange with a white neighbor, one of the poor whites who lived near their country house, in which a man asks for a kiss in trade for some popcorn. She is dared by her brother to accept the offer, already regretting her choice:

> She knows better, knows that kisses are for friends and other loved ones. She fears the history of this exchange. White man taking black girls, black women, the word they do not understand but hear the grown-ups use: white men raping black women. After eating the popcorn he assures her that he will tell as soon as they are home, that she will be punished. Rushing home, running through the dark, she hopes the punishment will wipe away the feeling of shame. (*Bone Black* 33)

As a child, the protagonist believes that the punishment would overtake the feeling that something wrong had happened. The lack of understanding of what "rape" means, but the knowledge that it happens with black women and girls and that it is performed by white men is enough to color this exchange as something negative. As a teenager, the narrator reports many cases in which whites abuse their position, such as the white man who drives naked through their neighborhood, or the ones who shout obscenities to women in the streets, unafraid of the consequences in their still segregated reality.

In the beginning of the process of integration, exceling black students were allowed to take classes with their white peers. As an attempt to create a bridge between black and white students, the students in the "smart classes", comprised completely of black girls at first: "[w]e are not surprised that black boys are not in the smart classes, even though we know many of them are smart. We know that white folks have this thing about black boys sitting in class with white girls" (*Bone Black* 155). Black boys are introduced little by little in these special classes, under heavy supervision, hinting the fear of the rape of white girls. The conversation regarding the integration of schools and the issues that arise from the measures taken to enforce this reality are seen as futile by the protagonist, who states:

> We are tired of the long hours spent discussing what can be done to make integration work. We discuss with them knowing all the while that they want us to do something, to change, to make ourselves carbon copies of them so that they can forget we are here, so that they can forget the injustice of their past. They are not prepared to change. (*Bone Black* 156)

When describing the life of segregated schools, the protagonist states: "[a]lthough black and white attend the same school, blacks sit with blacks and whites with whites. In the cafeteria there is no racial mixing. [...] School is a place where we came face to face with racism" (156). She is taught to distrust white men in all environments, even in school, as figures of authority might abuse their power to commit violence (sexual, psychological and physical) (*Bone Black* 164).

This mistrust of white men might, an act of self-preservation, may be traced back to slavery times, as black women were often raped and abused by slave owners, a kind of violence that continued in different forms until our times. One white boy in her class displays an interest in her as they work together in different projects, eventually becoming friends. Her parents approve of their friendship in school but do not want him to visit her at home and she is disappointed in her parents for maintaining these racial divisions inside their home. She is invited by the boy to have dinner in his house, something her parents reluctantly accept. The dinner, however, reveals his true intentions:

She can see that he has demanded that his parents prove they are not racists with actions, not just with words. She admires his parents, that they love him enough to act. She tells him later that she will not be this little experiment he uses to test his parents. Alone in his room, listening to records, she says no to his kiss. She says no she will not be used to test his parents' love. They are friends. He is not surprised that she can see through him. He tells her that she cannot be interested when it seems that she is only a way for him to announce his own rebellion. They take the long way home. A carload of white men seeing a black girl and a white boy together try to run them off the road. (*Bone Black* 165)

Race impacts the relation between these characters differently from the previous case. Here, her mistrust of whiteness is expanded when she perceives that though her classmate is not a racist in the terms that she has been taught to fear and avoid, he continues to be racist as he uses her to publicly demonstrate his acceptance of difference. In his perspective, she is simply a prop that is going to be used to prove a point to his parents. He does not demonstrate any remorse as she can astutely perceive that this relationship exists only to benefit him, as her desires will not be met, and that deep down she would never be fully accepted in his reality. Finally, the events concerning the ride back home showcase the racist hatred still felt by many during that time, corroborating the protagonist's education on whiteness, and that white man were dangerous and should not be trusted.

4.2.4 A Door Closing in a Room Without Air – Domestic Violence and Childhood

In one of the rare passages that precisely indicate the age of the protagonist, hooks is sixteen years old, and the subject of marriage is brought up one more time. Her distancing from the idea of marriage at this time was more concerned with the role of men in the matter than with the education of women. The narrator describes: "[w]henever she thought of marriage she thought of it for someone else, someone who would make a beautiful bride, a good wife. From her perspective, the problem with marriage was not the good wife, but the lack of good husband" (*Bone Black* 97). hooks stresses that the women had been prepared to be part of the patriarchal structure from the beginning of their lives, leaving aside in this argument their position of vulnerability in the patriarchal structure. Her rebellion is shown as she defies the education she is given, as her mother tries to teach her the importance of cooking and cleaning as skills a good wife would possess. When she tries to explain to her mother the reasons she does not want to be married, she states:

[...] Seems like, she says, stammering, marriage is for men, that women get nothing out of it, men get everything. She did not want the mother to feel as if she was saying unkind things about her marriage. She did not want the mother to know that it was precisely her marriage that made it seem like a trap, a door closing in a room without air. (*Bone Black* 98)

Her explanations are received with scorn by the family, grouped in a non-disclosing "they", responding to her claims with another set of criticism regarding her body and temperament, confirming that they too did not believe she was fit for marriage, especially because of her constant talking back:

They agreed with her when she said that marriage was not a part of her dreams, they said she was too thin, lacking the hips, breasts, and thighs that men were interested in. But more importantly she was too smart, men did not like smart women, men did not like women whose head was always in a book. And even more importantly men did not like a woman who talked back. She had been hit, whipped, punished again and again for talking back. They had said they were determined to break her – to silence her, to turn her in one of them. (*Bone Black* 98–99)

The description created by the ominous "they" focuses on the traits that are desirable in a wife according to a patriarchal order, designating that besides her undesirable body, what was more relevant in her character was her argumentative personality. Her intelligence figured as something that would hinder her prospects, since she could not easily be manipulated, or even, she did not display the capability to be submissive that was necessary to become a desirable wife and later a mother. This "they" also reports the ways in which the patriarchal order was imposed over her body, in the violence of a gendered education that was determined to "break her", to tame her sense of identity that insisted on existing outside the norm of sexist hierarchies.

hooks' understanding of the dynamics she could see in her house discouraged her from pursuing the life of a wife mother. Her intent of trying to hide these notions from her mother, however, demonstrate the extent to which she would like to protect this woman figure, focusing her attention on the role of men, and more specifically, on her father's role, in the creation of the imbalance she saw in married life. By observing her mother's behavior, in the presence of her father and otherwise, she concluded that marriage was not a positive experience for women. She describes:

She could not tell her mother how she became a different person as soon as her husband left the house in the morning, how she became energetic, noisy, silly, funny, fussy, strong, capable, tender, everything she was not when he was around. When he was around she became silent. She reminded her daughter of a dog sitting, standing obediently until the master, the head of the house, gave her orders o move, to do this to do that, to cook his food just so, to make sure the house was clean just so. (*Bone Black* 98)

The understanding that hooks had of this marriage, and finally of marriage at large, was that women existed to serve men in this arrangement, and that any traits that would display a more proactive personality, such as the ones that her mother suppressed when in the presence of her father, should not be fostered. For the protagonist, the price to pay for being a good wife and mother, suppressing one's identity and conforming to a passive role, was too much to be paid to participate in this social compromise. hooks also comments on her parents' sex life, stating that she never heard funny noises coming from their bedroom, which was located directly under her and her sister's. What she does hear, however, is the voice of a woman who must humiliate herself in order to obtain a little more financial support from her husband: "[s]he heard the plaintive pleading voice of the woman – she could not hear what she was asking for, but she knew that the schoolbooks, the bit of pocket money, the new dresses, the *everything* had to be paid with more than money, with more than sex" (*Bone Black* 98). The currency of exchange in this relationship, and for hooks in marriage in general, was submissiveness, something she could not afford to offer. The joylessness of married life for women was something hooks did not understand, and her puzzlement was conveyed in the following statements: "[w]hatever joy there was in marriage was something the women kept to themselves, a secret they did not share with one another or their daughters. She never asked where the joy was, when it appeared, why it had to be hidden. She was afraid of the answer" (*Bone Black* 98).

hooks recounts of an instance in which her talking back resulted in direct corrective violence perpetrated by her father, in addition to commenting on the mother's reaction, illustrating one more reason why she would not like to participate in married life:

> She answers her mother back one day in the father's presence. He slaps her hard enough to make her fall back, telling her Don't you ever let me hear you talking to your mother like that. She sees pride in her mother's face. She thinks about the way he speaks to her, ways that at this moment do not matter. He has taken a stand in her honor against the daughter. She has accepted it. This, the daughter thinks, must be a kind of marriage – and she hopes never to bear a daughter to sacrifice in the name of such love. (99)

The violence that is described here illustrates the conflicting relation that hooks develops with her parents, as the father only communicates with her through violence, at the same time he blames the mother for not educating the children better, which is her responsibility as a mother according to the patriarchal order.

The mother does not interfere, which is perceived by the protagonist as a sign of siding with the father who is enacting justice, vindicating the mother's grievances caused by this rebellious child. Marriage figures for hooks as a bad arrangement, that creates more pain than pleasure to all involved (and especially to women), if there is any pleasure to be found.

Her dismay with marriage also includes an episode of domestic violence as the violent kind of "education" that her father provides also extends to the mother, attacking her in a surge of rage and jealousy. The children watch this episode, fearfully trying to make sense of his actions:

> Out of nowhere he comes home from work angry. He reaches the porch yelling and screaming at the woman inside the house – yelling that she is his wife, he can do with her what he wants. They do not understand what is happening. He is pushing, hitting, telling her to shut up. She is pleading – crying. He does not want to hear, to listen. They catch his angry words like lightning bugs – store them in a jar to sort them out later. Words about other men, about phone calls, about how he had told her. They do not know what he has told her. They have never heard them talk in an angry way. (*Bone Black* 146)

The children, comprised in the non-descriptive "they" include the protagonist this time, emphasizing the impact of this act of violence in a collective manner. The language used to describe his anger, the words found in bits and pieces that escape their grasp are also illustrative of the incomprehension of this episode in their eyes. The mother is reduced to a position of passiveness, something that also is received as a shock by the witnesses, who do not recognize the woman if front of them as the proactive woman they are used to calling mother. The violence continues, as the narrator describes:

> Yelling, screaming, hitting: they stare at the red blood that trickles through the crying mouth. They cannot believe this pleading, crying woman, this woman who does not fight back, is the same person they know. The person they know is strong, gets things done, is a woman of ways and means, a woman of action. They do not know her still, paralyzed, waiting for the next blow, pleading. They do not know their mama afraid. Even if she does not hit him back, they want her to run, to run and to not stop running. She wants her to hit him with the table light, the ashtray, the one near her hand. She does not want to see her like this, not fighting back. (*Bone Black* 146–147)

Just like the children feel too afraid to take action while the father disciplines them, the mother does not fight back, reduced as she seems to a position of powerlessness. The desire of the collective "they" is for the mother to escape that situation, to encounter means to flee, not to avenge this violence. The personification of the discourse takes place as the narrator switches from the plural "they" to the singular "she", demonstrating the particular feelings of the protagonist as she witnesses this violence, resorting to more hypothetical violence as a way of defense. The presence of the witnesses is noticed, and more threatens of violence are heard, as the protagonist tries to resist the dominant power exerted by the father, focusing on the safety of the battered mother.

> He notices them, long enough to tell them to get out, go upstairs. She refuses to move. She cannot move. She cannot leave her mama alone. When he says What are you staring at, do you want some too? she is afraid enough to move. She will not take orders from him. She asks the woman if that is right to leave her alone. The woman – her mother – nods her head yes. She still stands still. It is his movement in her direction that sends her upstairs. She cannot believe all her sisters and her brother are not taking a stand, that they go to sleep. She cannot bear their betrayal. When the father is not looking she creeps down the steps. She wants the woman to know that she is not alone. She wants to bear witness. (*Bone Black* 147)

The defiance of the patriarchal order achieves its culminating point as the protagonist decides she will no longer take orders from the perpetrator of violence. The defense of her mother exemplifies the kind of solidarity among women that would ultimately foster the combative attitude necessary to oppose the violence imposed by the patriarchal order that ultimately wishes to maintain women, and all those who do not belong to the patriarchal order, in a position of submissiveness. The protagonist wants to assure the wellbeing of her mother, risking her own wellbeing in this effort. The mother's nod may be seen as a form of reciprocity of this solidarity, as she also seems to want to protect her daughter from the violent father. The protagonist's feelings toward her siblings reproach their lack of attitude when facing this violence, as they abstain from interfering, trying to protect themselves, proving in the eyes of the protagonist that those who are neutral in situations of oppression involuntarily side with the oppressor. The violence continues as the daughter refuses to go to her room, and hiding on the steps of the stairs watches as the men continues to threaten the mother:

> She sees that the man has a gun. She hears him tell the woman that he will kill her. She sits in her place on the stair and demands to know of herself is she able to come to the rescue, is she willing to fight, is she ready to die. Her body shakes with the answers. She is fighting back her tears. When he leaves the room she comes to ask the woman if she is alright, it there is anything that she can do. The woman's voice is full of tenderness and hurt. She is in her role of mother. She tells her daughter to go upstairs and go to sleep, that everything will be all right. The daughter does not believe her. Her eyes are pleading. (*Bone Black* 148 – 149)

The daughter shows allegiance to the mother, demonstrating her concern, risking her life to protect her mother's. The role of the mother is fulfilled as she tries to protect the daughter from the possibility of violence, yet, her position as a victimized wife is clear for the protagonist. She knows that nothing will be all right, and that the mother and possibly herself are in imminent danger. The father returns and tells the protagonist to go upstairs once again, but she refuses to acknowledge him as a figure of authority. "He turns to the woman, tells her to leave, tells her to take the daughter with her" (*Bone Black* 149). The protagonist's presence and resistance possibly saves their lives. While the visibly shocked mother prepares to leave the house the protagonist is able to perceive the economic and material influences that determine the submissiveness of the woman in this marriage, as she is the one who must leave the house and is allowed to take only her clothes with her.

> The woman does not protest. She moves like a robot, hurriedly throwing things into suitcases, boxes. She says nothing to the man. He is still screaming, muttering. When she tries to say to him he is wrong, so wrong, she is more angry, threatening. All the neat drawers are emptied out on the bed, all the precious belongings that can be carried, stuffed, are to be taken. There is sorrow in every gesture, sorrow and pain –like dust collecting on everything, so thick she can gather it in her hands. She is seeing that the man owns everything, that the woman has only her clothes, her shoes, and other personal belongings. She is seeing the woman can be told to go, can be sent away in the silent, long hours of the night. [...] The gun is pointed at love. He lays it on the table. He wants his wife to finish her packing, to go. (*Bone Black* 149)

The protagonist believes that what she is witnessing in not the possible killing of a woman, but actually the "death of love", as she believes that had there been love between her parents it would have stopped everything. This interpretation, which might be seen in a sentimental light, exposes that the love that created this relationship is no longer present, as he does not respect her fully, and perceives her as less important, or even less human than he is. Seeing her as his property, the woman is cast-off from his house as he pleases. The mother tells the daughter that she does not need to go, that this "is not her fight". The daughter sees in this another motherly gesture, wishing to protect the daughter from the uncertainty of what is to come. She does not consider that the daughter, and even the rest of the children, may be in more danger staying with this violent father. The mother's brother comes to pick her up, not questioning the father, not questioning the victimized woman, simply accepting that he is exercising his prerogative as the head of the house. "She cannot bear the silent agreement that the man is right, that he has done what men are able to do. She cannot take the bits and pieces of her mother's heart and put them together" (*Bone Black* 150). The protagonist cannot perceive how this form of masculinity is not fought against, as it seems that everyone concurs in the violence that is taking place in front of their eyes. Her inability to help her mother also shows the limitations inside the patriarchal order, that even though she perceives that something is not correct, that there are structural forms of oppression that are difficult to be challenged by subjects who are already being impaired by them, that risking her life might not be enough, and that she cannot ultimately help her mother as she wished her mother would help her.

The next vignette seems to take place some time later, with the mother back in the house. The readership is not given access to how the things were resolved, or how long it took for the mother to be back, neither the family repercussions of not having the mother at home. What is clear now is that the protagonist does not forgive the mother for resuming her position of wife. This is seen through the description of frequent disagreements between mother and daughter. More importantly, the factor that creates the fights between them is that the mother sides with the father in matters related to the protagonist, something that she attributes to the idea of marriage, the couple supporting each other's decisions. The protagonist's rebellion becomes even more evident, as she feels betrayed by this mother figure:

> She is hurting me. This is my dream of her – that she will stand between me and all that hurts me, that she will protect me at all costs. It is only a dream. In some way I understand that it has to do with marriage, that to be the wife to the husband she must be willing to sacrifice even her daughters for his good. For the mother it is not simple. She is always torn. She works hard to fulfill his needs, our needs. When they are not the same she must maneuver, manipulate, choose. She has chosen. She has decided in his favor. She is a religious woman. She has been told that a man should obey god, and that a woman should obey man, that children should obey their fathers and mothers, particularly their mothers. I will not obey. (*Bone Black* 151)

The acceptance of her role of wife, confirmed and sustained by religion, sets this mother and daughter against each other. The protagonist does not fail to see that to ultimately continue to be part of this structure, possibly for the benefit of the children, this mother must sacrifice not only her daughter but her own needs. The mother is put in an impossible position, in which her needs will probably never be fully met. Later, while physically punishing the daughter, the rebellious nature of the girl is seen in a new level, as she does not allow her mother the hit her:

> She says that she punishes me for my own good. I do not know what it is I have done this time. I know that she is ready with her switches, that I am to stand still while she lashes out again and again. In my mind there is the memory of the woman sitting still while she is being hit, punished. In my mind I am remembering how much I want that woman to fight back. Before I can think clearly my hands reach out, grab the switches, are raised as if to hit her back. For a moment she is stunned, unbelieving. She is shocked. She

tells me that I never *ever* as long as I live raise my hand against my mother. She is even more shocked. Enraged, she lashes out again. This time I am still. This time I cry. I see the hurt in her eyes when I say I do not have a mother. I am ready to be punished. My desire was to stop the pain, not to hurt. I am ashamed and torn. (*Bone Black* 152)

In *All About Love* hooks comments on the dynamics of abuse, stating: "[m]ost psychologically and/or physically abused children have been taught by parenting adults that love can coexist with abuse. And in extreme cases that abuse is an expression of love" (9), something that is seen in the beginning of this vignette. Remembering her mother and her vulnerability, the protagonist decides to take action and stop the violence, disrupting the patriarchal order (backed by the religious discourse). Though the daughter does not wish to hurt the mother, the attitude of standing up for herself signals that she is no longer passive, that she does not want to perpetuate the order under which she was raised. The violence of daughter towards mother takes place in discourse, as she unmakes the sanctity of the order by claiming she does not have a mother, a decision she regrets when seeing the reaction of the mother. The mother then threatens to report these events to the father, complying one more time with the patriarchal order, something that corroborates the feeling of betrayal she has towards the battered mother she continues to pity. The relation between the mother and the father seems to be resumed without much complication, something the protagonist does not believe to be true, thinking that the violence continues to take place. "Although they act as if everything between them is the same, that life is as it was. It is only a game. They pretend. There is no pain in the pretense. Everything is hidden" (*Bone Black* 152).

One of the outcomes of the violence experienced by the protagonist is the occurrence of repeated nightmares related to the night of the violence. Dream discourse is used one more time in the narrative to illustrate the extent of the impact of the presence of violence in the life of the protagonist. In her sleep the protagonist revisits the violence as if she were watching a film regarding not the act but the consequences of it: this time the man does kill the woman and also kills their daughter:

In her sleep is the place of remembering. It is the place where there is no pretense. She is dreaming, always the same dream. A movie is showing. It is a tragic story of jealousy and lost love. It is called *Crime of Passion*. In the movie a man has killed his wife because he believes she has lovers. He has killed the daughter because she witnesses the death of the wife. When they go to trial all the remaining family come to speak on behalf of the man. At his job he is calm and quiet, a hardworking man, a family man. Neighbors come to testify that the dead woman was young and restless, that the daughter was wild and rebellious. Everyone sympathizes with the man. His story is so clean and white. Like flags waving, they are a signal of peace, of surrender. They are a gesture to the man that he can go on with life. (*Bone Black* 153)

The inclusion of this dream in the narrative may be seen as the representation of PTSD symptoms experienced by the protagonist, who deals with the traumatic memory of violence in her sleep, similarly to Martine in *Breath, Eyes, Memory*. The memory of the violence would return repeatedly and uncontrollably, as the trauma remained unresolved. The protagonist's dream may be seen as a critique of the upholding of the sexist and violent patriarchal order, in which men are capable to be forgiven of their violent crimes by most members of the society, since they seem to be simply acting according to their masculine "passion", fulfilling their role as disciplining husbands and fathers, applying a justified kind of justice over deviating women. Society, including here even the members of the family of the victims, seems to comply to this order by upholding him as a calm, quiet, and hardworking family man, ignoring the crimes that completely contradict this characterization. Ultimately, men will not be punished for being agents of violence in a patriarchal society.

4.2.5 They Need to be Valued

The positive relationships that hooks had inside her family were few but were enough to salvage this period of her life, guaranteeing the development of a positive sense of identity that would embrace everything that was seen as wrong by most of her relatives. She finds solace in the company of the oldest generation of her family, focusing on her grandparents as sources of understanding and positive nurturing. This confession comes early in the narrative, as the protagonist recalls the words of a religious song learned in school, in which the lives of children are cherished and appreciated, something she hardly ever experiences in her harsh reality:

> They imagine that Jesus even if he is white understands children, respects them. When they learn the words to a song "Jesus loves the little children of the world, red and yellow, black and white, they are precious in his sight," they sing over and over again to the grown-ups, hoping they will hear and understand that children are precious, that they need to be valued. No one black or white seems to understand this except for a few old men. (*Bone Black* 32)

These empathetic men were few in hooks' life, but made an impact in her childhood. Her own father was exactly the opposite of this kind of masculinity, only displaying feelings of rage and frustration, arousing fear in all other members of the family. These other men were capable of demonstrating different, tender feelings, displaying emotion, showing interest, listening, and answering questions without judgment. The narrator stresses:

> TO HER CHILD mind old men were the only men of feeling. They did not come at once smelling like alcohol and sweet cologne. They approached one like butterflies, moving light and beautifully, staying still for only a moment. She found it easy to be friends with them. They talked to her as if they understood one another, as if they were the same – nothing standing between them, not age, not sex. They were the brown-skinned men with serious faces who were the deacons of the church, the right-hand men of god. They were the men who wept when they felt his love, who wept when the preacher spoke of the good and faithful servant. They pulled handkerchiefs out of their pockets and poured tears in them, as if they were pouring milk into a cup. She wanted to drink those tears that like milk would nourish her and help her grow. (*Bone Black* 64)

The respect and admiration the protagonist feels toward these members of her community derive from their openness to demonstrate their feelings, as well as from a position of equality that unmade any hierarchies that could exist between them. By describing the tears of these men as nourishing, she is pointing to the empowering aspect of showing vulnerability, of accepting one's own humanity, instead of aiming at unattainable models of masculine virility that are fundamentally created through violence inflicted upon others and upon the self.

Upon seeing an old man in the church, with a humped back, she asks her family the reason why he did not walk straight, something that received only a disapproving look back as an answer from her family. During service she would hold his hand and watch as he shed his tears. He told the little girl that he needed his hand back to work, to build his house, and invited her to visit him whenever she wanted. She could see the loneliness in his life, a feeling that she shared, and told him that she always held the right hand because she knew all his loneliness was stored there, something he automatically also saw about her. She was able to provide him with conversation and attention, and he provided her with respect and interest. When recounting about their afternoons, the protagonist stresses how they could talk without judgment, something she did not often experience (*Bone Black* 66).

Another older man that fits the description of a caring adult is her grandfather Daddy Gus, her mother's father. The protagonist describes their relationship in very positive terms, as he seems to provide her with the attention and sensibility that she often lacks. She states: "[w]ith him all the broken pieces of my heart get mended, put together again bit by bit" (85). Their relationship is special because he seems to notice the oppression that the protagonist faces in her own house. "He knows that I am a wounded animal, that they pour salt on the open sores just to hear me moan" (*Bone Black* 85). The terms used by the protagonist to describe her relationship with her family indicate the cruelty she feels her immediate family is capable of imposing upon her, and it seems that this older man is also capable of noticing it, offering comfort by his understanding. One of the main characteristics that differentiates him from the other masculine figures in her life is that he is not a man who uses harsh words, never displaying rage as a form of affirmation. The protagonist thus concludes that "I need his presence in my life to learn that all men are not terrible, are not to be feared" (*Bone Black* 85).

Daddy Gus is the one who provides the protagonist with the healing metaphor that will help her survive the difficult years of her childhood and adolescence, a primordial cave: "[h]is voice comes from a secret place of knowing, a hidden cave where the healers go to hear the messages from the beloved" (86). His tenderness guides her to this place of healing, a place in which she can restore from the negative aspects experienced in her life. This metaphor comes again in dream form, similarly to Saru's metaphor of the storyteller, conveying its message to the protagonist:

> In my dream we run away together, hand in hand. We go to the cave. To enter we must first remove all out clothes, we must wash, we must rub our body with a red mud. We cover ourselves so completely that we are no longer recognizable as grandfather and granddaughter. We enter without family ties or memory. The cave is covered with paintings that describe the way each animal has come to know that inside all of us there is a place for healing, that we have only to discover it. (*Bone Black* 86)

The description of the rituals that are necessary to enter the cave and finally reach the place of self-healing may be seen as the necessary steps to achieve self-awareness, leaving behind the markers that attach the self to a specific condition of being: memory, family ties, and body. By covering their bodies to the point of no recognition, the metaphor suggests an erasing of the self in order to have access to a better place. The cave does not guard the knowledge, but the ways in which each animal was able to have access to the place of healing, conjuring a kind of experience that is ultimately collective, and that is based on the sharing of stories. According to the narrator, upon entering the cave the first reaction of the animals is weeping, as they are finally able to really see, and the burden is too much to bear. This also happens to the protagonist and her grandfather. Then they make a fire:

> In the fire are all the lost spirits that show us ways to live in the world. I do not yet have a language with which to speak with them. He knows. He speaks. I am the silent one, the one who bears witness. In the dream we can leave the cave in quiet. Just as we reach the outside

he begins talking to me without opening his mouth. He places his voice inside my head telling me that knowledge of the cave can be given to anyone, only they must be seeking, that until I can tell a seeker from someone who is just curious I must not speak about it. (*Bone Black* 86)

The spirits of the animals may be seen as the diversity that the protagonist is able to see around her, different ways that have found acceptance in their own diversity. The grandfather's power to communicate with them stand for someone who has left judgment aside, that is able to accept the difference in others because he has accepted the difference in himself. By telling the secret of the cave to his troubled granddaughter, what he is really sharing is the knowledge of acceptance, the knowledge that she is able to thrive despite all the harsh and violent reactions of the outside world to her difference. She now is capable of sharing this wisdom with other seekers of true understanding.

This positive view that was found in the dreams is corroborated in their usual interactions, as the protagonist describes lovingly the ways in which this male provided her a sense of security, as well as a sense of wonder, filling her days with the imaginative input that was often obscured by the difficult life she experienced in her home. Daddy Gus is a collector of odd small objects, believing that each object is imbued with history. He also seems to be a writer of diaries, something that would ultimately inspire the protagonist:

Every object has a history. He teaches me to listen to the stories things tell, to appreciate their history. He has many notebooks, little black notebooks filled with faded yellow paper. I understand from him that the notebooks are a place for the storage of memory. He writes with a secret pencil; the pages seem covered in ash, the ash left by the fire we have visited. This fire he says now burns inside us. (*Bone Black* 87)

The description of the souls of the small objects contributes to the fostering of a perspicacious mind, which will constantly inquire its reality, as well as to the instilling of the notion that everything that surrounds her has a particular history. This positive perspective counteracts much of the notion that was imposed by her parents of not discussing things, of not focusing on, or even assessing the processes by which reality came to be. Finally, Daddy Gus provides an education on remembering the path of all things, material or otherwise. By connecting the fire of her dreams with the ashes found in the notebooks, the metaphor of healing is made concrete in the act of writing, a principle that would serve the protagonist as a coping mechanism in the working through of her family-induced trauma. Like the protagonist, Daddy Gus was also the target of critiques by this family, as she points out: "[t]he people in this house think of him as a coward, a small man shrinking into his chair like a shadow. They make fun of him, of his clothes, of his habits. They think all his treasures are junk" (*Bone Black* 89). His sensitive manner is not accepted by the family either, just like they cannot make sense of the protagonist's. Most of this criticism comes not only for the kind of life that Daddy Gus lives, but also from his not having wanted to go to war, for having refused to participate in the killing of other human beings. He tells the protagonist that all his sons have gone to war and lost part of themselves there. Yet, old age grants him the privilege to act as he pleases, something the protagonist is able to notice.

> THERE IS MUCH to celebrate about being old. I want to be old as soon as possible for I see the way the old ones live – free. They are free to be different – unique – distinct from one another. None of them are alike. Some of them were already on their way to being old when I was born. I do now know them young. I do not have to forgive them past mistakes. They have not caused me any sorrow. My grandfather tells me that all he ever wanted was for the world to leave him be, that it won't let you be when you are a young man. The world demands that you do work for it, make families, provide, take no time to listen to your own heart beating. (*Bone Black* 88)

The right to be different, to cultivate one's own idea of the self, is the privilege that the protagonist wants to obtain, something that seems to be granted on the basis of seniority. When the grandfather tells the protagonist that the world makes demands, and that his wish was simply to have the freedom to live his life the way he wanted, the protagonist encounters a perfect description of her own desires. However, the demands to conform to the necessities of the world, which fall mainly under a capitalist description of society (work, multiply, provide) are the forces that keep one away from the desired self. It is also interesting to notice that her sympathy for the older generations is something that derives from a clean slate, as they have not harmed her in any way, and therefore she does not need to pardon any wrong doing, does not need to overcome any grievance, constituting an exchange between generations that exists in a relation of love.

Another loving elder in her life is Big Mama, her great-grandmother on her father's side. Her description as a loving character happens during the protagonist's early childhood, demonstrating once again the importance of the older generations in her identity construction. Like Daddy Gus, Big Mama treats children with kindness, and though the protagonist does not describe a particular experience like the one with her grandfather, she certainly provides a sense of reassurance and respect when dealing with the younger members of the family. The narrator states: "BIG MAMA – TO us she is special, unique, one of a kind. We do not know that there are other big mamas in the world. She is short and fat. We stand and look straight into her eyes even though we are children" (*Bone Black*

25). The comment that the children can look at her straight in the eye might be interpreted in two different ways: the children are the same height as Big Mama, or that she is does not care for the hierarchies that the protagonist is submitted to at home, allowing the children to connect to her on the same level that she connects to them. Given her old age, it may also be inferred that Big Mama was possibly a slave during her youth, a hypothesis that is supported by the narrator: "[w]e know she is old because she is our father's mother's mother, because she does not read or write, because she chews tobacco and smokes a pipe. We know that women did these things in the old-old days" (*Bone Black* 25). As the subjects of slavery and racialization are not openly discussed in her home, the child's impression is limited by simply saying she learned doing these things in a distant past. Her desire to please the grandchildren is also perceived by the adults: "[t]he grown-ups say she lets us have our way. They are not eager to let us go and stay in her house. We come back spoiled" (*Bone Black* 26). Her description might be read as a stereotypical Mammy, an overly sweet black woman, fat and short, who would please all the children, spoiling them with her kindness. Food is an idea that is also associated with this character, as she is the bearer of sweets, always available inside the pocket of her apron, and also a cook, mastering the wood-burning stove. "At Big Mama's the kitchen is our home" (*Bone Black* 26), a part of the house that in her own home she does not remember very fondly, as it is the place of house work for her mother, and later would be a place for punishment when the protagonist does not eat. While baking a cake to their father, the children drop it on the ground while removing it from the stove. Instead of castigating them she provides comfort, demonstrating the care and love that the protagonist often lacked at home:

> Big Mama says there is plenty of time in life to bake cakes, tells us about not crying over spilled milk. We love this telling. We would rather not cry, it is the punishment we fear usually when we spill things that makes us cry. Big Mama never punishes us. She always talks soothingly, quietly, laughing in between her words. We love her because she does not hide anything from us. Her pocketbook is not a secret woman world we cannot enter. We can go all through it searching for dimes and pennies. She tells us stories of our father as a young man. She is the one who gives us caring memories of him as a young boy who loved her, who waited on her hand and foot as we do, who searched her apron pockets as we do. (*Bone Black* 27)

Big Mama is responsible for providing the protagonist and her siblings the positive memories they would have about their father. He too was once a loving child, a representation that differs from the adult that these children know. By telling the stories of his childhood, Big Mama reconstructs a positive side to this character that the children do not have access to.

Saru, the grandmother that is conjured in the first dream of the narrative, also figures as an important member of the older generation that has created a positive impact on the protagonist's life. Saru's Native-American ancestry is firstly described as the protagonist revisits a box with old photographs:

> She is the woman in the black-and-white photo wearing a blue and green cotton dress with buttons like tiny pearls, wearing no shoes with her hair hanging, jet black, long and straight. This is what the camera does. Her hair has not been combed. It has not been plaited into the two braids that identify this woman as the grandmother I have always known. It is she who explains that picture taking is no innocent act – that it is a dangerously subtle way we drive our souls into extinction. If this is not so why is it that the photographers always manage to arrive just when the tribe is dying out, just when the traditional practices lose power, just when people are blinded by sorrow. (*Bone Black* 46 – 47)

Saru's mistrust of the photographs and photographers declares a clear position of resistance to the colonialist invasion, seeing in the act of registering the images the marker of the extinction of her own traditions. The reference to dying out tribes, and the weakening of the traditional practices also determines the impact of colonialism on her life and her ancestors. Seeing herself dressed as a cowgirl when a child in one of the photographs, the protagonist talks about the joined plight of Native-Americans and African-Americans, as her opponents are not the Indians, as the traditional western narrative would state, but their common perceived antagonist, white men:

> I want never to grow up, to be a cowgirl forever riding in my skirt, with matching vest and hats, with my pointed boots and my one gun. I can defend myself against any enemy. I can shoot straight. I do not kill Indians – they are family. I protect us from the enemy white men. I shoot straight. (*Bone Black* 47– 48)

Saru recounts of the dynamics of miscegenation that often took place, revealing some part of her past. She states that lighter-skinned black men frequently wanted to marry indigenous women, revealing that maybe her father was one of these black men, who would prefer "[...] not a white-skinned bride, but a woman with skin the color of warm honey, with straight jet black hair, blacker than white folk's hair" (*Bone Black* 49). According to Saru these kinds of unions were common in the past, possibly a way to "lighten" the race without running the risk of committing a crime. Black men would often be chastised for claiming Native-American heritage, something that Saru feels grieved about since forcibly denying a part of their ancestral past would cause immense harm to the person, hinting that Saru herself might have had to conceal her origins at some point. While discussing her ancestral past, Saru comments on the influence of school in the protagonist's life, as it clearly comprises and disseminates visions of racial minorities that are biased, favoring the colonial project:

> SARU CALLS THEM the People of the First Snow, I call them Indians. I tell her we learn at school that they are Indians, that like the Africans, they were called savages in our books. She tells me that we go to school to learn the white men's way, to learn to deny parts of ourselves. (*Bone Black* 52)

Saru contends that she has learned much from her mother, knowledges that are not part of the "white men's way", survival skills that would help them thrive. Learning to hunt, trapping small animals for food, learning to keep worms for fishing, as well as learning to grow things, dealing directly with the earth, skills that she tries to pass onto the protagonist. From her, the girl learns that it is possible not to buy everything from a store, and that the equality of the sexes is real: "[t]he sight of her eldest daughter whirling a chicken in the air without blinking, without feeling moved by its cries and scattered feathers convinces me that women are the equals of men" (*Bone Black* 58). She also teaches the protagonist that believing in god has nothing to do with going to church, demonstrating that her spirituality was also not compromised by the "white men's way". The protagonist describes Saru as follows: "[s]he was a woman of spirit, a woman of strong language, a fighter. She tells me that she has inherited this fighting spirit from her mother, that I may have a little of it but it is too early to tell" (*Bone Black* 53). When describing her infancy, Saru mentions the people who arrived from Africa remembering their original homes across the sea, referring to a time in which the connection between the experience of the middle passage and their present history was still fresh on their minds (*Bone Black* 49–50). The erasure of memory in the case of the Africans, the ways in which they "stopped talking" about their ancestral land is attributed to the white men, an oblique reference to colonialism. By separating the peoples from different lands and languages, depriving these subjects from the ability to share their past, colonialism successfully unmade the connection between lived history and memory. On the other hand, Saru lays claim to her indigenous background as a form of resistance to the indignities of slavery, refusing to deny this part of her ancestral past, finding in it a locus for identity construction. Saru is also aware of the need to share the oral history of her Native-American side, maintaining her past alive through the act of actively partaking it with the younger members of her family.

Later, the protagonist inserts another dream in her narrative, this one dealing with a story told by Saru about a magic woman who lived inside smoke as a way of never being captured, being able to turn the smoke into other things for her benefit and protection. She uses the smoke to turn herself into a male, so she can become a warrior. The protagonist sees the warrior in her sleep and finds that the male warrior has her face: "I stare into his eyes as if I am looking into a mirror" (*Bone Black* 51). The magic warrior is Saru, and when she turns back into a woman, she no longer has the protagonist's face. Troubled by this dream, the protagonist reaches for the wisdom of the grandmother:

> When I tell Saru my dream, of the young warrior who wears my face in battle, she says that this is the face of my destiny, that I am to be a warrior. I do not understand. I do not intend to fight in wars or battles. She says that there are many battlegrounds in life, that I will live the truth of the dream in time. (*Bone Black* 51)

This dream foresees the protagonist's future, in which her battles are not fought in a war field, but in different arenas. The author may be hinting that her fight for social justice was already predicted, that the idea of being a woman who could face the adversities of life, and that would thrive even in the presence of conflict, was already present in her life since her childhood. Ultimately Saru's wisdom would help the protagonist understand the dream, creating a model of femininity that was not limited by the colonial heteronormative patriarchal archetype, fostering the creation of self-esteem, as well as corroborating the combative traits that compose the protagonist's identity.

Finally, the trope of the quilt resurfaces, as the readership is granted access to the place in which the quilt the protagonist received from her mother's hope chest is made. Saru finally redresses the motif of marriage in a positive light, as she tells the protagonist that the specific quilt she selected and would later inherit was one of the first she created as a young bride, and that ultimately quilting is a way of teaching a young woman the virtue of patience:

> When she is not fighting she is quietly making quilts. Sewing the small pieces of fabric together ease her mind. [...] In the sewing room she pulls back the mattress of a feather mattress made firm by layers and layers of quilts that rest under it. I select the quilt of my choice, the Star of David pattern. She tells me that it was one of the first quilts she made as a young bride. I imagine each part of the star, each different bit of cotton, has been stitched with the intensity of her love and will to make this marriage work, make it complete and fulfilling like the quilt. With my hands I trace the pattern. She tells me a woman learns patience making quilts. (*Bone Black* 54)

Besides the elderly generation in her family, the protagonist also finds solace in painting, as she discovers how her creativity finds a place to be explored through this media. At the care of Mr. Harold, the art teacher, the protagonist feels confident to investigate this form of expression. Although he is white, their relation is very positive, as he seems to be a person capable of understanding things across the color line:

In this integrated high school he is one of the few white teachers who do not keep black kids at arm's length, who is not afraid. He cares. He is the only one who seems to understand that the whites and their hatreds are the problem, and not us. He does not deny us. (*Bone Black* 170)

While in art class, the protagonist enjoys mixing paint and water, creating different hues and colors, before experimenting them in paper. For her, painting grants her another opportunity to visit the cave of her grandfather's dream, trying to remember the painting in the cave wall. She believes that if she is able to remember the animals painted in the cave walls, she will be able to remember the secret of living, the thing she had left back in the cave. The discovery of painting and pigments also provides the author with the name of her memoir, *Bone Black*, as she discovers that "Bone black is a black carbonaceous substance obtained by the calcifying of bones in closed vessels. Burning bones, that's what it makes me think about – flesh on fire, turning black, turning into ash" (*Bone Black* 170). The flesh on fire that is being elicited here may be a reference to the lynching of African-Americans during the 20th century, who were frequently burned after being brutalized and tortured. Black also figures as an important color for the protagonist since is signifies attaining womanhood, "Mr. Harold laughs at me when I tell him that all my life I have heard my mother say black is a woman's color – a color denied me because I am a child" (*Bone Black* 170). Mr. Harold does exactly the opposite: "[h]e does not deny me the color black. He urges me to stay with it, but to add more color, to do more with it" (*Bone Black* 170).

She does so, beginning by painting the cave as if in an expedition to this sacred place. By adding red to the black, she honors the hearts of seekers, human or animal. Then she paints the fire, adding red, blue, yellow and green. At the bottom of the fire, the color black stands for the ashes the fire will become, "[t]his is the remain of all the animals who have given their life in sacrifice to keep the spirit moving, burning bright" (*Bone Black* 170). Next she wants to paint a world covered in grey mist, representing the confused and undistinguished reality she finds once out of the cave. Then she tries to paint the animals, something she does not promptly succeed in:

I try and try but cannot get them right. Mr. Harold looks at me from his desk and says no as he sees me about to rip the paper, to throw it away. He shakes his head no. He has told me many times to keep at it, to look at it, to rethink what it is I am trying to do. Without remembering all the animals I leave watercolor behind; I am on to acrylic, to painting on canvas. (*Bone Black* 171)

The metaphor for the need to remember what happened in the past in order to live the present and to project the future is recuperated here, through the act of painting. Though the protagonist might not remember all the animals she saw in the cave, to keep looking at her attempts to remember is a form of learning from them anyway. When she moves to new materials, the protagonist paints in red, yellow and brown " [...] the wilderness [her] spirit roams in" (*Bone Black* 171), exploring life outside of the cave, trying to encounter the healing she found in the fire in places outside the mythical cave.

Finally, in the last vignettes of the memoir, the protagonist gets involved with religion as a way of coping with her harsh reality. She approaches the catholic church in the campus ministry. Her interests, however, might seem political at first, as she confides: "[l]eaders in the crusade for christ in our town do radical political work as well for they dare to cross the barriers between white and black" (*Bone Black* 172). The protagonist confesses though that she becomes slowly involved because her faith is not as strong as the others', since she does not believe in sin at all. She gives as example of a ritual performed to cure her asthma, and though it has not clinically resolved her problems "[i]t is the power of that night that makes all other nights of healing possible" (*Bone Black* 173). She experiences a retreat with people of the faith, something her mother allows since she has become "disconsolate", a word she heard during mass and that she believes describes her perfectly.

> I am not crazy, I tell them. I am disconsolate. I show them in the dictionary that it means dejected, deprived of consolation. Whatever it is, they are sick of it. They are waiting for it to go away. They do not understand that I am also waiting for it to go away. (*Bone Black* 173)

What the protagonist seems to be describing is her state of depression, something that neither she, nor her parents are able to address by themselves. hooks comments in *Sisters of the Yam*, though not specifically referring to her own experience, that depression might reach life-threatening levels, enlisting many factors that contribute to it, stressing that the matter of depression is not restricted to adult age. All the factors that are present in her comment may all be found in her memoir:

> Unreconciled grief, sadness, and feeling that life has lost meaning are all states of being that lead black women into life-threatening depression. Loss is no respecter of age. Very young children suffer debilitating depression. This is all the more likely if they are living in an abusive, dysfunctional family. For some grown black women, the depressions we face can be traced back to childhood roots. Some of us hold our pain through years and years, letting it trouble our health. (*Sisters of the Yam* 122)

The feelings of inadequacy in school and in life in general are factors that heavily contribute to her disconsolation, and the religious retreat offers the protagonist a chance to distance herself from these stressors, experiencing the possibility of a positive network of influence besides her grandparents. "Here among the faithful I can reveal that I am anguished in spirit. They understand the primacy of the spirit" (*Bone Black* 173). The protagonist confesses:

> I am glad to be at the retreat, to escape the tensions of home, the feeling that I stand on the edge of a cliff about to fall off. I know that many people come to god to be rescued, to be taken from the cliff and placed on solid ground. I have not been rescued. For comfort I read over and over the story of John the Baptist wandering in the wilderness. I too linger in the wilderness desperate to find my way. (*Bone Black* 176)

Although she does not experience the feeling of salvation in this religious experiment, she is able to feel comforted in this environment. The identification with John the Baptist is also relevant, since she is able to relate with feeling lost in a world, feeling inadequate and adrift. The imagery of the standing on the edge of a cliff is suggestive of suicidal ideation, as the protagonist's disconsolation might take her to the edge of life. The protagonist describes this space of loneliness she inhabits, expressing her most vulnerable feelings:

> LONELINESS BRINGS ME to the edge of what I know. My soul is dark like the inner world of the cave – bone black. I have been drowning in blackness. Like quicksand it sucks me in and keeps me there in the space of all my pain. I never say out loud that I could die in this space of loneliness, of outsiderness. I never say out loud I want to kill myself – to go away from all this. I never tell anyone how much I want to belong. (*Bone Black* 181)

Later, when hearing a talk given by a catholic priest as the opening session of the retreat, the protagonist experiences something different, the feeling that both are alone in the room and he is talking directly to her. In his speech he approaches the matter of loneliness. The protagonist feels she is seen completely, like no other time in her life:

> For the first time in my life I hear someone say that there is nothing wrong with feeling alone, that he, too, has been at the edge, has felt the fear of drowning, of being moved toward death without consciously contemplating suicide. [...] When I weep and sob over the slate grey clothing he tells me that the young woman standing on the cliff, alone and afraid to live, is only suspended in a moment of hesitation, that she will overcome her fear and leap into life – that she will bring with her the treasures that are her being: the beauty, the courage, the wisdom. He tells me to let that young woman into my heart, to begin to love her so that she can live and live and go on living. (*Bone Black* 177)

This religious experience does not concern divine intervention, but the realization that she is not alone in her loneliness, that this is an experience that happens with other people, and more importantly that there is a possibility for redemption. Though her relationship with Daddy Gus offers some consolation, he continues to be teased by an insensitive family. The experience of hearing the priest talk about her most intimate problems, even those she cannot precisely name like the suicidal ideation, functions like the fire in the cave she had previously explored with her grandfather, the priest's words are what ultimately saves her. The priest is even capable of reverting the cliff metaphor, as she is urged to jump into life with all her positive attributes. Ultimately, the protagonist learns from the priest that she must accept and love her inner self in order to live and keep on living. Later, the priest sends one of his students to spend time with the protagonist, bringing to her a copy of Rilke's *Letters to a Young Poet*, a collection of ten letters in which Rilke advises his friend Franz Xaver Kappus about his poetry over the course of six years. The most important message in these letters is for the young poet to stop looking at criticism, and to trust his inner feelings, to believe in himself.

Writing becomes the strategy to survive for the protagonist, as she finds identification in the words of Rilke. Commenting on the power of writing in *Sisters of the Yam*, hooks connects it to her spiritual experience, as it offers the opportunity to recover life, to give it new meaning, and to experience all that has been denied to her. Citing the work of William Goyen, hooks concludes:

> Writing was always a sanctuary for me in my wounded childhood, a place of confession, where nothing had to be hidden or kept secret. It has always been one of the healing places in my life. At the end of William Goyen's essay "Recovering," he states, "It is clear that writing – recovering life – for me is a spiritual task." Like Goyen, I believe that writing is "the work of the spirit." (*Sisters of the Yam* 197)

In words she is able to find the comfort that she so desperately needs, and most importantly, she is able to fully realize that her experience is shared by other people who are also deemed to be too sensitive, to feel too much. When she affirms that she has been seen, she reiterates that she has finally been perceived as a whole, not as someone who is always missing a piece. Ultimately, she is also able to see herself completely in the dark cave of her inner self. " [...] I tell myself stories, write poems, record my dreams. In my journal I write – I belong in this place of words. This is my home. This dark, bone black inner cave where I am making a world for myself" (Bone Black 182–183). Daddy Gus' wisdom is added to that of Rilke, as the protagonist finds her way to belong in the world of words. The protagonist, thus, evades the expectations of belonging to the larger world that is promised by the canonical *Bildungsroman*, meeting instead the possibility of belonging to a particular world of the private self. In becoming a writer she finds a way to be part of something, even if this something is not associated to the values of the society that surrounds her, ultimately refusing to fit in. hooks is also able to subvert the promise of achieving adulthood that is preconized in the canonical *Bildungsroman* at the end of *Bone Black*, only playing pretend, projecting her future self as someone who will ultimately be able to fulfill her wishes, to became a woman in her own terms, permanently deferring the conclusion:

> When she is older she will wear black every day. She wants to know how soon it will be, how soon will she be able to wear a black dress. They say never if you do not shut up talking about it. She cannot wait to be a woman. She cannot wait to wear the color black. She is looking in the mirror, playing pretend. She is a woman wearing a black dress. She is not mourning. She has learned to put all the broken bits and pieces of her heart back together again. She is a woman. She is dressed in black. She has been told all her life Black is a woman's color. (*Bone Black* 180)

Finally, hooks comments in *Talking Back* that by writing her autobiography she did not succeed in killing the Gloria of her childhood, she ultimately saved her by reclaiming the tortuous past in writing, making sense of it:

> In the end I did not feel as though I had killed the Gloria of my childhood. Instead I had rescued her. She was no longer the enemy within, the little girl who had to be annihilated for the woman to come into being. In writing about her, I reclaimed that part of myself I had long ago rejected, left uncared for, just as she had often felt alone and uncared for as a child. Remembering was part of a cycle of reunion, a joining of fragments, "the bits and pieces of my heart" that the narrative made whole again. (*Talking Back* 268)

The author also comments on the redemptive power of writing in *Remembered Rapture – The Writer at Work*, concluding that writing was a way of knowing:

> That woundedness that I was once so ashamed to recognize became for me a place of recovery, the dark deeps into which I could enter to find both the source of that pain and the means to heal. Only in fully knowing the wound could I discover ways to attend to it. Writing was a way of knowing. (*Remembered Rapture 16*)

5 Conclusion

"*There is really nothing more to say – except why. But since* why *is difficult to handle, one must take refuge in* how" (*The Bluest Eye* 4). This sentence belongs to *The Bluest Eye*, in which Toni Morrison vividly explores the matters of child abuse, poverty, violence, racism and discrimination, all the while dealing with the slow descent into madness that the abused protagonist went through, after being raped by her own father and carrying the baby to term. The facts that constitute Pecola Breedlove's predicament were, hence, the *how* of all that happened. The *why*, on the other hand, as the author strikingly puts it, "is difficult to handle". Though this novel is not directly the object of analysis here, it is one that has always been present when considering the narratives under scrutiny here. Morrison's words convey the discomfort and the necessity of discussion that underlies the experience of dealing with matters of violence and trauma ("the unspeakable things unspoken"), especially when it is related to younger vulnerable subjects. This work exists much in this spirit, as it tries to tackle the *how* of violence and trauma present in coming-of-age narratives, since the *why* of this presence continues to be a challenging interrogation.

The present work intends to demonstrate the ways in which trauma and violence are present in the literature produced by African-American and Afro-Caribbean women in the United States, broadening the comprehension regarding those topics in relation to the unveiling of inequalities, as well as the denouncing of how structural violence and trauma are present in the coming-of-age stories that were analyzed. The present work also intends to demonstrate the ways in which the different kinds of resistence to violence, in addition to showcasing the strategies found in the texts related to the attempt of overcoming trauma that derives from the experience of violence. The investigation regarding the presence of trauma and violence in coming-of-age stories and the aspects of subversion of the traditional *Bildungsroman* genre also contribute to a larger understanding related to the representation of both these conditions, and the development of the specific genre itself, as it is adapted and reinvented to encompass the different concerns voiced by the authors, all of them dealing with the complex intersectionality of gender, race, and class.

In *Lucy*, Kincaid explores the ways in which colonialism continues to be a force that strongly shapes reality, demonstrating as well the (conscious) obliviousness of her white counterparts to the protagonist's claims of violation. The matters related to identity construction in this novel are also revealing of the colonial education that continues to exist in our times, as the character of the mother, the school, and even the politics of the island insist in upholding a system of

https://doi.org/10.1515/9783110752755-007

beliefs that does not correspond to Lucy's needs as a young black woman born in the Caribbean. The encounters with whiteness in the United States are illustrative of the rage that builds up in Lucy from being constantly misunderstood, or from not being acknowledged as a full human being. The trauma and violence that are reported in Lucy's story are finally related to the cultural trauma of a colonized subject in our times, whose identity becomes the locus for the historical violence.

Danticat's *Breath, Eyes, Memory* deals with questions of tradition and violence. Practices that have been carried out for generations are responsible for the creation of trauma and violence in the lives of contemporary people. Danticat addresses these issues by describing the ways in which these practices need to be re-examined by all those involved in its culture of perpetuation. She approaches the question of overcoming trauma through the reassessment of what is meant to be kept and what must be eliminated from culture, while she interrogates her own role as a woman in the fostering of a safe environment for women and girls to develop healthy relations with the world and with themselves. The author also explores the ways in which trauma affects the family structure, as this mother-daughter pair must confront the ways in which violence has affected their lives in order to achieve the healing of their psychological wounds.

Morrison's *God Help The Child* cements the idea of the pervasiveness of violence in the lives of children, addressing matters of sexual exploitation, abuse, trauma and racism/colorism. The question of colorism is deeply explored by Morrison, as she depicts discrimination from multiple perspectives, demonstrating the ways in which this specific kind of violence creates division among between black subjects, as well as reshapes the hierarchies of discrimination created by white supremacy. Bride demonstrates the ways in which the idea of discrimination may be reversed, as she takes control of her own narrative, investing in the creation of a self that acknowledges race and uses it as its most powerful asset. The addition of magic realism as a trope to depict the reversal of development triggered by the fear of abandonment felt by the protagonist demonstrates how the *Bildungsroman* genre is liable of being negotiated by authors that differ from the canonical expectations placed on this traditional model. The presence of violence and trauma in different contexts, being told from different perspectives in non-linear narratives, showcases the plasticity of literature, and its ability to deal with issues that defy representation, resulting in a complex composite that displays how prevalent these issues are around us.

Finally, hooks' *Bone Black* investigates the ways poverty is a structural form of violence, shaping the reality of growing up, as well as interrogating the role of family ties in the fostering of a healthy environment for the development of a child. The author also explores the confrontation with racism in the segregated

South, sharing with the readership the oblique education on race that she received from her family, and exposing the ways in which racism is constructed. The author also unveils the silences that are imposed on the bodies of girls, and the African-American community at large, demonstrating the need for the unmaking of the structures that inhibit these subjects from confronting silencing. The gendered education that hooks experienced is demonstrative of the ways gender roles are frequently a form of oppression, since their imposition takes place through disciplinary action that more often than not is enacted through violence. By placing the path for achieving a sense of self in the oldest generation of her family, hooks also elicits the importance of memory and the sharing of stories as forms of healing, as the word becomes for her a place of belonging.

The contributions of studies regarding subalternity as well as the black-feminist epistemologies allowed for an intersectional understanding of the different kinds of oppression that generate trauma and violence in the lives of African diasporic subjects in the United States and in the Caribbean. Only by recognizing the different kinds of oppression that co-create the violent experience of these subjects it will be possible to de-construct the systemic and structural forces that continue to diminish the life opportunities of all those involved in the process of colonization. In the selected works the decolonization of the *Bildungsroman* genre takes place through the negotiation of the conventions that are subverted by the authors, showcasing different paths of growth, ultimately removing them from the invizibilization fashioned by the colonial process that continues to create the conditions of living and existing in our times. The Sociology of Absences comes to the fore in this process, as the authors are capable of making evident through their writings the distinctions that have been created by the colonial process, addressing how the invizibilization of the oppressions are created and perpetuated, seeing clearly the abysmal line that separates their experiences as African diasporic women in the Americas from the so-called mainstream understanding of Americanness. As a form of addressing these issues, the present investigation tries to work under the principles of the ecology of recognition ("Para além do Pensamento Abissal" 24), to identify the ways difference and hierarchy are constitutive of each other.

By negotiating the *Bildungsroman*, the authors effectively become agents who intervene in the social realities they literally inhabit or perceive to exist, reiterating the understanding that literature, especially when created by the margins, becomes a form of cultural translation (seen here in the previously discussed approaches provided by Santos, Bhabha, and Butler), since the authors use their narratives to create a better understanding regarding the ways in which different social struggles inform their lives, exposing this previously concealed reality to a larger and more diverse public, and fostering mutual intelligibility between all the agents that are involved in this process. Ultimately, counter-hegemony becomes possible in the articulation of non-hegemonic knowledges (literature from the margins) in conjunction with their agents (the authors and readers). They offer their narratives as tools in the creation of social practices of non-discrimination and the deconstruction of oppressive paradigms. The knowledges provided by literature regarding the experience of trauma and violence ultimately serve as instruments in the unmaking of violence and trauma at large for the readership. As stated in the introduction, the present work intends to be conceived as a form of emancipatory research, engaged with social change and the raising of awareness regarding the presence of trauma and violence responsible for the generation and perpetuation of social inequalities.

Literature becomes a place in which trauma and violence come out of the realm of invisibility, becoming a tool for the voicing of concerns that might not be fully realized otherwise. As imagination is capable of seeing and delivering more than what exists, it is possible to claim that it also becomes the context in which alternative perspectives emerge, promising ways of seeing both the *how* and perhaps even the *why*. The representation of trauma and violence, in its (im)possibility, finds its way through the flexibility of the narrated word that reimagines the disrupting events, either fictional or autobiographical, trying to create order out of chaos.

There are, however, many questions that are left unanswered, as the complexities that were approached here are manifold, and the contexts that have been explored could be seen from different angles, in addition to the multiplicity of comparisons to other literary works that could have been made.

Some questions that can be pursued in future works regarding the topics developed here are: is there a difference in representations of violence and trauma in the work of diasporic male authors? Or LGBTQ+ authors? What is there to be said about empathy in literature towards the perpetrators of violence? What are the other ways literature may effectively serve as tool in the overcoming of trauma? What other genres may better voice concerns regarding violence and trauma? The present work does not claim to be exhaustive in its scope, as it deals solely with four different titles that by no means account for the plurality of the African-American and Afro-Caribbean experience of women writers. It intends, however, to demonstrate the richness that can be found in literature, and more specifically its capacity to adapt and absorb themes and to reshape forms, in order to encompass the developments of the human condition, either unitarily or collectively. Most importantly, it intends to reaffirm literature as a space for resistance, and for the fostering of a more egalitarian reality.

6 Works Cited

Abruna, Laura Niesen. "Jamaica Kincaid's Writing and the Maternal-Colonial Matrix."
  Caribbean Women Writers: Fiction in English, edited by Mary Condé and Thorunn
  Lonsdale, Macmillan Press, 1999, pp. 172–183.
Andrews, William. "African-American Autobiography Criticism: Retrospect and Prospect."
  American Autobiography – Retrospect and Prospect, edited by John Paul Eakin. Madison:
  University of Wisconsin Press, 1991, pp. 195–215.
Angelou, Maya. *I Know Why the Caged Bird Sings* in *The collected Autobiographies of Maya
  Angelou.* New York: The Modern Library, 2004.
Ashcroft, Bill, et al. *Key Concepts in Post-Colonial Studies.* London: Routledge, 2001.
Bauman, Zygmunt. *Liquid Times – Living in an Age of Uncertainty.* Cambridge and New York:
  Polity Press, 2007.
Bhabha, Homi. *The Location of Culture.* London and New York: Routledge, 1994.
Bolaki, Stella. *Unsettling the Bildungsroman – Reading Contemporary American Ethnic
  Women's Fiction.* New York: Rodopi Press, 2011.
Bonetti, Kay. "Interview with Jamaica Kincaid." *The Missouri Review,* Fall 2002, www.missou
  rireview.com/article/interview-with-jamaica-kincaid/. Accessed on 3 Oct. 2017.
Bourdieu, Pierre. *Masculine Domination.* Stanford: Stanford University Press, 2001.
Braziel, Jana Evans. "Daffodils, Rhizomes, Migrations: Narrative Coming of Age in the
  Diasporic Writings of Edwidge Danticat and Jamaica Kincaid." *Meridian,* vol. 3 2, Indiana
  University Press, 2003, pp. 110–31, www.jstor.org/stable/40338577. Accessed on 2 Mar.
  2018.
Broeck, Sabine. "Trauma, Agency and Kitsch and the Excesses of the Real: Beloved Within
  the Field of Critical Response." *America in the Couse of Human Events*, edited by Josef
  Jarab, Marcel Arbeit, and Jenel Virden, VU University Press, 2006, pp. 201–15.
Butler, Judith. *Precarious Life – The Powers of Mourning and Violence,* London and New York:
  Verso, 2004.
Butler, Judith. *Bodies That Matter,* London and New York: Routledge, 2011.
Butler, Judith and Athena Athanasiou. "Recognition and Survival, or Surviving Recognition."
  Dispossession – The Performative in the Political, Cambridge: Polity Press, 2013,
  pp. 75–91.
Caldeira, Isabel. "Who Has the Right to Claim America?" *America Where? Transatlantic Views
  of the United States in the Twenty-First Century,* eds. Isabel Caldeira, Maria José Canelo
  and Irene Ramalho Santos, Amsterdam: Peter Lang, 2002, pp. 173–196.
Caldeira, Isabel. "Toni Morrison and Edwidge Danticat – Writers as Citizens of the African
  Diaspora or 'The Margins as a Space of Radical Openness.'" *The Routledge Companion
  of Inter-American Studies,* edited by Wilfried Raussert. New York: Routledge, 2017,
  pp. 207–18.
Caldeira, Isabel. "'What Moves at the Margins' as Vozes Insurretas de Toni Morrison, bell
  hooks e Ntozake Shange." *The Edge of Many Circles – Homenagem a Irene Ramalho
  Santos,* orgs. Isabel Caldeira, Graça Capinha, and Jacinta Matos, vol. I, Coimbra:
  Imprensa da Universidade de Coimbra, 2017, pp. 139–160.
Caldeira, Isabel. "Mourning for Citizenship in Toni Morrison's Fiction." 28th Annual
  Conference of the American Literature Association, 25 to 28 May 2017, Boston, USA.

https://doi.org/10.1515/9783110752755-008

Caruth, Cathy. "Trauma and Experience." *The Holocaust: Theoretical Readings,* eds. Neil Levi e Michael Rothberg, Edinburgh: Edinburgh University Press, 2003.

Ceceña, Ana Esther, et al. *El Gran Caribe. Umbral de la geopolítica mundial.* La Habana: Ciencias Sociales, 2011.

Collins, Patricia Hill. "Foreword" *Emerging Intersections : Race, Class, and Gender in Theory, Policy, and Practice.* edited by Bonnie Thornton Dill and Ruth Enid Zambrana New Brunswick, N.J. :Rutgers University Press, 2009.

Combahee River Collective. "A Black Feminist Statement." *But Some Of Us Are Brave: All The Women Are White, All The Blacks Are Men: Black Women Studies,* ed. Gloria T. Hull, Patricia Bell Scott and Barbara Smith. New York: The Feminist Press, 1982, pp. 13–22.

Commonwealth of Nations. "Antigua and Barbuda: History", 2018, thecommonwealth.org/our-member-countries/antigua-and-barbuda/history. Accessed on 12 Dec. 2018.

Coronil, Fernando. "Beyond Occidentalism: Toward Nonimperial Geohistorical Categories." *Cultural Anthropology,* no. 11, Blackwell Publishing, 1996, pp. 51–87.

Crenshaw, Kimberlé. "Demarginalizing the Intersection of Race and Sex: A Black Feminist Critique of Antidiscrimination Doctrine, Feminist Theory and Antiracist Politics." *University of Chicago Legal Forum,* vol. 1989 issue 1, 1989, pp. 139–167, chicagounbound.uchicago.edu/uclf/vol1989/iss1/8. Accessed on 18 Sep. 2015.

Crenshaw, Kimberlé. "Mapping the margins: intersectionality, identity politics, and violence against women of color." *Stanford Law Review,* vol. 43 no. 6, Stanford Law Review, 1991, pp. 1241–99.

Cudjoe, Selwyn. "Identity and Caribbean Literature: A lecture delivered to the Japanese Black Studies Association at Nara Women's College, Nara, Japan.", 2001, www.trinicenter.com/Cudjoe/2001/June/24062001b.htm. Accessed on 18 Sep. 2015.

Danticat, Edwidge. "We are Ugly, but We are Here." *The Caribbean Writer,* vol. 10, 1996, faculty.webster.edu/corbetre/haiti/literature/danticat-ugly.htm . Accessed on 11 Sep. 2018.

Danticat, Edwidge. *Breath, Eyes, Memory,* New York: Randon House Vintage Books, 1999.

Danticat, Edwidge. *Create Dangerously : The Immigrant Artist at Work.* New York: Vintage Books, 2010.

Danticat, Edwidge. "NEED TO KNOW". Interview with author Edwidge Danticat, 2011, www.youtube.com/watch?v=_laWVbxEhoM. Accessed on 8 Sep. 2015.

Danticat, Edwidge. "The Long Legacy of Occupation in Haiti." *The New Yorker,* 2015, www.newyorker.com/news/news-desk/haiti-us-occupation-hundred-year-anniversary. Accessed on 30 Sep. 2015.

Danticat, Edwidge. *The Art of Death – Writing the Final Story,* Minneapolis: Graywolf Press, 2017.

Dash, J. Michael. *The Other America – Caribbean Literature in a New World Context,* Charlottesville and London: University Press of Virginia, 1998.

Davis, Angela. *Women, Race and Class,* New York: Fist Vintage Books Edition, 1983.

Davis, Angela. *The Meaning of Freedom and Other Difficult Dialogues,* San Francisco: City Lights Books, 2012.

DeGruy, Joy. *Post Traumatic Slave Syndrome – America's Legacy of Enduring Injury and Healing,* Portland: DeGruy Publication Inc, 2005.

Deleuze, Gilles and Félix Guattari. "What is a Minor Literature?", *Kafka: Toward a Minor Literature,* Minneapolis: University of Minnesota Press, 1986, pp. 16–26.

Deloria, Philip. "American Indians, American Studies, and the ASA", *American Quarterly* vol. 55 no. 4, 2003, muse.jhu.edu/login?auth=0&type=summary&url=/journals/ american_quarterly/v055/55.4deloria.html. Accessed on 4 Mar.

Dill, Bonnie Thornton and Ruth Enid Zambrana, eds. *Emerging Intersections: Race, Class, and Gender in Theory, Policy, and Practice.* Foreword by Patricia Hill Collins. Rutgers University Press, 2009.

Dreiding, Michelle. "Inaugurating Ambivalence – Toni Morrison's *God Help the Child.*" *Variations*, vol. 24, Bern: Peter Lang, 2016, pp. 129–140.

DuBois, W. E, B. *The Souls of Black Folk,* New York: Dover Thrift Editions, 1994.

Easton, Alison. "Subjects in Time – Slavery and African-American Women's Autobiographies." *Feminism and Autobiography – Texts, Theories, Methods,* eds. Tess Cosslett, Celia Lury and Penny Summerfield. London and New York: Routledge, 2000, pp. 169–182.

Feng, Pin-Chia. "'Ou libéré!' Trauma and Memory in Edwidge Danticat's Breath, Eyes, Memory." *Canadian Review of Comparative Literature/Revue Canadienne de Littérature Comparée*, September – December 2003, Canadian Comparative Literature Association, 2003, pp. 738–52.

Ferguson, Moira and Jamaica Kincaid. "A Lot of Memory: An Interview with Jamaica Kincaid." *The Kenyon Review, New Series,* vol. 16 no. 1, Kenyon College, 1994, pp. 163–188, www. jstor.org/stable/4337017. Accessed on 19 Jan. 2018.

Figley, Charles, et al. "Beyond the 'Victim' Secondary Traumatic Stress." *Beyond Trauma: Cultural and Societal Dynamics,* eds. Rolf Kleber, Charles Figley and Berthold Gersons, Ontario: University of Waterloo, 1995, pp. 75–98.

Figueiredo, D. H. and Frank Argote-Freyre. *A Brief History of the Caribbean.* New York: Infobase Publishing, 2008.

Fishkin, Shelley. "Crossroads of Cultures: The Transnational Turn in American Studies. Presidential Address to the American Studies Association." *American Quarterly*, vol. 57 no. 1, March 2005, pp. 17–57.

Foran, Sheila. "Toni Morrison Reflects on What it Means to be a Citizen in a Changing World.", 2011, today.uconn.edu/blog/2011/04/toni-morrison-reflects-on-what-it-means-to-be-a-citizen-in-a-changing-world. Accessed on 17 Apr. 2017.

François, Irline. "The Daffodil Gap: Jamaica Kincaid's Lucy." *Bloom's Modern Critical Views: Jamaica Kincaid, New Edition,* New York: Infobase Publishing, 2008, pp. 81–96.

Galtung, Johan. "Violence, Peace, and Peace Research.", *Journal of Peace Research, vol.* 6 no. 3, 1969, pp. 167–191.

Galtung, Johan. "Cultural Violence." *Journal of Peace Research, vol.* 27 no. 3, 1990, pp. 291–305.

Garis, Leslie. "Through West Indian Eyes." *The New York Times, 1990,* www.nytimes.com/ 1990/10/07/magazine/through-west-indian-eyes.html. Accessed on 12 Mar. 2016.

Gates, Henry Louis Jr. "Writing 'race' and the difference it Makes." *Critical Inquiry*, vol. 12 no. 1, Chicago: The University of Chicago Press, 1985, pp. 1–20.

Gay, Roxane. "God Help the Child by Toni Morrison review – 'incredibly powerful'." *The Guardian*, 2015, www.theguardian.com/books/2015/apr/29/god-help-the-child-toni-morri son-review-novel. Accessed on 30 Oct. 2018.

Gay, Roxane. *Hunger – A Memoir of (My) Body,* London: Corsair, 2017.

Ghansah, Rachel Kaadzi. "The Radical Vision of Toni Morrison." *The York Times Magazine, 2015*, www.nytimes.com/2015/04/12/magazine/the-radical-vision-of-toni-morrison.html. Accessed on 17 Sep. 2018.

Gilbert, Susan. "I Know Why the Caged Bird Sings: Paths to Escape." *Bloom's Biocritiques Maya Angelou*, ed. Harold Bloom, Broomal: Chelsea House Publishers, 2002, pp. 73–84.

Gilmore, Leigh. *The Limits of Autobiography – Trauma and Testimony.* Ithaca: Cornell University Press, 2011.

Grafenstein, Von Johanna. *Haití.* Ciudad de Mexico: Alianza Editorial-Instituto José María Luis Mora, 1988.

Grayson, Erik. "The Most Important Meal: Food and Meaning in Jamaica Kincaid's Lucy." *Journal of the Georgia Philological Association*, 2006, pp. 212–227, www.academia.edu/363586/_The_Most_Important_Meal_Food_and_Meaning_in_Jamaica_Kincaids_Lucy_. Accessed on 26 Jan. 2018.

Grillo, Trina. "Anti-Essentialism and Intersectionality: Tools to Dismantle the Master's House." *Theorizing Feminism,* ed. Elizabeth Hackett and Sally Haslanger, Oxford University Press, 2006, pp. 30–40.

Hall, Stuart. "The West and The Rest: Discourse and Power" in *Formations of Modernity* Oxford: Polity in association with Open University, 1992.

Harris-Perry, Melissa. *Sister Citizen – Shame, Stereotypes, and Black Women in America.* New Heaven and London: Yale University Press, 2011.

Hirsch, Marianne. "The Generation of Postmemory." *Poetics Today,* vol. 29 issue 1, Duke University Press, 2009, pp. 103–22, read.dukeupress.edu/poetics-today/article/29/1/103/20954/The-Generation-of-Postmemory. Accessed on 03 Nov. 2018.

hooks, bell. *Feminist Theory From Margin to Center.* Boston: South End Press, 1984.

hooks, bell. *Yearning – Race, Gender, and Cultural Politics.* Boston: South End Press, 1990.

hooks, bell. *Bone Black – Memories of Girlhood.* New York: Henry Holt and Company, 1996.

hooks, bell. *All About Love – New Visions.* New York: HarperCollins Publishers, 2000.

hooks, bell. *Salvation – Black People and Love.* New York: HarperCollins Publishers, 2001.

hooks, bell. *Remembered Rapture – The writer at Work.* New York: Owl Books, 2013.

hooks, bell. *Talking Back – Thinking Feminist, Thinking Black.* New York and London: Routledge, 2015.

hooks, bell. *Sisters of the Yam: Black Women and Self-Recovery.* New York and London: Routledge, 2015.

Hull, Gloria T. Scott, Patricia Bell and Smith, Barbara, eds., *All the Women Are White, All the Blacks Are Men, but Some of Us Are Brave, Black Women's Studies* :Beverly Guy-Sheftall The Journal of Negro History, 1982.

Ibarrola-Armendariz, Airtor. "The Language of Wounds and Scars in Edwidge Danticat's *The Dew Breaker:* a case in trauma symptoms and the recovery process." *Journal of English Studies,* vol. 8, Universidad de la Rioja, 2010, pp. 23–56.

hooks, bell. "Too Huge a Theme for Too Slight a Treatment: Toni Morrison's *God Help the Child." Taking Stock to Look Ahead – Forty Years of English Studies in Spain,* eds. María Ferrández San Miguel and Claus Peter Neumann, Zaragoza: Prensas Universitarias de Zaragoza, 2018, pp. 75–84.

Iqbal, Razia. "God Help The Child by Toni Morrison, book review: Pain and trauma live just under the skin." *The Independent, 2015,* www.independent.co.uk/arts-entertainment/

books/reviews/god-help-the-child-by-toni-morrison-book-review-pain-and-trauma-live-just-under-the-skin-10164870.html. Accessed on 30 Oct. 2018.

Jablonski, Nina. *Living Color – The Biological and Social Meaning of Skin Color.* Berkeley and Los Angeles: University of California Press, 2012.

Keita, Fatoumata. "Conjuring Aesthetic Blackness/ Abjection and Trauma in Toni Morrison's God Help the Child." *Africology: The Journal of Pan African Studies,* vol. 11 no. 3, 2018, pp. 43–55.

Kendi, Ibram. *Stamped From the Beginning – The Definitive History of Racist Ideas in America.* New York: Nation Books, 2016.

Kidd, David and Emanuele Castano. "Reading Literary Fiction Improves Theory of Mind." *Science,* 2013, pp. 377–80, science.sciencemag.org/content/342/6156/377. Accessed on 31 Jan. 2017.

Kidd, David and Emanuele Castano. "Different Stories: How Levels of Familiarity With Literary and Genre Fiction Relate to Mentalizing." *Psychology of Aesthetics, Creativity, and the Arts,* 2016, pp. 474–86, dx.doi.org/10.1037/aca0000069. Accessed on 1 Feb. 2017.

Kincaid, Jamaica. *At the Bottom of the River.* New York: Farrar Straus Giroux, 1983.

Kincaid, Jamaica. (1990) *Lucy.* Farrar Straus Giroux: New York.

Kincaid, Jamaica. (1991) "On Seeing England for the First Time" in *Transition* No 51. 32–40. Indiana University Press – W.E.B. DuBois Institute. Available at http://www.jstor.org/sta ble/2935076?origin=JSTOR-pdf on 23/01/2018.

Kincaid, Jamaica. *Annie John.* London: Vintage Books, 1997.

Kincaid, Jamaica. *My Brother.* New York: Farrar Straus Giroux, 1998.

Kincaid, Jamaica. *My Garden Book.* New York: Farrar Straus Giroux, 1999.

Kincaid, Jamaica. *Talk Stories.* New York: Farrar Straus Giroux, 2001.

Kincaid, Jamaica. *See Now Then.* New York: Farrar Straus Giroux, 2013.

LeSeur, Geta. *Ten Is the Age of Darkness: The Black Bildungsroman.* Columbia: University of Missouri Press, 1995.

Lorde, Audre. *Sister Outsider.* New York: Crossing Press, 2007.

Lugones, Maria. "The Coloniality of Gender." *Worlds and Knowledges Otherwise, vol. 2,* Duke University, Spring 2008, pp. 1–17.

Lupton, Mary Jane. *Maya Angelou – A Critical Companion.* Westport: Greenwood Press, 1998.

Martins, Catarina. "The West and the Women of the Rest" in T*he Edge of Many Circles* Volume II, 2017.

Marquis, Claudia. "'Not At Home In Her Own Skin': Jamaica Kincaid, History and Selfhood." *Caliban French Journal of English Studies, no.* 21, 2007, pp. 101–14, caliban.revues.org/ 1883#text. Accessed on 24 Oct. 2017.

Moffitt, Robert and Peter Gottschalk. "Ethnic and Racial Differences in Welfare Receipt in the United States." *America Becoming: Racial Trends and Their Consequences,* eds. Neil J. Smelser, William Julius Wilson and Faith Mitchel, vol. II, National Research Council, 2001, pp. 153–73.

Moi, Toril. *Sexual/Textual Politics.* New York: Methuen & Co., 1985.

Morrison, Toni. "What the Black Women Thinks About the Woman's Lib." *The New York Times,* 1971, www.nytimes.com/1971/08/22/archives/what-the-black-woman-thinks-about-womens-lib-the-black-woman-and.html?_r=1. Accessed on 25 Oct. 2016.

Morrison, Toni. "Unspeakable Things Unspoken – The African-American Presence in American Literature." *The Tanner Lectures in Human Value,* University of Michigan, 1988, tannerlectures.utah.edu/_documents/a-to-z/m/morrison90.pdf. Accessed on 7 Sep. 2017.

Morrison, Toni. *The Bluest Eye.* London: Vintage Books, 1999.

Morrison, Toni. "Nobel Lecture December 7, 1993." *Nobel Lectures, Literature 1991–1995,* ed. Sture Allén, World Scientific Publishing Co., 2007, www.nobelprize.org/nobel_prizes/liter ature/laureates/1993/morrison-lecture.html. Accessed on 4 May 2015.

Morrison, Toni. *Burn This Book: Writers Speak Out on the Power of the World.* New York: Harper, 2009.

Morrison, Toni. *God Help the Child.* London: Chatto and Windus, 2015.

Mukherjee, Kusumita. "Politics of Selfhood and Magic Realism in Morrison's *The Bluest Eye* and *God Help the Child.*" *The Criterion: An International Journal in English,* vol. 8, issue III, 2017, pp. 497–504.

Myrdal, Gunnar. *An American Dilemma: The Negro Problem and Modern Democracy.* London and New York: Harper & Brothers, 1944.

Page, Yolanda. *Encyclopedia of African American Women Writers.* Westport: Greenwood Press, 2007.

Pease, Donald. "Re-thinking American Studies after US Exceptionalism." *American Literary History* vol. 21 issue 1, 2009, pp. 19–27.

Pinkston, Victoria Paige "'Our voices will not be silenced': Edwidge Danticat, Haiti, and the Silences of History." *Undergraduate Honors Theses,* paper 427, College of William and Mary, 2011, scholarworks.wm.edu/honorstheses/427/?utm_source=publish.wm.edu% 2Fhonorstheses%2F427&utm_medium=PDF&utm_campaign=PDFCoverPages. Accessed on 14 Aug. 2018.

Puga, Rogério. *O Bildungsroman: (Romance de Formação) Perspectivas.* Lisboa: Cetaps – Center for English, Translation and Anglo-Portuguese Studies/Institute of Modern Languages Research, 2016.

Queeley, Andrea. "Remembering the Wretched: Narratives of Return as a Practice of Freedom." *The Journal of Pan African Studies*, vol. 4 no.7, 2011, pp. 109–25.

Quijano, Aníbal and Immanuel Wallerstein. "Americanity as a Concept, or the Americas in the Modern World System." *International Social Science Journal* 134, November 1992, pp. 549–57.

Radway, Janice. "What's in a Name? Presidential Address to the American Studies Association." *American Quarterly,* vol. 51 issue 1, Baltimore: The Johns Hopkins University Press, March 1999, pp. 1–32.

Ramalho, Maria Irene. "Who Owns American Studies? Old and New Approaches to Understanding the United States of America." *Op. Cit.: A Journal of Anglo-American Studies* 2, 2013, pp. 1–21.

Ramírez, Manuela López. "'What You Do To Children Matters'- Toxic Motherhood In Toni Morrison's *God Help The Child.*" *The Grove – Working Papers in English Studies,* Universidad de Jaén, 2015, pp. 107–19, revistaselectronicas.ujaen.es/index.php/grove/ article/view/2700. Accessed on 30 Oct. 2018.

Ramírez, Manuela López. "'Childhood Cuts Festered and Never Scabbed Over': Child Abuse in Toni Morrison's *God Help the Child." Alicante Journal of English Studies,* vol. 29, 2016, pp. 145–64.

Ramírez, Manuela López. "'Racialized Beauty': The Ugly Duckling in Toni Morrison's *God Help the Child.*" *Complutense Journal of English Studies,* no. 25, Madrid: Ediciones Complutenses, 2017, pp. 173–89.

Raynaud, Claudine. "Coming of Age in the African American Novel." *The Cambridge Companion to the African American Novel,* ed. Mearyemma Graham, Boston: Cambridge University Press, 2004, pp. 106–21.

Ribeiro, António Sousa and Maria Irene Ramalho. "Dos estudos literários as estudos culturais?" *Revista Crítica de Ciências Sociais,* vol. 52/53, Coimbra: Centro de Estudos Sociais, 1998/1999, pp. 61–84.

Ribeiro, António Sousa. *Representações da Violência.* Coimbra: Almedina, 2013.

Rich, Adrienne. "Notes Toward a Politics of Location." *Blood, Bread, and Poetry – Selected Prose 1979–1985,* New York: W. W. Norton and Company, 1986, pp. 210–31.

Rishoi, Christy. *From Girl to Woman: American Women's Coming-of-age Narratives.* Albany: State University of New York, 2003.

Rogoziński, Jan. *A Brief History of the Caribbean – From the Arawak and the Carib to the Present.* New York: Facts On File, 1999.

Rosello, Mireille. "Marassa with a difference: Danticat's *'Breath, Eyes, Memory'.*" *Edwidge Danticat: a reader's guide,* ed. Martin Munro, Charlottesville: University of Virginia Press, 2010, pp. 117–29.

Roth, Julia. "Entangled Inequalities as Intersectionalities: Towards an Epistemic Sensibilization." *DesiguAldades.net Working Paper Series no. 42,* Berlin: DesiguALdades.net Reseach Network on Interdependent Inequalities in Latin America, 2013.

Roth, Julia. *Occidental Readings, Decolonial Practices: A Selection on Gender, Genre, And Coloniality In The Americas.* Trier: WVT Wissenschaftlicher Verlag Trier, 2014.

Sader, Emir and Rodrigo Nobile. *Latinoamericana – Enciclopédia Contemporânea da América Latina e do Caribe.* São Paulo: Boitempo, 2006.

Salván, Paula Martín. "Secrets, Lies And Non-Events: The Production Of Causality And Self-Deconstruction In Toni Morrison's *God Help the Child.*" *THEORY NOW: Journal Of Literature, Critique And Thought,* vol. 1 no. 1, July-December 2018, pp. 65–80.

Santos, Boaventura de Sousa. "A Crítica Da Razão Indolente: Contra O Desperdício Da Experiência." *Para Um Novo Senso Comum: A Ciência, O Direito E A Política Na Transição Paradgmática,* São Paulo: Cortez, 2002.

Santos, Boaventura de Sousa. "Para uma sociologia das ausências e uma sociologia das emergências." *Revista Crítica de Ciência Sociais,* vol. 63, Coimbra: Centro de Estudos Sociais, 2002, pp. 237–80.

Santos, Boaventura de Sousa. "A Critique of Lazy Reason: Against the Waste of Experience" *The Modern World-System In The Longue Durée,* ed. Immanuel Wallerstein, Boulder: Paradigm, 2004, pp. 103–25.

Santos, Boaventura de Sousa. "The World Social Forum, a Users Manual.", 2004, www.ces. uc.pt/bss/documentos/fsm_eng.pdf. Accessed on 15 Feb. 2017.

Santos, Boaventura de Sousa. "Para Além do Pensamento Abissal: Das Linhas Globais a Uma Ecologia de Saberes." *Revista Crítica de Ciências Sociais,* vol. 78, Coimbra: Centro de Estudos Sociais, 2007, pp. 3–46.

Santos, Boaventura de Sousa. *Para Descolonizar Occidente: Más Allá Del Pensamiento Abismal.* Buenos Aires: Consejo Latinoamericano de Ciencias Sociales, 2010.

Santos, Boaventura de Sousa. "Public Sphere and Epistemologies of the South" in *Africa Development*, vol. XXXVII, no. 1, 2012, pp. 43–67, www.boaventuradesousasantos.pt/media/Public%20Sphere_AfricaDevelopment2012.pdf. Accessed on 12 Jan. 2017.

Santos, Boaventura de Sousa, João Arriscado Nunes and Maria Paula Meneses. "Introduction: Opening Up the Canon of Knowledge and Recognition of Difference." *Another Knowledge is Possible – Beyond Northern Epistemologies,* ed. Boaventura de Sousa Santos, New York and London: Verso, 2008, pp. VIX–LXII.

Santos, Maria Irene Ramalho. "Difference and Hierarchy Revisited by Feminism." *Anglo Saxonica* Serie III, No. 6, 2013.

Sarthou, Sharrón Eve. "Unsilencing Défilés Daughters: Overcoming Silence in Edwidge Danticat's *Breath, Eyes, Memory* and *Krik? Krak!*" *The Global South – Special Issue: The Caribbean and Globalization,* ed. Adetayo Alabi, Indiana University Press/Journals, vol. 4 no. 2, Fall 2010, pp. 99–123.

Schinkel, Willem. *Aspects of Violence: A Critical Theory.* New York: Palgrave Macmillan, 2010.

Seligmann-Silva, Márcio. "Narrar o trauma – a questão dos testemunhos de catástrofes históricas" *Psicologia Clínica,* vol. 20 no. 1, Rio de Janeiro: Pontifícia Universidade Católica, 2008, pp. 65–82.

Shaw, Denise. "Textual Healing: Giving Voice to Historical and Personal Experiences in the Collective Works of Edwidge Danticat." *The Hollins Critic,* vol. 44 no. 1, Virginia Hollins University, 2007, pp. 1–13.

Sielke, Sabine. *Reading Rape – The Rhetoric of Sexual Violence in American Literature and Culture 1770–1990.* Princeton University Press, 2002.

Smith, Barbara. "Toward a Black Feminist Criticism." *The Radical Teacher,* University of Illinois Press, 1978, pp. 20–7.

Smith, Valerie. *Not Just Race, Not Just Gender: Black Feminist Readings.* New York: Routledge, 1998.

Smith, Valerie. *Toni Morrison – Writing the Moral Imagination.* Oxford: Wiley-Blackwell, 2012.

Spivak, Gayatri Chakravorti. *The Spivak Reader,* eds. Donna Landry and Gerald MacLean, New York: Routledge, 1996.

Taylor, Ula. "The Historical Evolution of Black Feminist Theory and Praxis." *Journal of Black Studies*, vol. 29 no. 2, Sage Publications, 1998, pp. 234–53.

Treisman, Deborah. "Edwidge Danticat On Her Caribbean Immigrant Experience." *The New Yorker, 2018,* www.newyorker.com/books/this-week-in-fiction/fiction-this-week-edwidge-danticat-2018-05-14. Accessed on 4 Jun. 2018.

Truth, Sojourner. "Ain't I a Woman?" *Great Speeches by African Americans,* ed. James Daley. New York: Dover Publications, 2006, pp. 11–2.

Updike, John. "Dreamy Wilderness – Unmastered Women in Colonial Virginia." *The New Yorker,* 3 Nov. 2008, www.newyorker.com/magazine/2008/11/03/dreamy-wilderness. Accessed on 19 Sep. 2018.

Vega-González, Susana. "The Dialetics of Belonging in bell hooks' *Bone Black: Memories of Girlhood." Journal of English Studies,* vol. 3, Universidad de la Rioja, 2002, pp. 237–48, publicaciones.unirioja.es/ojs/index.php/jes/article/view/79. Accessed on 15 Nov. 2018.

Warrior, Robert. "Native American Scholarship and the Transnational Turn" *Cultural Studies Review, vol. 15* no. 2, 2011, pp. 119–30, dx.doi.org/10.5130/csr.v15i2.2041. Accessed on 2 Dec. 2017.

Washington, Mary Helen. "Disturbing the Peace: What Happens to American Studies If You Put African American Studies at the Center?" *American Quarterly,* vol. 50 no. 1, 1998, pp. 1–23, http://xroads.virginia.edu/~drbr2/washington.html. Accessed on 4 Mar. 2016.

Wordsworth, William. "I Wandered Lonely as a Cloud.", 1807, www.poetryfoundation.org/poems/45521/i-wandered-lonely-as-a-cloud. Accessed on 14 Feb. 2018.

Index

https://doi.org/10.1515/9783110752755-009

www.ingramcontent.com/pod-product-compliance
Lightning Source LLC
Chambersburg PA
CBHW021918190326
41519CB00009B/837